Springer Textbooks in Earth Sciences, Geography and Environment

The Springer Textbooks series publishes a broad portfolio of textbooks on Earth Sciences, Geography and Environmental Science. Springer textbooks provide comprehensive introductions as well as in-depth knowledge for advanced studies. A clear, reader-friendly layout and features such as end-of-chapter summaries, work examples, exercises, and glossaries help the reader to access the subject. Springer textbooks are essential for students, researchers and applied scientists.

More information about this series at http://www.springer.com/series/15201

Lars Petter Røed

Atmospheres and Oceans on Computers

Fundamental Numerical Methods for Geophysical Fluid Dynamics

Lars Petter Røed
Department of Geosciences
University of Oslo
Oslo, Norway

and

Norwegian Meteorological Institute
Oslo, Norway

ISSN 2510-1307 ISSN 2510-1315 (electronic)
Springer Textbooks in Earth Sciences, Geography and Environment
ISBN 978-3-030-06733-5 ISBN 978-3-319-93864-6 (eBook)
https://doi.org/10.1007/978-3-319-93864-6

Softcover re-print of the Hardcover 1st edition 2019

This Springer imprint is published by the registered company Springer Nature Switzerland AG
The registered company address is: Gewerbestrasse 11, 6330 Cham, Switzerland

Preface

The central purpose of this book is to give the reader an insight into the fundamental issues involved in putting oceans and atmospheres on computers. In short, it describes the numerical solution of the partial differential equations (PDEs) governing the motion of atmospheres and oceans, and geophysical fluids in general. The presentation is geared toward students and researchers who aim to understand the physical content of the equations used in the various models and gain experience of the numerical methods used to solve these equations. Much of the formulation is general, hence applicable to any model. Moreover, it is hoped that this book will be useful to students who wish to become forecasters. In this capacity, they will no doubt find themselves analyzing results generated by model simulations and comparing them with observations. In doing so, and knowing something about the model's inner workings, they will be in a better position to provide feedback to model developers on what is realistic and what is not so realistic about the results.

Apart from a brief discussion of spectral methods, commonly applied when solving atmospheric problems, this book focuses on finite difference methods. Present research and development has evolved beyond finite difference methods (e.g., finite element methods), but the simplicity of the finite difference method is useful in a pedagogical treatment of the subject. Moreover, the finite difference methods are still the most commonly applied approach to atmospheric and ocean science.

It is assumed that the reader has little or no prior knowledge of or experience in the use of finite difference methods, so these are explained in some detail. Furthermore, details are provided on how to develop a set of instructions to enable the computer to do the arithmetic.[1] In this way, the full highly nonlinear PDEs that govern the motion of geophysical fluids may be solved. Hence, putting atmospheres and oceans on computers enables an understanding of their global circulation, and the possibility of predicting it. Moreover, it provides an insight into the nonlinear processes that govern their behavior. It is therefore hoped that this book will be a useful source of information for forecasters and researchers alike in their daily work.

[1]This set of instructions is commonly referred to as the model code.

What This Book Is and What It Is Not

Certain topics in the modelling of atmospheres and oceans may appear complex and esoteric, or even magical, to some students. This may lead to a growing distance between those scientists who analyze results from model simulations and observations and those who deal with the model's inner workings. In an endeavor to reduce this gap, the present treatment focuses on the fundamental issues involved in putting atmospheres and oceans on the computer. For this reason, no fully fledged models are presented. On the contrary, the treatment is limited to some of the fundamental balance equations inherent in models of geophysical fluids, such as the advection, diffusion, and rotating shallow water equations. Moreover, we shall discuss numerical methods that work and others that do not, giving reasons in the latter case.

To familiarize the reader with what it takes to develop a numerical model of atmospheres and oceans, and to develop a code that works, we present computer problems to be solved by the reader at the end of each chapter. These computer problems are always associated with the content of the given chapter. They also contain some analytical work to give the reader a better appreciation of the physics involved in each problem. The reader is therefore encouraged to solve these computer problems as they go along.

Some of the models used to forecast atmospheric or oceanic weather are so-called limited area models (LAM). As their name suggests, LAM models cover a limited geographical area, and therefore open boundaries are present at which there are no natural boundary conditions to replace the governing equations (Røed and Cooper 1986). This means that the governing equations are valid at such boundaries, and conditions must be developed that avoid deteriorating the solution in the interior of the LAM domain. More often than not the LAM model is embedded in a global model, or at least one that covers a much larger geographical area, and under these circumstances, the LAM model is nested. Since nested LAM models are commonly used in day-to-day weather forecasting operations, for both atmospheres and oceans, a chapter is included on how to develop sound conditions or nesting techniques at open boundaries (Chap. 7), and there is a section on two-way nesting (Sect. 10.5).

When a model code is developed to emulate the motion of the atmosphere or the ocean, the aim is often to provide a tool that helps us to understand a certain process, or to provide a model to forecast the geophysical fluid's future states. Hence, the model code must provide results that imitate the behavior of the geophysical fluid in question. Consequently, a model developer must be well versed in physics as well as numerics. The former is needed to judge whether the solution provided by the model code is indeed an approximation of the physical reality, and the latter to ascertain whether the model code yields a stable and true solution of the governing PDEs. The book therefore differs from a book on geophysical fluid dynamics. At the same time, it also differs from a book on numerical methods to solve PDEs. We thus include the physics required to understand why some numerical methods work and some do not.

Organization

This book is organized into eleven chapters and one appendix. It is assumed that the book will not be read cover to cover. Thus, each chapter is more or less independent from the rest. To enhance this independence, each chapter starts with a brief introduction and ends with a summary and a few remarks. This organization allows the reader to read the book in a somewhat arbitrary manner, and it is hoped that this will enhance the book's readability and accessibility, both as a monograph and as a reference for students and researchers. Moreover, one or more computer problems are suggested at the end of each chapter. Students and readers are encouraged to do these problems, and the exercises to test their progress. Equally important, it is hoped that by performing the computer problems students will obtain some hands-on experience in programming and visualization of results. By solving the problems, it is also hoped that students will gain an appreciation of the inner workings of an numerical weather prediction (NWP) or an numerical ocean weather prediction (NOWP) model.

To motivate the reader, and for later reference, Chap. 1 starts by describing how the diffusion, advection, and shallow water equations relate to the full equations governing the motion of the atmosphere and ocean. This requires a recapitulation of the governing equations (Sect. 1.1) and the boundary conditions (Sect. 1.2). The chapter continues by discussing the basic approximations commonly made in meteorology and oceanography (Sects. 1.3 and 1.4), which lead to simplified systems of equations. For instance, Sect. 1.5 shows how to derive the shallow water equations, and Sect. 1.6 the quasi-geostrophic equations. The assumptions and approximations needed to obtain them are highlighted. Readers who are well versed in these matters may skip this chapter without loss of continuity.

The diffusion, advection, and shallow water equations belong to a class of equations known as partial differential equations (PDEs). Consequently, Chap. 2 provides a rather detailed account of the way various types of PDEs relate to the advection–diffusion equations and the shallow water equations, and exposes the different physics inherent in elliptic (Sect. 2.1), parabolic (Sect. 2.2), and hyperbolic (Sect. 2.3) PDEs. At the heart of the finite difference methods are Taylor series. Consequently, Sects. 2.5 and 2.6 are devoted to the basics of Taylor series and how to use them to construct finite difference approximations of the various derivatives entering the governing equations, respectively. This involves a discussion of truncation errors and accuracy (Sect. 2.7). It is convenient at this stage to introduce the reader to the notation used throughout the book (Sect. 2.8). The following sections describe some important and useful mathematical concepts such as orthogonal functions (Sect. 2.9), Fourier series (Sect. 2.10), and Fourier transforms (Sect. 2.11). Again readers who are well acquainted with the above may skip this chapter without loss of continuity, except for the part on notation.

Most of the problems associated with geophysical phenomena are time marching problems, which is the subject of Chap. 3. The focus is not on how to solve them, but rather on some of the properties, e.g., conservation of mechanical energy, inherent in the diffusion (Sect. 3.2), advection (Sect. 3.3), and shallow water equations (Sect. 3.4). These are properties that should carry over when solving them numerically, and are therefore useful in assessing model performance. Checking whether these properties are adhered to is part of the quality assurance procedure outlined in some detail in Chap. 11, at the end of this book.

The bulk of the book is contained in the next three chapters, i.e., Chaps. 4, 5, and 6. They lay out various schemes whereby, respectively, the one-dimensional versions of the diffusion, advection, and shallow water equations may be solved numerically by finite difference methods, e.g., the forward-in-time centered-in-space (FTCS) scheme and the centered-in-time centered-in-space (CTCS) leap-frog scheme.

Chapter 4 introduces important and general concepts belonging to the world of finite difference methods, e.g., necessary and sufficient conditions for numerical stability (Sects. 4.3 and 4.7), consistent and convergent schemes (Sect. 4.9), explicit and implicit schemes (Sect. 4.8), and numerical dissipation (Sect. 4.5). Section 4.4 offers a comprehensive exposition of von Neumann's stability analysis, used to investigate under which conditions a given scheme is stable. This is then used to investigate the stability of the diffusion equation (Sect. 4.5). Finally, Chap. 4 also includes a section on how to solve an elliptic problem, using Gauss elimination as a direct elliptic solver (Sect. 4.11).

Chapter 5 continues by exploring the one-dimensional advection equations. Besides introducing various schemes, e.g., the leap-frog scheme (Sect. 5.3), the upstream scheme (Sect. 5.9), and the Lax–Wendroff scheme (Sect. 5.11), by which it may be solved, this chapter also introduces pivotal concepts such as numerical dispersion (Sect. 5.5), computational modes (Sect. 5.7), and numerical diffusion (Sect. 5.15). Section 5.16 presents methods and schemes whereby these unwanted properties may be lessened or avoided. Section 5.12 offers an explanation of yet another scheme, the so-called semi-Lagrangian scheme. This is becoming more and more popular in today's models of atmospheres and oceans, and paves the way for introducing the quasi-Lagrangian schemes used to solve the shallow water equations in Chap. 6. Among the computer problems suggested at the end, the reader is encouraged to complete Computer Problem 5.19.1, which gives insight into how dispersion, diffusion, instabilities, computational modes, etc., tend to show up in the results produced by a numerical model.

Chapter 6 treats the shallow water equations, both in linear and nonlinear form. After presenting the nonlinear, rotating shallow water equations, Sect. 6.2 discusses the various assumptions and approximations required to derive the linear shallow water equations. Section 6.3 is devoted to analytic solutions of the linear equations, such as inertia-gravity waves, coastal Kelvin waves, and Rossby waves, and introduces the potential vorticity as a first integral of the shallow water equations. These waves are observed phenomena and manifest themselves, for instance, as tide and storm surges in the ocean. In contrast, the energy contained in the inertia-gravity waves of the atmosphere is so small that it is regarded as noise. The chapter therefore includes a section on the semi-implicit method commonly applied in atmospheric models (Sect. 6.8). Also discussed are staggered grids (Sect. 6.5), originally introduced by Mesinger and Arakawa (1976) to avoid specifying more boundary conditions than allowed. The remaining sections provide details about various useful schemes for solving the shallow water equations. Since the shallow water equations, like the advection equation, are hyperbolic systems, many of the schemes that work for the advection equation also work for the shallow water equations, e.g., the leap-frog scheme, but since they involve more than one dependent variable they also open the way to new schemes, such as the forward–backward scheme and the quasi-Lagrangian scheme. Computer Problems 6.11.1 and 6.11.2 are particularly recommended here.

Chapter 7 is about open boundaries. In a numerical forecasting model, the size of the grid cells determines its resolution. Since computers have a limited capacity, there will always be a trade-off between grid size and geographical coverage. This implies the existence of a computational boundary where the governing equations are still valid, but where the computation stops. At these boundaries, referred to as open boundaries, a condition must still be constructed to replace the governing equations. This is the theme of Sects. 7.3 through 7.6. However, most of today's limited area forecasting models (LAM) are embedded in a global or basin scale model. The open boundary is then an interface between the LAM and the global model, which has a coarser grid size than the LAM. The solution within the limited area must therefore be nested together with the solution of the coarser mesh model without deteriorating the solution within the limited area. When information is passed on solely from the larger scale model to the limited area model, the technique is referred to as one-way nesting. An example of such a technique, widely used in meteorology, is presented in Sect. 7.5. Recently, two-way nesting techniques have also been suggested, in which information goes both ways. Details of the two-way nesting technique are postponed to Sect. 7.10.

It is common in today's models of atmospheres and oceans to replace the orthogonal Cartesian coordinate system by a non-orthogonal system, in which the vertical coordinate is replaced by a generalized vertical coordinate such as pressure, isopycnal, or terrain-following coordinates. This is the theme of Chap. 8. Section 8.1 explains in general terms how to transform from a Cartesian coordinate system to a new one. In Sect. 8.2, these "rules" are applied to the governing equation of a non-Boussinesq hydrostatic fluid to gain insight into how the equations transform. Finally, Sect. 8.3 provides an example in which the equation is transformed to a particular terrain-following σ-coordinate system.

Throughout this book, we stick to the mantra: "Make things as simple as possible, but no simpler" (Albert Einstein, 1879–1955). Thus, when developing numerical models to solve the diffusion, advection, and shallow water equations (Chaps. 4 through 6), they are simplified to one-dimensional equations. In reality, all numerical models of the atmosphere and oceans are three-dimensional in space. Hence, to gain insight into how to treat equations with more than one independent variable, Chap. 9 investigates the effect of including more than one-dimension in space. Section 9.1 studies the diffusion equation, and Sects. 9.2 and 9.3 the advection equation and the shallow water equations, respectively. Special consideration is given to the effect on numerical stability.

Chap. 10 presents several topics of a more advanced nature. For instance, Sect. 10.1 shows how to construct higher order spatial schemes. The advection equation is used as a sample equation, and special attention is given to the effect on stability, dispersion, and computational modes. Section 10.2 shows how to treat advection and diffusion when they are combined into an advection–diffusion equation, looking particularly at the effects on the stability of the scheme. Another topic discussed in Sect. 10.3 is nonlinearity, which is an important aspect of the motion of atmospheres and oceans. Again, focus is on the effect of nonlinearity on the stability. We discuss in particular the concept of nonlinear instability. Section 10.4 presents filtering methods like the Shapiro filter, used to avoid nonlinearity causing the model to blow up. Applying such a filter is analogous to adding an artificial eddy viscosity (diffusion term), but filtering has wider applications, because it may also be used to smooth out stable, but noisy results. Moreover, filtering is also an inherent part of the two-way nesting technique presented in Sect. 10.5, which is becoming more and more popular in atmosphere and ocean modelling.

Today's models are complex, consisting of more than a thousand executable statements. Only rarely are the codes making up numerical atmosphere and ocean models written by a single person. They are normally written by a group of people and over many years. Thus, an important aspect of maintaining a numerical model is to ensure its quality. The last chapter of this book is therefore devoted to procedures for assessing the performance of a model (Chap. 11). In particular, the notions of tuned, transportable, and robust models are discussed in Sect. 11.2. These pave the way for the introduction of the detailed quality assurance procedures presented in Sect. 11.3. The crucial elements of these procedures are *model verification*, *sensitivity analysis*, and *model validation*.

Finally, we stress that the use of numerical methods to solve atmospheric and oceanographic problems is to a large degree a "hands-on-experience". It is no use learning how to turn the governing equations into sound finite difference equations without learning how to develop a computer code (a set of instructions) to solve them, how to run them on the computer, and finally how to visualize the results. The first part concerns what is called *programming*. It involves writing the instructions, or "code", in some "language" that the computer understands. The programming language used today in most atmosphere and ocean modelling is FORTRAN. The reason is simply that, when compiled, there is no other language that provides the same speed. To give insight into the fundamentals of the language, a brief introduction to FORTRAN programming is given in Appendix A. To gain familiarity with programming in FORTRAN, the reader is encouraged to solve as many as possible of the problems at the end of each chapter in the sections called "Computer Problems". The reader is also encouraged to solve the exercises at the end of each chapter.

Some Historical Notes

In view of the nonlinear character of the equations they had to solve, it is hardly surprising that scientists and forecasters working in meteorology and oceanography so quickly embraced electronic computers as a tool. It all started in 1946 when the mathematician John von Neumann, then a well-known professor at Princeton University, approached the meteorologist Carl-Gustav Rossby[2] to organize a "Conference on Meteorology". The idea was to acquaint the meteorological community with the Electronic Numerical Integrator and Computer (ENIAC) and the Institute of Advanced Study (IAS) machines, and to solicit their advice and support in designing research strategies. The outcome of the conference was the Princeton Meteorological Project (1947–1953) which was managed by Dr. Jule G. Charney.[3] Among the participants were two young Norwegians, Arnt Eliassen[4] and Ragnar Fjørtoft.[5] The project successfully produced a 24-hour numerical weather prediction in less than 2 hours. This very first attempt to produce a numerical weather forecast was published in 1950 by Charney et al. (1950).

The Meteorology Project marked the beginning of the scientific field known as Numerical Weather Prediction (NWP). An important basis for the rapid development of NWP in the 1950s and 1960s was the deterministic paradigm stated by Vilhelm Bjerknes[6] at the turn of the century. In his famous 1904 paper Bjerknes, stated that (Bjerknes 1904):

> If it is true, as most scientists think, that the atmospheric state at any time can be developed from its earlier state using physical laws, then it follows that a necessary and sufficient condition for a rational solution to the problem of weather forecasting is a sufficiently accurate knowledge of the present atmospheric state, and a sufficiently accurate knowledge of the equations that govern the development of the atmosphere from one state to the next.

[2]Carl-Gustaf Arvid Rossby (1898–1957) was a Swedish-born American meteorologist. He was the first to explain the large-scale motions of the atmosphere in terms of fluid mechanics. He identified and characterized both the jet stream and the long waves in the westerlies that were later named Rossby waves.

[3]Jule Gregory Charney (1917–1981) was an American meteorologist. As part of his Ph.D. work, he developed a set of equations for calculating the large-scale motions of planetary-scale waves (the quasi-geostrophic vorticity equations). He gave the first convincing physical explanation for the development of midlatitude cyclones, known as the baroclinic instability theory.

[4]Arnt Eliassen (1915–2000) was a Norwegian meteorologist who pioneered the use of numerical analysis and computers for weather forecasting. His early work was done at the Institute for Advanced Study in Princeton, New Jersey, together with John von Neumann. He received the Carl-Gustaf Rossby Research Medal in 1964 and the prestigious Balzan Prize in 1996 "for his fundamental contributions to dynamic meteorology that have influenced and stimulated progress in this science during the past fifty years".

[5]Ragnar Fjørtoft (1913–1998) was an internationally recognized Norwegian meteorologist. He was part of the Princeton team that performed the first successful numerical weather prediction using the ENIAC electronic computer in 1950. He was also a professor of meteorology at the University of Copenhagen and director of the Norwegian Meteorological Institute.

[6]Vilhelm Friman Koren Bjerknes (1862–1951) was a Norwegian physicist and meteorologist who did much to develop the modern practice of weather forecasting.

Another important basis was the later attempt by Lewis Fry Richardson[7] to compute a 6-hour weather forecast by hand. He did this by first casting the governing equations into finite difference form using numerical methods (Richardson 1922). Afterwards, while serving with the Quaker ambulance unit in northern France during World War I, he solved the finite difference equations using only pen and paper.

Although Bjerknes did not mention it, the statement quoted above is also true when forecasting oceanic "weather", that is, the growth and fate of meanders, jets, and eddies, the latter being the ocean's high- and low-pressure systems. Oceanic lows and highs are, however, much smaller than their atmospheric counterparts. Hence, the power and capacity of the early computers was insufficient to make oceanic forecasts in the 1950s and early 1960s. However, as computer power and capacity grew,[8] computers and numerical methods were also more commonly used to solve the equations governing oceanic motion. The first published results from numerical experiments using a basin scale ocean model appeared in Bryan and Cox (1967).[9] This started the scientific field of Numerical Ocean Weather Prediction (NOWP) in the late 1960s. In fact, the capacity and power of today's computers are suitable for forecasting oceanic weather, at least for limited areas. Furthermore, it is interesting to note that NWP and NOWP are among the major scientific fields pushing computer technology to its very limits.

The atmosphere and the ocean form a coupled system exchanging momentum, heat, and moisture. This was recognized early on in climate modelling, and the very first climate models were thus coupled atmosphere–ocean models (an excellent review can be found in Edwards 2011). With the ever-growing capacity of computers, coupled atmosphere–ocean models are also developed with NWP/NOWP in mind (see, e.g., Warner et al. 2010). In this undertaking, sea ice and wave models are also taken into account. Thus, in the not too distant future, fully coupled models will probably be the standard prediction tool for making weather forecasts of atmospheres and oceans in one sweep. In light of this, anyone aspiring to become a meteorologist, an oceanographer, or a climatologist must have a sound knowledge of and insight into the fundamental methods used to develop reliable numerical methods for solving oceanographic and meteorological problems.

---

[7]Lewis Fry Richardson (1881–1953) was an English mathematician, physicist, meteorologist, psychologist, and pacifist who pioneered modern mathematical techniques of weather forecasting.

[8]Growth in computer power and capacity has been almost exponential since the 1940s.

[9]Kirk Bryan (Jr.) (b. 1929) is an American oceanographer who is considered to be the founder of numerical ocean modelling. Starting in the 1960s at the Geophysical Fluid Dynamics Laboratory, Bryan worked with a series of colleagues to develop numerical schemes for solving the equations of motion describing flow on a sphere. His work on these schemes led to the so-called Bryan–Cox code, used for many early simulations, and it also led to the Modular Ocean Model currently used by many numerical oceanographers and climate scientists.

Readers who are interested in a full account of the history, personalities, and ideas behind the successes of NWP and NOWP are encouraged to read the excellent book published recently by the mathematicians Ian Roulstone and John Norbury (Roulstone and Nordbury 2013).

Caveats and Concerns

It should be emphasized that the numerical solutions returned by the computer depend on the method used. The computer is very good at producing numbers based on the instructions it receives. The numbers may look reasonable, even when they provide a false solution. When developing a model code, or making use of an existing model code of the atmosphere and/or ocean, it is important to be able to judge whether the given numerical method yields the "true" solution to the PDEs. The essence of this book is about acquiring just such knowledge.

As already mentioned, most processes in the ocean and atmosphere are highly nonlinear, so more and more research in NWP and NOWP relies on sometimes hugely complex computer codes. It is a growing concern that many of these codes are written and amended by scientists who are not necessarily skilled programmers. This concern is corroborated in a statistical survey by Hannay et al. (2009), which states "the knowledge required to develop and use scientific software is primarily acquired from peers and through self-study, rather than from formal education and training". Even though the numerical methods may be sound, the codes themselves may be rather poorly written from a skilled programmer's point of view. Only rarely do these codes undergo rigorous testing. Hence, model codes may inadvertently contain errors that are potentially damaging to the results.

Another serious concern is that computers always produce results in terms of numbers. These numbers may look reasonable, but in reality be totally false due to the use of unsound numerical methods. Such results may even lead to wrong conclusions. In fact, there are examples in the literature where a numerical solution is interpreted as a new physical phenomenon, which is later shown to be a pure artifact produced by an incorrect numerical method. It is therefore important to understand why some methods are sound and some unsound for a specific problem. Likewise, it is important to acquire knowledge of the "quality" of the computations. In light of these concerns, a chapter is added at the end of the book to give insights into quality assurance procedures that can be used to assess the quality of a specific model (Chap. 11).

Many of the numerical methods employed today have been developed historically to solve atmospheric problems. Nevertheless, these methods also work when solving oceanographic problems, simply because the *dynamics* of the two systems are very similar. In a numerical context, there is therefore no need to treat meteorology and oceanography separately, in particular regarding the more fundamental methods. A further rationale is, as already mentioned, that atmospheres and oceans are inherently coupled. However, it should be kept in mind though that there are also numerical methods and techniques unique to each of them. In particular this concerns methods relevant to *thermodynamics*. However, the treatment of these issues goes beyond the scope of this book.

Oslo, Norway
March 2017

Lars Petter Røed

References

Bjerknes V (1904) Das Problem der Vettervorhersage, betrachtet vom Standpunkte der Mechanik und der Physik. Meteor Zeitschr 21:1–7

Bryan K, Cox MD (1967) A numerical investigation of the oceanic general circulation. Tellus 19:54–80, https://dx.doi.org/10.1111/j.2153-3490.1967.tb01459.x

Charney JG, Fjørtoft R, von Neumann J (1950) Numerical integration of the barotropic vorticity equation. Tellus 2:237–254

Edwards PN (2011) History of climate modeling. Wiley Interdisciplinary Rev Clim Change 2 (1):128–139, https://dx.doi.org/10.1002/wcc.95

Hannay JE, MacLeod C, Singer J, Langtangen HP, Pfahl D, Wilson G (2009), How do scientists develop and use scientific software? In: SECSE '09: Proceedings of the 2009 ICSE workshop on software engineering for computational science and engineering, pp 1–8, IEEE Computer Society, Washington, DC, USA, http://dx.doi.org/10.1109/SECSE.2009.5069155

Mesinger F, Arakawa A (1976) Numerical methods used in atmospheric models, GARP Publication Series No. 17, 64 pp, World Meteorological Organization, Geneva, Switzerland

Richardson LF (1922) Weather prediction by numerical process, 250 pp, Cambridge University Press, UK

Røed LP, Cooper CK (1986) Open boundary conditions in numerical ocean models. In: O'Brien J (ed) Advanced physical oceanographic numerical modelling, (Series C: Mathematical and Physical Sciences), vol 186, pp 411–436. D. Reidel Publishing Co., Dordrecht

Roulstone I, Nordbury J (2013) Invisible in the storm: the role of mathematics in understanding weather, 376 pp, Princeton University Press, New Jersey

Warner JC, Armstrong B, He R, Zambon JB (2010) Development of a coupled ocean–atmosphere–wave–sediment transport (COAWST) modeling system. Ocean Model 35 (3):230–244, http://dx.doi.org/10.1016/j.ocemod.2010.07.010

Acknowledgements

During my years as a scientist and oceanographer, the Department of Geosciences at the University of Oslo, Norway (MetOs) and the Norwegian Meteorological Institute (MET Norway) have provided me with a friendly and rewarding environment for modeling the ocean, from both the forecasting and the climate perspective. Without the many, sometimes heated, discussions with colleagues at these two institutes, this book would not have become what it is. My early years as a scientist were, however, spent at the Geophysical Institute, University of Bergen, where I was taken on as a fellow by Prof. Martin Mork (1933–2017). I would like to take this opportunity to express my appreciation and sincere thanks to him for introducing me to, and guiding me through, the fascinating world of oceanography and meteorology which have captivated me ever since.

This book draws its inspiration from lectures delivered by Prof. Dr. James J. O'Brien (1935–2016) at the Florida State University. In fact, some of the material covered in this book is based on my notes when I followed his lectures as far back as 1980. I am indebted to Jim for his lasting friendship and boundless enthusiasm for this book project. I wish he could have lived to see it through.

The material gathered for this book is based on the lecture notes I have compiled and amended over the years while teaching masters students the fundamental numerical methods necessary to put atmospheres and oceans on computers. I am indebted to the many students and colleagues who have pointed out misprints and errors in earlier versions of my lecture notes over the years. I would also like to thank Dr. Thor Erik Nordeng for helping me to compile material regarding the atmosphere. In this respect, my gratitude is also extended to Prof. Arne M. Bratseth (1949–2004). Some of the material covered in this book is influenced by his lecture notes on "Numerical Atmosphere Models". I am also grateful to Mr. Gunnar Wollan, University of Oslo, for letting me include a modified and shortened version of his notes entitled "A Fortran 2003 introduction by examples" as an appendix.

This book is also influenced by the many creative scientists and researchers within the international geophysical fluid dynamics and fluid mechanics communities that I have had the opportunity to meet and befriend. I am also indebted to the mathematical professors and physicists I encountered during my undergraduate and graduate studies in mathematics, physics, and fluid mechanics. Hopefully, their many insights into the way general mathematical statements lead to deep physical understanding are reflected in these pages. In this respect, I extend special gratitude to Einar Høiland, Enok Palm, Eivind Riis, and Vetle Lauvstad.

I would also like to take this opportunity to thank Stephen N. Lyle[10] for painstakingly going through the manuscript and correcting my English. It certainly improved the readability of the text and took it to a new level. In addition, he helped in many instances to clarify passages that might otherwise have puzzled the reader. My appreciation also goes to two anonymous reviewers who reviewed the original manuscript, and helped me to improve the text.

Last, but not least, I would like to thank my wife for her enduring patience and support over the years. Without it, this book could never have been completed.

[10]http://www.stephenlyle.org/.

Contents

Governing Equations and Approximations

1

The purpose of this chapter is to familiarize the reader with the equations governing the motion of atmospheres and the oceans. It introduces the well-known Boussinesq and hydrostatic approximations, and outlines the way the shallow water equations and the quasi-geostrophic equations follow from these approximations. The importance of boundary conditions is also emphasized. Readers who are familiar with these equations, approximations, and conditions may skip this chapter without loss of continuity.

1.1 Governing Equations

In the atmosphere and ocean, and in geophysical fluids in general, e.g., seas and lakes, the most prominent dependent variables are the three components u, v, and w of the three-dimensional velocity vector[1] $\mathbf{v}$, pressure p, density ρ, and (potential) temperature[2] θ. In addition, humidity q and cloud liquid water content q_{L} must be included among the prominent variables in the atmosphere, while the salinity s must be included in the ocean. To determine the unknowns, an equal number of equations are needed. Since they determine the time evolution of the motion of atmospheres and oceans, they are normally referred to as the *governing equations*.

Among the above variables, only the velocity is a vector. The remaining variables, commonly referred to as the state variables, are all scalars. The state variables, except density and pressure, are all examples of what are referred to as tracers. Other examples of tracers are any dissolved chemical component or substance. Tracers that influence the motion through the density, such as salinity, temperature, and humidity,

[1]Velocity is normally referred to as wind in the atmosphere and current in the ocean, seas, and lakes.
[2]In the following, bold upright fonts, e.g., $\mathbf{u}$, denote vectors, while bold special italic fonts, e.g., $\boldsymbol{\mathcal{U}}$, denote tensors.

L. P. Røed, *Atmospheres and Oceans on Computers*,
Springer Textbooks in Earth Sciences, Geography and Environment,
https://doi.org/10.1007/978-3-319-93864-6_1

are commonly referred to as *active* tracers. Tracers that do not influence the motion, like for instance dissolved chemical components, are *passive* tracers.

As is common in the mathematical formulation of physical problems, the governing equations are developed by exploiting conservation principles, in our case the conservation of mass, momentum, internal energy, and tracer content. Regarding atmospheres and oceans, and geophysical fluids in general, the governing equations in their non-Boussinesq form are (Gill 1982; Griffies 2004)

$$\partial_t \rho + \nabla \cdot (\rho \mathbf{v}) = 0 \,, \tag{1.1}$$

$$\partial_t (\rho \mathbf{v}) + \nabla \cdot (\rho \mathbf{v}\mathbf{v}) = -2\rho \boldsymbol{\Omega} \times \mathbf{v} - \nabla p + \rho \mathbf{g} - \nabla \cdot (\rho \mathscr{F}_{\mathrm{M}}) \,, \tag{1.2}$$

$$\partial_t (\rho C_i) + \nabla \cdot (\rho C_i \mathbf{v}) = -\nabla \cdot (\rho \mathbf{F}_i) + \rho S_i \,, \quad i = 1, 2, \ldots \,, \tag{1.3}$$

$$\rho = \rho(p, \theta, \gamma) \,. \tag{1.4}$$

Here, C_i represents the concentration of any tracer including temperature, humidity, and/or salinity, $\boldsymbol{\Omega}$ is the Earth's rotation rate, $\mathbf{g}$ is the gravitational acceleration, S_i is the tracer source, if any, and the tensor $\mathscr{F}_{\mathrm{M}}$ and the vector $\mathbf{F}_i$ represent fluxes due to turbulent mixing of momentum and tracers, respectively. As usual, ∂_t, ∂_x, ∂_y, and ∂_z indicate partial derivatives with respect to the subscript. Thus, $\partial_t \rho$ denotes the time tendency or time rate of change of the density. Finally, ∇ is the three-dimensional del-operator defined by

$$\nabla = \mathbf{i}\partial_x + \mathbf{j}\partial_y + \mathbf{k}\partial_z \,. \tag{1.5}$$

Of the above equations, (1.1) is the conservation of mass, (1.2) is the conservation of momentum, and (1.3) is the conservation of the tracer. The equation of state (1.4) relates density and pressure to the active tracers alone, so that γ is replaced by some measure of humidity in the atmosphere and by the salinity in the ocean.

In the (dry) atmosphere, the equation of state is linear and follows the ideal gas law, that is,

$$p = \rho R \theta \,, \tag{1.6}$$

where R is the gas constant[3] and θ is the potential temperature.[4] In contrast, the equation of state for the ocean is highly nonlinear and cannot be expressed in a formal, closed form (Fofonoff 1985; McDougall et al. 2013).

1.2 Boundary and Initial Conditions

To solve (1.1)–(1.4), conditions at the spatial boundaries of the domain need to be specified. The number of conditions necessary depends on the number of integration constants (Sect. 2.4). Such conditions are referred to as *boundary conditions*. In addition, the state of the ocean and/or atmosphere at some given time also needs to

[3] $R = 287.04\,\mathrm{J\,kg^{-1}\,K^{-1}}$.

[4] The potential temperature of a parcel of fluid at pressure p is the temperature that the parcel would acquire if adiabatically brought to a standard reference pressure p_0.

be specified (Bjerknes 1904). The latter boundary condition is commonly referred to as the *initial conditions*, a nomenclature adopted here.

The boundary conditions are of two main types: *dynamic boundary conditions* and *kinematic boundary conditions*. Some of the bounding surfaces of the volume containing the ocean or the atmosphere are material surfaces, e.g., the interface between the atmosphere and the ocean. Since a material surface is one that consists of the same particles at all times, the kinematic boundary condition states that any particle at the material surface at some instant of time will remain there forever. It thus requires that the particle move with the surface. The dynamic boundary conditions associated with a material surface require that the pressure and fluxes are continuous across the surface, so that no acceleration takes place there.

As an example, consider a system consisting of the atmosphere on top of the ocean. Let $\eta = \eta(x, y, t)$ denote the deviation of the atmosphere/ocean interface away from its equilibrium level at $z = 0$, and let $z = -D$, where $D = D(x, y)$ is the equilibrium depth of the ocean. Then, the kinematic boundary condition at the interface is

$$w = \partial_t \eta + \mathbf{u} \cdot \nabla_{\mathrm{H}} \eta \,, \quad \text{at } z = \eta \,. \tag{1.7}$$

Here $\mathbf{u}$ and w are, respectively, the horizontal and vertical components of the three-dimensional velocity $\mathbf{v}$, while $\nabla_{\mathrm{H}} = \mathbf{i}\partial_x + \mathbf{j}\partial_y$ is the horizontal component of the three-dimensional del-operator (1.5). This condition ensures that a particle at the surface will stay there and move with it. The dynamic condition at the interface is

$$p_{\mathrm{A}} = p_{\mathrm{O}} \,, \quad \text{at } z = \eta \,, \tag{1.8}$$

where p_{A} denotes the atmospheric pressure, and p_{O} the oceanic pressure. Consequently, to avoid heat accumulating at the interface, the heat flux must be continuous there as well. If not, pressure differences would be avoidable and an acceleration would ensue.

The kinematic boundary condition at the bottom of the ocean is similar to (1.7), that is,

$$w = -\mathbf{u} \cdot \nabla_{\mathrm{H}} D \,, \quad \text{at } z = -D \,, \tag{1.9}$$

where it is assumed that the bottom $z = -D$ is a material surface, and that D is time independent. Hence, (1.9) dictates that the bottom is impermeable, so that there is no flow across it. This is ensured by requiring $\mathbf{n} \cdot \mathbf{v} = 0$ at $z = -D$, where $\mathbf{n}$ is the unit vector perpendicular to the bottom, which leads to (1.9).

1.3 Hydrostatic Approximation

In the atmosphere and ocean, the horizontal scales of the dominant motions are large compared to the vertical scale. As a consequence the vertical acceleration, $\mathrm{D}w/\mathrm{d}t$, is small in comparison to, e.g., the gravitational acceleration ρg. Consequently, when describing the large-scale motion, the vertical momentum equation is assumed to be reduced to a balance between the gravitational acceleration and the vertical pressure gradient. When solving this demoted system, the model is said to be *hydrostatic*, implying that the motion satisfies the *hydrostatic approximation*.

To illustrate the hydrostatic approximation, consider first the vertical component of the momentum equation in full[5]:

$$\partial_t(\rho w) + \nabla \cdot (\rho \mathbf{v} w) = -\partial_z p - \rho g - \nabla \cdot (\rho \mathbf{F}_{\mathrm{M}}^{\mathrm{V}}) \; , \tag{1.10}$$

where $\mathbf{F}_{\mathrm{M}}^{\mathrm{V}}$ is the vertical vector component of the mixing tensor $\mathscr{F}_{\mathrm{M}}$. The hydrostatic assumption implies that the terms on the left-hand side of (1.10) are small compared to the gravitational acceleration and may safely be neglected. Since in most cases, the dominant energetic part of the motion consists of long waves in shallow water, the horizontal scale is significantly longer than the vertical scale. The exceptions are cases that include steep topography and/or strong convection. Under these circumstances, one has to revert to non-hydrostatic equations or parameterizations. Furthermore, since the vertical motion is small compared to the horizontal motion, the friction term may also be neglected. Thus, (1.10) reduces to

$$\partial_z p = -\rho g \; , \tag{1.11}$$

which is the *hydrostatic equation*. The word "hydrostatic" is used since a fluid at rest in the gravitational field satisfies precisely this equation. Moreover, since a fluid at rest is static, it is referred to as hydrostatic.

When the hydrostatic approximation is valid, the momentum equation is normally split into its vertical and horizontal components. The vertical component is the hydrostatic equation (1.11), while the horizontal component is

$$\partial_t(\rho \mathbf{u}) + \nabla_{\mathrm{H}} \cdot (\rho \mathbf{u}\mathbf{u}) + \partial_z(\rho w \mathbf{u}) + \rho f \mathbf{k} \times \mathbf{u} = -\nabla_{\mathrm{H}} p + \partial_z \boldsymbol{\tau} - \nabla_{\mathrm{H}} \cdot (\rho \mathscr{F}_{\mathrm{M}}^{\mathrm{H}}) \; , \tag{1.12}$$

where $f = 2\Omega \sin\phi$ is the Coriolis parameter, ϕ is the latitude, and Ω is the Earth's rotation rate. $\mathscr{F}_{\mathrm{M}}^{\mathrm{H}}$ is the horizontal component of the three-dimensional flux tensor $\mathscr{F}_{\mathrm{M}}$ due to turbulent mixing, while $\boldsymbol{\tau}$ is the vertical component also commonly referred to as the vertical shear stress. Moreover, the third term on the left-hand side and the second term on the right-hand side of (1.12) are singled out and separated from their horizontal counterparts.

The tracer equation

$$\partial_t(\rho C_i) + \nabla_{\mathrm{H}} \cdot (\rho \mathbf{u} C_i) + \partial_z(\rho w C_i) = -\partial_z(\rho F^{\mathrm{V}}) - \nabla_{\mathrm{H}} \cdot (\rho \mathbf{F}_i^{\mathrm{H}}) + \rho S_i \; , \quad i = 1, 2, \ldots \; , \tag{1.13}$$

is left unchanged, but as in the momentum equation the third term on the left-hand side, associated with the vertical acceleration, is singled out and written separately. Similarly, the first term on the right-hand side associated with the vertical component of the turbulent mixing (F^{V}) is also separated from the second term on the right-hand side associated with the horizontal component of the turbulent mixing ($\mathbf{F}_i^{\mathrm{H}}$).

[5] In a Cartesian coordinate system fixed to the Earth's surface, the vertical component of the Coriolis force is small compared to the gravitational pull. The former is therefore dropped in (1.10).

1.4 Boussinesq Approximation

Another common approximation, used particularly in ocean models, is the *Boussinesq approximation*. The fundamental basis for this approximation is that in many cases, the atmosphere and the ocean behave as if they were incompressible fluids. Under these circumstances, any parcel of fluid conserves its volume, even if the parcel is heated. Thus, for the Boussinesq approximation to be valid, the change in density for any parcel of fluid must be small compared to the density itself, that is,

$$\frac{1}{\rho}\frac{\mathrm{D}\rho}{\mathrm{d}t} = \frac{\mathrm{D}\ln\rho}{\mathrm{d}t} \approx 0\,, \tag{1.14}$$

where the operator $\mathrm{D}/\mathrm{d}t$ is the material derivative,[6] defined by

$$\frac{\mathrm{D}}{\mathrm{d}t} = \partial_t + \mathbf{v}\cdot\nabla\,. \tag{1.15}$$

Under the Boussinesq approximation, the approximation (1.14) is taken as an equality. The mass conservation equation (1.1) then reduces to

$$\nabla\cdot\mathbf{v} = 0\,. \tag{1.16}$$

Thus, under the Boussinesq approximation, volume is conserved rather than mass (e.g., Griffies 2004, p. 17).

The Boussinesq approximation is widely used in ocean models, but rarely in atmospheric models. The reason is that application of this approximation effectively filters out the acoustic waves, while retaining pressure changes in response to density changes. In the ocean, the phase speed of acoustic waves is about $1\,500\,\mathrm{m\,s^{-1}}$, compared with about $300\,\mathrm{m\,s^{-1}}$ in the atmosphere. From a numerical point of view, it is advantageous to filter out the acoustic waves in the ocean. Because they are fast, they put a severe restriction on the time step to avoid the model blowing up (Sect. 4.3). Including the acoustic waves in an ocean model therefore dramatically increases the wall clock time spent to perform even relatively short time integrations.

Employing the Boussinesq approximation in ocean models has one major disadvantage though. One particularly pertinent example is the expected change in sea level, or ocean volume, due to global warming. If the ocean is uniformly heated, the equation of state implies that the density will decrease. For a non-Boussinesq ocean, which is mass conserving, the response to the decrease in density is to expand its volume. Hence the sea level rises. In contrast, a Boussinesq ocean, which conserves volume, responds to heating by a decrease in the density achieved by losing mass. Obviously the latter is highly unrealistic.

In summary, use of the Boussinesq approximation implies that the fluid is volume conserving rather than mass conserving. The implication regarding the governing equations is that one may treat the density as a constant everywhere, except when it appears together with the gravitational acceleration. This greatly simplifies them.

[6] Also referred to as the individual derivative.

For instance, combining the Boussinesq approximation and the hydrostatic approximation leads to

$$\nabla_{\mathrm{H}} \cdot \mathbf{u} + \partial_z w = 0\,, \tag{1.17}$$

$$\partial_z p = -\rho g\,, \tag{1.18}$$

$$\partial_t \mathbf{u} + \nabla_{\mathrm{H}} \cdot (\mathbf{u}\mathbf{u}) + \partial_z (w\mathbf{u}) + f\mathbf{k} \times \mathbf{u} = -\rho_0^{-1} \nabla_{\mathrm{H}} p + \rho_0^{-1} \partial_z \boldsymbol{\tau} - \nabla_{\mathrm{H}} \cdot (\mathscr{F}_{\mathrm{M}}^{\mathrm{H}})\,, \tag{1.19}$$

$$\partial_t C_i + \nabla_{\mathrm{H}} \cdot (C_i \mathbf{u}) + \partial_z (C_i w) = -\partial_z F^{\mathrm{V}} - \nabla_{\mathrm{H}} \cdot \mathbf{F}_i^{\mathrm{H}} + S_i\,, \tag{1.20}$$

together with the equation of state (1.4). When applying the hydrostatic and Boussinesq approximation, the time rate of change of the vertical velocity and the density disappears. Like pressure, they are reduced to so-called *diagnostic variables*. In contrast, the time rates of change of the horizontal velocity $\mathbf{u}$ and the tracers C_i are kept as so-called *prognostic variables*, and their time evolution is determined by the *prognostic equations* (1.19) and (1.20), respectively.

1.5 Shallow Water Equations

A very common reduced set of equations in meteorology and oceanography is the so-called shallow water equations. To arrive at this set, the hydrostatic and Boussinesq approximations are assumed. Hence, the starting point is (1.17)–(1.20) together with the equation of state (1.4). It is also assumed that the density is uniform in time and space wherever it appears. Thus, it is assumed that $\rho = \rho_0$, where ρ_0 is a constant. In doing so, the equation of state (1.4) and the tracer equations (1.20) are obsolete. Thus, applying these assumptions leads to the simplified set

$$\nabla_{\mathrm{H}} \cdot \mathbf{u} + \partial_z w = 0\,, \tag{1.21}$$

$$\partial_z p + \rho_0 g = 0\,, \tag{1.22}$$

$$\partial_t \mathbf{u} + \nabla_{\mathrm{H}} \cdot (\mathbf{u}\mathbf{u}) + \partial_z (w\mathbf{u}) + f\mathbf{k} \times \mathbf{u} = -\rho_0^{-1} \nabla_{\mathrm{H}} p + \rho_0^{-1} \partial_z \boldsymbol{\tau} - \nabla_{\mathrm{H}} \cdot \mathscr{F}_{\mathrm{M}}^{\mathrm{H}}\,. \tag{1.23}$$

Since the density is assumed to be uniform (constant), (1.22) may be integrated from an arbitrary height/depth z to a reference surface $z = \eta(x, y, t)$, which leads to

$$p = p_{\mathrm{s}} + g\rho_0(\eta - z)\,, \tag{1.24}$$

where p_{s} is the pressure at the reference surface. In the atmosphere, this reference surface is normally the top of the atmosphere where the pressure is zero. In the ocean, it is the interface between the atmosphere and the ocean. Thus, in the ocean η is identical to the deviation of the sea surface from its equilibrium level $z = 0$, while in the atmosphere it is equivalent to the height of the atmosphere above the level $z = 0$. The horizontal pressure gradient entering (1.19) is therefore given by

$$\rho_0^{-1} \nabla_{\mathrm{H}} p = g \nabla_{\mathrm{H}} \eta\,, \tag{1.25}$$

which says that the pressure gradient is independent of the depth z.

Using (1.25), the set (1.21)–(1.23) may be simplified even further. Integrating (1.21) and (1.23) from the bottom to the top yields

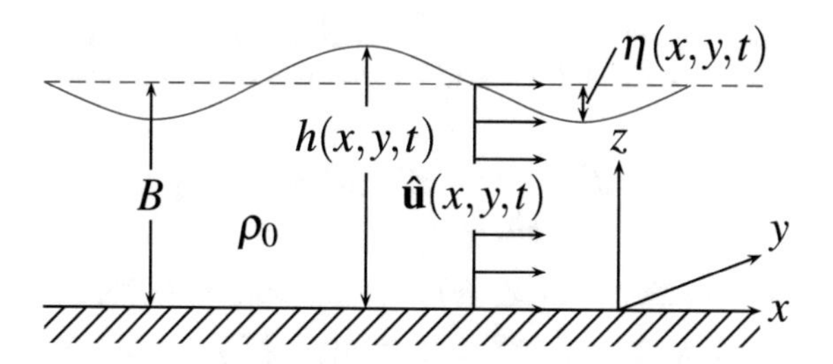

Fig. 1.1 One-layer model of the atmosphere or ocean. Here $h = \eta + B$ is the total depth of a fluid column, where $B = B(x, y)$ is the equilibrium depth and $\eta = \eta(x, y, t)$ is the deviation of the top reference surface away from its (level) equilibrium position. The mean velocity of the fluid column is $\hat{\mathbf{u}}$, while the uniform density of the layer is ρ_0

$$\partial_t \mathbf{U} + \nabla_{\mathrm{H}} \cdot \left(\frac{\mathbf{U}\mathbf{U}}{h} \right) + f\mathbf{k} \times \mathbf{U} = -gh\nabla_{\mathrm{H}}(h + B) + \rho_0^{-1}(\boldsymbol{\tau}_{\mathrm{s}} - \boldsymbol{\tau}_{\mathrm{b}}) + \mathbf{X}\,, \tag{1.26}$$

$$\partial_t h + \nabla_{\mathrm{H}} \cdot \mathbf{U} = 0\,, \tag{1.27}$$

where

$$\mathbf{U} = \int_B^\eta \mathbf{u}\, \mathrm{d}z \tag{1.28}$$

is the volume flux through a fluid column. As shown in Fig. 1.1, $h = \eta + B$ is the total height of a fluid column, where B denotes the ground in the atmosphere and the bottom in the ocean. Furthermore, the kinematic boundary conditions (1.7) and (1.9) and the dynamic boundary condition[7] $p = 0$ at $z = \eta$ are used to derive (1.26) and (1.27). Moreover, $\boldsymbol{\tau}_{\mathrm{s}}$ and $\boldsymbol{\tau}_{\mathrm{b}}$ are the turbulent vertical momentum fluxes at the top and bottom of the fluid column, respectively, and $\mathbf{X}$ is what is left of the horizontal momentum fluxes when integrated vertically from bottom to top. Since η is independent of z, the quantity $\partial_t \eta$ is replaced by $\partial_t h$ in (1.27). Finally, the term arising from the approximation

$$\int_B^\eta \nabla_{\mathrm{H}} \cdot (\mathbf{u}\mathbf{u})\mathrm{d}z \approx \nabla_{\mathrm{H}} \cdot \left(\frac{\mathbf{U}\mathbf{U}}{h} \right) \tag{1.29}$$

is absorbed into the last term $\mathbf{X}$ on the right-hand side of (1.26). The set consisting of (1.26) and (1.27) is commonly referred to as the *shallow water equations*. This form of the shallow water equations is referred to as the flux form.

[7]In the ocean, $p_{\mathrm{s}} = p_{\mathrm{a}}$ at $z = \eta$, where p_{a} is the atmospheric pressure at the atmosphere/ocean interface. Setting $p_{\mathrm{a}} = 0$ neglects the effect of the atmospheric pressure on the oceanic motion. It is important to include atmospheric pressure forcing when forecasting storm surges.

The shallow water equations may also be written in non-flux form. Defining the depth average velocity by $\hat{\mathbf{u}} = \mathbf{U}/h$, and replacing $\mathbf{U}$ by $\hat{\mathbf{u}}$ in (1.26) and (1.27), leads to

$$\partial_t(h\hat{\mathbf{u}}) + \nabla_{\mathrm{H}} \cdot (h\hat{\mathbf{u}}\hat{\mathbf{u}}) + f\mathbf{k} \times h\hat{\mathbf{u}} = -gh\nabla_{\mathrm{H}}(h+B) + \rho_0^{-1}(\tau_{\mathrm{s}} - \tau_{\mathrm{b}}) + \mathbf{X}\,, \quad (1.30)$$

$$\partial_t h + \nabla_{\mathrm{H}} \cdot (h\hat{\mathbf{u}}) = 0\,. \quad (1.31)$$

Moreover, the acceleration terms $\partial_t(h\hat{\mathbf{u}}) + \nabla_{\mathrm{H}} \cdot (h\hat{\mathbf{u}}\hat{\mathbf{u}})$ in (1.30) may be written

$$\begin{aligned} \partial_t(h\hat{\mathbf{u}}) + \nabla_{\mathrm{H}} \cdot (h\hat{\mathbf{u}}\hat{\mathbf{u}}) &= h\left(\partial_t\hat{\mathbf{u}} + \hat{\mathbf{u}} \cdot \nabla_{\mathrm{H}}\hat{\mathbf{u}}\right) + \hat{\mathbf{u}}\left[\partial_t h + \nabla_{\mathrm{H}} \cdot (h\hat{\mathbf{u}})\right] \\ &= h\left(\partial_t\hat{\mathbf{u}} + \hat{\mathbf{u}} \cdot \nabla_{\mathrm{H}}\hat{\mathbf{u}}\right)\,, \end{aligned} \quad (1.32)$$

where use is made of (1.31) to arrive at the last equals sign. Thus substituting (1.32) into (1.30) leads to

$$\partial_t\hat{\mathbf{u}} + \hat{\mathbf{u}} \cdot \nabla_{\mathrm{H}}\hat{\mathbf{u}} + f\mathbf{k} \times \hat{\mathbf{u}} = -g\nabla_{\mathrm{H}}(h+B) + \frac{\tau_{\mathrm{s}} - \tau_{\mathrm{b}}}{\rho_0 h} + \frac{\mathbf{X}}{h}\,, \quad (1.33)$$

$$\partial_t h + \nabla_{\mathrm{H}} \cdot \left(h\hat{\mathbf{u}}\right) = 0\,. \quad (1.34)$$

When the shallow water equations are written in their non-flux form (1.33) and (1.34), the mass conservation equation (1.34) becomes nonlinear. This is in contrast to the linear mass conservation equation (1.27) written in flux form.

1.6 Quasi-geostrophic Equations

Another common set of reduced equations are the quasi-geostrophic equations. For details, the reader is referred to Pedlosky (1979) or Stern (1975). The derivation here essentially follows the latter.

The starting point is the shallow water equations in non-flux form, namely (1.33) and (1.34). Neglecting the forcing terms on the right-hand side of (1.33), they take the form

$$\frac{\mathrm{D_H}\mathbf{u}}{\mathrm{d}t} + f\mathbf{k} \times \mathbf{u} = -g\nabla_{\mathrm{H}}(h+B)\,, \quad (1.35)$$

$$\frac{1}{h}\frac{\mathrm{D_H}h}{\mathrm{d}t} + \nabla_{\mathrm{H}} \cdot \mathbf{u} = 0\,, \quad (1.36)$$

where the circumflexes are dropped for clarity and the operator

$$\frac{\mathrm{D_H}}{\mathrm{d}t} = \partial_t + \mathbf{u} \cdot \nabla_{\mathrm{H}} \quad (1.37)$$

is the two-dimensional version of (1.15). Furthermore, it is assumed that the acceleration $\mathrm{D_H}\mathbf{u}/\mathrm{d}t$ is small compared to the Coriolis acceleration. Defining the *Rossby number* by

$$R \equiv \frac{|\mathrm{D_H}\mathbf{u}/\mathrm{d}t|}{|f\mathbf{k} \times \mathbf{u}|}\,, \quad (1.38)$$

this is equivalent to assuming that $R \ll 1$.[8] Under these circumstances, (1.35) reduces to

$$f\mathbf{k} \times \mathbf{u} = -g\nabla_{\mathrm{H}} h \,, \tag{1.39}$$

where the bottom gradient term $\nabla_{\mathrm{H}} B$ is assumed to be of $\mathcal{O}(R)$ as well. Equation (1.39) describes a balance between the Coriolis term and the pressure term, a balance referred to as the *geostrophic balance*, and (1.39) is therefore referred to as the *geostrophic equation* or the *thermal wind equation*. Solving for the horizontal velocity component leads to

$$\mathbf{u} = \frac{g}{f}\mathbf{k} \times \nabla_{\mathrm{H}} h \,. \tag{1.40}$$

To arrive at the geostrophic equation, terms of $\mathcal{O}(R)$ or higher are neglected. Therefore, (1.40) is viewed as a first approximation to an expansion in terms of the Rossby number R. It is therefore alternatively written

$$\mathbf{u} = \frac{g}{f}\mathbf{k} \times \nabla_{\mathrm{H}} h + \mathcal{O}(R) \,. \tag{1.41}$$

Obviously, (1.39) provides no information about the space–time variations in either the velocity field or the pressure field. To obtain information on dynamics to $\mathcal{O}(R)$ one has to look elsewhere, for instance, the relevant asymptotic form of the vorticity equation. The vorticity equation is derived from the nonlinear shallow water equations (1.35) and (1.36). Operating on (1.35) using the operator $\mathbf{k} \cdot \nabla_{\mathrm{H}} \times$, and substituting for $\nabla_{\mathrm{H}} \cdot \mathbf{u}$ from (1.36), leads to

$$\frac{\mathrm{D_H}}{\mathrm{d}t}\left(\frac{\zeta + f}{h}\right) = 0 \,, \tag{1.42}$$

where ζ is the *relative vorticity* defined by

$$\zeta \equiv \mathbf{k} \cdot \nabla_{\mathrm{H}} \times \mathbf{u} \,, \tag{1.43}$$

and $\zeta + f$ is the absolute vorticity. The quantity $(\zeta + f)/h$ is the *potential vorticity* of a barotropic fluid, since the density is assumed to be constant.[9] If L is a typical lateral (horizontal) scale of $\mathbf{u}$ and U a typical speed, then $|\mathbf{u} \cdot \nabla_{\mathrm{H}}\mathbf{u}| \sim U^2/L$. In view of (1.38), it follows that a *necessary* condition for $R \ll 1$ to be satisfied is

$$\frac{U}{fL} \ll 1 \,. \tag{1.44}$$

The condition is not *sufficient* since the remaining acceleration term in $\mathrm{D_H}\mathbf{u}/\mathrm{d}t$ is the local time rate of change $\partial_t \mathbf{u}$, which might be comparable to the Coriolis term $f\mathbf{k} \times \mathbf{u}$. However, requiring the initial condition to satisfy (1.39), the smallness of $\partial_t \mathbf{u}$ compared to the Coriolis acceleration depends on the smallness of $\mathbf{u} \cdot \nabla_{\mathrm{H}}\mathbf{u}$. Under these circumstances, (1.44) may be safely regarded as being equivalent to requiring $R \ll 1$.

[8] $R \ll 1$ is commonly referred to as the quasi-geostrophy condition [see (1.38)].

[9] The potential vorticity may also be derived in a similar fashion for a baroclinic fluid, but the mathematical expression is different (e.g., Griffies 2004, p. 70).

To proceed, we note, using (1.43), that

$$\frac{|\zeta|}{f} = \frac{U}{fL} . \tag{1.45}$$

Hence, (1.44) requires the relative vorticity to be smaller by a factor of R than the planetary vorticity f. Furthermore, the variation in the layer thickness h, obtained from (1.39), is

$$h - H_m \sim \frac{fL}{g}U \quad \text{or} \quad \frac{h - H_m}{H_m} \sim RF^2 , \tag{1.46}$$

where H_m is the mean layer thickness,

$$F \equiv \left(\frac{L^2}{L_R}\right)^{1/2} , \tag{1.47}$$

and

$$L_R = gH_m/f^2 \tag{1.48}$$

is the Rossby deformation radius. Assuming $F \sim \mathcal{O}(1)$ or less, which is equivalent to assuming that L is not large compared to Rossby's deformation radius, the layer thickness variation in (1.46) is small to the same order as the ratio ζ/f of the relative vorticity ζ to the planetary vorticity f.

The potential vorticity (1.42) may be rewritten to yield

$$\frac{\zeta + f}{h} = \frac{f}{H_m} \frac{1 + \zeta/f}{1 + \dfrac{h - H_m}{H_m}} , \tag{1.49}$$

and using the above, it leads to

$$\frac{\zeta + f}{h} = \frac{f}{H_m}\left[2 + \frac{\zeta}{f} - \frac{h}{H_m} + \mathcal{O}(R^2)\right] . \tag{1.50}$$

Substituting this into the potential vorticity equation (1.42) leads to

$$(\partial_t + \mathbf{u} \cdot \nabla_H)\left(\frac{\zeta}{f} - \frac{h}{H_m}\right) + \mathcal{O}(R^3) = 0 . \tag{1.51}$$

Since the (dimensionless) magnitudes of the acceleration terms $\partial_t \mathbf{u}$ and $\mathbf{u} \cdot \nabla_H \mathbf{u}$ are $\mathcal{O}(R)$, the leading term in (1.51) is $\mathcal{O}(R^2)$. Thus, the fractional errors in the asymptotic vorticity equation

$$(\partial_t + \mathbf{u} \cdot \nabla_H)\left(\frac{\zeta}{f} - \frac{h}{H_m}\right) = 0 , \tag{1.52}$$

and in the asymptotic momentum equation (1.39), are both of $\mathcal{O}(R)$. Hence, substituting $\mathbf{u}$ from (1.40) wherever the latter appears in (1.52), the resulting differential equation for the layer thickness (or pressure) h is also asymptotic when $R \ll 1$. It is therefore permissible to evaluate the velocity and the relative vorticity in (1.52) using the geostrophic equation (1.40). This leads to

$$\zeta = \frac{g}{f}\nabla_H^2 h , \tag{1.53}$$

which yields, when substituted in (1.52),

$$\left[\partial_t + \frac{g}{f}(\mathbf{k} \times \nabla_{\mathrm{H}} h) \cdot \nabla_{\mathrm{H}}\right]\left(\nabla_{\mathrm{H}}^2 h - L_{\mathrm{R}}^{-1} h\right) = 0 . \tag{1.54}$$

This equation together with the geostrophic equation (1.39) are commonly referred to as the *quasi-geostrophic equations* (QG equations). Solving the quasi-geostrophic vorticity equation (1.54) then provides the layer thickness h at an arbitrary time $t > 0$ for a specified initial distribution. Recall that the latter must be in geostrophic balance. Thereafter the velocity $\mathbf{u}$ is computed by solving the geostrophic equation (1.40). The resulting solution is then almost geostrophic, but not quite, which is why the name quasi-geostrophic is used. Recall that it is only under very strict conditions that these equations are valid. For instance, since $|\nabla_{\mathrm{H}} B|$ is assumed to be of $\mathcal{O}(R)$, the ground or bottom slopes must be very gentle.

1.7 Summary and Remarks

In this chapter, we have discussed the full three-dimensional differential equations that govern the motions of atmospheres and oceans. These are the starting point for constructing the numerical models used today to forecast atmospheric and oceanic weather, to make climate projections, and to study the circulation of seas and lakes. In short, these are the equations that govern all fluids on Earth, when described in a coordinate system fixed relative to the Earth.

Section 1.1 presents the governing equations in all their complexity and nonlinearity. Nevertheless, from a mathematical point of view, they are just a set of equations belonging to the class of partial differential equations. Section 1.2 followed up by discussing boundary and initial conditions. It should be emphasized that any PDE is valid only within a domain, whether it is global or covers a limited geographical area. It is the conditions at the boundary of the domain that dictate what the solutions are. When solving a PDE, it is integrated in time and space to yield a certain number of unknown integration constants. The only way to determine these constants is via the boundary conditions. Thus, it is absolutely necessary that the number of boundary conditions should match the number of integration constants. If not, the existence and uniqueness of the solution are not ensured.

Furthermore, Sects. 1.4 and 1.3 discussed the hydrostatic and Boussinesq approximations leading to the development of conceptually simpler models, such as the shallow water equations (Sect. 1.5) and the quasi-geostrophic equations (Sect. 1.6). It should be kept in mind that each step in the hierarchy of approximations removes or filters out a certain class of phenomena or processes. The advantage of such procedures is that they allow us to isolate effects on different space–time scales, and make more efficient models on the computer.

In this regard, there appears to be a continuing awareness in the scientific community that numerical prediction models do not provide perfect simulations of reality. The reason is that the resolution of all important space–time scales remains far beyond our current capabilities. In addition, some of these unresolved processes actually control the evolution of the atmosphere and oceans, and our scientific understanding is still notably incomplete in some cases, e.g., regarding turbulence, clouds, and oceanic convection. Despite the capacities of present-day supercomputers, they cannot be represented in full detail in the models, not even in limited area models. This is compounded by the fact that contemporary models are complex, nonlinear, and inherently chaotic systems (Lorenz 1963).

Consequently, simpler models, such as those based on the shallow water and quasi-geostrophic equations, have not disappeared, despite continuously increasing computer capacities. On the contrary, a stronger emphasis has been given to the concept of a "hierarchy of models" as the only way to provide a link between theoretical understanding and the complexity of realistic models (Held 2005). It is still necessary to consult models that are either conceptually simpler, or limited to a number of processes or a specific region, to assist in the use and interpretation of complex models, therefore enabling a deeper understanding of the processes at work or a more relevant comparison with observation. In a numerical context, the simpler models are extremely useful because they provide solutions against which the numerical solutions may be assessed and checked (Chap. 11).

References

Bjerknes V (1904) Das Problem der Vettervorhersage, betrachtet vom Standpunkte der Mechanik und der Physik. Meteor Z 21:1–7

Fofonoff NP (1985) Physical properties of seawater: a new salinity scale and equation of state for seawater. J. Geophys. Res. 90(C2):3332–3342. https://doi.org/10.1029/JC090iC02p03332

Gill AE (1982) Atmosphere–ocean dynamics. International geophysics series, vol 30. Academic, New York

Griffies SM (2004) Fundamentals of ocean climate models. Princeton University Press, Princeton. ISBN 0-691-11892-2

Held IM (2005) The gap between simulation and understanding in climate modelling. Bull Am Meteorol Soc 86:1609–1614

Lorenz EN (1963) Deterministic non-periodic flow. J Atmos Sci 20:130–141

McDougall TJ, Feistel R, Pawlowicz R (2013) Thermodynamics of seawater. In: Siedler G, Griffies SM, Gould J, Church JA (eds) Ocean circulation and climate. A 21st century perspective. International geophysics, vol 103. Elsevier, San Diego, pp 141–158. https://doi.org/10.1016/B978-0-12-391851-2.00006-4

Pedlosky J (1979) Geophysical fluid dynamics. Springer, Berlin

Stern M (1975) Ocean circulation physics. International geophysics series, vol 19. Academic, New York

Preliminaries

2

In this chapter, we present a general analysis of *partial differential equations* (henceforth, PDEs). These are of importance here because the governing equations of the atmosphere and the ocean, including the hierarchy of simpler equations outlined in Chap. 1, belong to this class. We shall see that a PDE has a different character depending on the physics it describes. As we shall show in Chaps. 4–6, this has to be taken into account when deciding which numerical method should be used to solve the PDE.

PDEs differ from ordinary differential equations in that there are at least two independent variables and, more often than not, more than one dependent variable. This is the norm when considering the equations governing the motion of geophysical fluids such as atmospheres and oceans. In most cases, and in particular for linear systems, the number of dependent variables can be reduced to one at the expense of increasing the order of the PDE, as illustrated in Sect. 2.3. Thus, the present analysis is limited to a single PDE that has two independent variables.

A PDE containing two independent variables has the general form

$$\overline{a}\partial_{x'}^2\phi + 2\overline{b}\partial_{x'}\partial_{y'}\phi + \overline{c}\partial_{y'}^2\phi + 2\overline{d}\partial_{x'}\phi + 2\overline{e}\partial_{y'}\phi + \overline{f}\phi = \overline{g}\,, \tag{2.1}$$

where $\partial_{x'}$, $\partial_{y'}$ denote differentiation with respect to the independent variables x', y', and $\phi = \phi(x', y')$ is the dependent variable. In general, the coefficients $\overline{a}, \overline{b}, \ldots, \overline{g}$ are functions of the independent variables, that is, $\overline{a} = \overline{a}(x', y')$, etc. Note that x' and y' represent any independent variable, for instance time or one of the spatial variables, while ϕ represents any dependent variable, e.g., velocity, pressure, density, salinity, or humidity.

We shall also introduce some basic mathematics underlying two of the most important numerical methods used to solve atmospheric and oceanic problems, namely *finite difference methods* and *spectral methods*. This mathematics includes Taylor series expansions (Sect. 2.5), orthogonal functions (Sect. 2.9), Fourier series (Sect. 2.10), and Fourier transforms (Sect. 2.11). Orthogonal functions, Fourier series, and Fourier transforms are the essential mathematical tools for developing

L. P. Røed, *Atmospheres and Oceans on Computers*,
Springer Textbooks in Earth Sciences, Geography and Environment,
https://doi.org/10.1007/978-3-319-93864-6_2

spectral methods. Taylor series is the tool used to develop *finite difference approximations*, which lie at the heart of the finite difference methods. Intrinsic to the establishment of finite difference approximations are the so-called truncation errors (Sect. 2.7). Finally, some convenient notation is introduced (Sect. 2.8).

2.1 Elliptic Equations

If $\overline{b}^2 - \overline{ac} < 0$, then the roots of (2.1) are imaginary, distinct, and complex conjugates. The corresponding PDE is *elliptic*. The classic example is *Poisson's equation*:

$$\nabla_{\mathrm{H}}^2 \phi = \partial_x^2 \phi + \partial_y^2 \phi = g(x, y) , \tag{2.2}$$

where ∇_{H} is the two-dimensional part of the three-dimensional del operator. To arrive at this equation, let $x' = x$, $y' = y$, $\overline{a} = \overline{c} = 1$, $\overline{g} = g$, and $\overline{b} = \overline{d} = \overline{e} = \overline{f} = 0$ in (2.1). Other examples are the *Helmholtz equation*, viz.,

$$\nabla_{\mathrm{H}}^2 \phi + f(x, y)\phi = g(x, y) , \tag{2.3}$$

and the *Laplace equation*, viz.,

$$\nabla_{\mathrm{H}}^2 \phi = 0 . \tag{2.4}$$

2.2 Parabolic Equations

If $\overline{b}^2 - \overline{ac} = 0$, the corresponding PDE is *parabolic*. The classic example is the *diffusion equation* or heat conduction equation:

$$\partial_t \phi = \kappa \partial_x^2 \phi , \tag{2.5}$$

where κ is the diffusion coefficient (heat capacity). To obtain (2.5) from (2.1), let $x' = x$, $y' = t$, $\overline{a} = 1$, $\overline{b} = \overline{c} = \overline{d} = \overline{f} = \overline{g} = 0$, and $\overline{e} = 1/2$. Observe that (2.5) is a simplified, one-dimensional version of the full three-dimensional tracer equation (1.3), where the advection term and also the source and sink terms are neglected. Under the latter circumstances, the full three-dimensional tracer equation of a Boussinesq fluid (1.20) takes the form

$$\partial_t C_i = \nabla \cdot (\mathscr{K} \cdot \nabla C_i) , \tag{2.6}$$

where the diffusive tracer flux $\mathbf{F}_i$ is parameterized as $\mathbf{F}_i = -\mathscr{K} \cdot \nabla C_i$, with $\mathscr{K}$ a matrix (dyade) describing the conductive efficiency of the medium with respect to the tracer C_i (see Sect. 3.1). Thus, $\mathscr{K} = \kappa_{mn} \mathbf{i}_m \mathbf{j}_n$, $m, n = 1, 2, 3$. To retrieve (2.5), simply let $\kappa_{11} = \kappa$ and $\kappa_{mn} = 0$ for $m \neq 1$ and $n \neq 1$, and assume that κ is constant.

If the atmosphere/ocean is at rest ($\mathbf{v} = 0$) and there are no sources or sinks, then (1.3) reduces to (2.6). This implies that the diffusion balance is of fundamental importance when solving atmospheric and oceanographic problems.

2.3 Hyperbolic Equations

If $\overline{b}^2 - \overline{ac} > 0$, then the roots of (2.1) are real and distinct. The corresponding PDE is then *hyperbolic*. The classic example is the wave equation, viz.,

$$\partial_t^2 \phi - c_0^2 \partial_x^2 \phi = 0 . \tag{2.7}$$

In order to derive (2.7) from (2.1), let $x' = t$, $y' = x$, $\overline{a} = 1$, $\overline{b} = 0$, $\overline{c} = -c_0^2$, and $\overline{d} = \overline{e} = \overline{f} = \overline{g} = 0$. Then $\overline{b}^2 - \overline{ac} = -(-c_0^2) = c_0^2$, which is indeed positive.

Define Φ by

$$\Phi = \partial_t \phi - c_0 \partial_x \phi . \tag{2.8}$$

Then $\partial_t \Phi = \partial_t^2 \phi - c_0 \partial_t(\partial_x \phi)$ and $c_0 \partial_x \Phi = c_0 \partial_t(\partial_x \phi) - c_0^2 \partial_x^2 \phi$, so using (2.7), we obtain

$$\partial_t \Phi + c_0 \partial_x \Phi = 0 . \tag{2.9}$$

Since c_0 is a constant, it finally takes the form

$$\partial_t \Phi + \partial_x (c_0 \Phi) = 0 . \tag{2.10}$$

Equation (2.9) is commonly referred to as the *advection equation*, while (2.10) is the flux form of this equation. Replacing ρ by Φ and letting $\mathbf{v} = c_0 \mathbf{i}$ in (1.1), (2.10) is equivalent to the one-dimensional version of the continuity equation. It is also a one-dimensional version of (1.3) with suitable replacements. This indicates that advection is an important process in atmospheric and oceanic circulation. Hence, it is crucial to solve the advection equation correctly numerically. Moreover, this shows that the equations governing atmospheres and oceans viewed as a time marching problem are inherently hyperbolic.

This can be illustrated by studying the linear version[1] of the two-dimensional shallow water equations (1.33) and (1.34), that is,

$$\partial_t u - f v = -g \partial_x h , \tag{2.11}$$

$$\partial_t v + f u = -g \partial_y h , \tag{2.12}$$

$$\partial_t h + H_\mathrm{m} \partial_x u = -H_\mathrm{m} \partial_y v . \tag{2.13}$$

Here u and v are the components of the horizontal velocity $\mathbf{u}$ along the horizontal axes x and y, respectively, $h - H_\mathrm{m}$ is the deviation of the height of the fluid column from its equilibrium height or depth H_m (assumed constant), f is the Coriolis parameter (assumed constant), and g is the gravitational acceleration.[2] Mathematically speaking, (2.11)–(2.13) are three first-order PDEs containing the three dependent variables u, v, and h, and the three independent variables t, x, and y. Differentiating (2.11) with respect to time and then adding (2.12) multiplied by f to the results leads to

$$(\partial_t^2 + f^2) u = -g f \partial_y h - g \partial_t (\partial_x h) . \tag{2.14}$$

[1]For details on how to linearize the shallow water equations, see Sect. 6.2.

[2]The geopotential is $\phi = gh$, but since g is constant, we may use h as the variable in the shallow water equations.

Similarly, differentiating (2.12) with respect to time and then adding (2.11) multiplied by $-f$ yields

$$(\partial_t^2 + f^2)v = gf\partial_x h - g\partial_t(\partial_y h) . \tag{2.15}$$

Applying the operator $(\partial_t^2 + f^2)$ to (2.13) leads to

$$(\partial_t^2 + f^2)\partial_t h + H_{\mathrm{m}}(\partial_t^2 + f^2)\partial_x u + H_{\mathrm{m}}(\partial_t^2 + f^2)\partial_y v = 0 . \tag{2.16}$$

Finally, differentiating (2.14) with respect to x and (2.15) with respect to y, then substituting the results into (2.16), the latter takes the form

$$(\partial_t^2 + f^2 - c_0^2\nabla_{\mathrm{H}}^2)\partial_t h = 0 , \tag{2.17}$$

where $c_0 = \sqrt{gH_{\mathrm{m}}}$. Let $h = 0$ at time $t = 0$. Integration in time t leads to

$$(\partial_t^2 + f^2 - c_0^2\nabla_{\mathrm{H}}^2)h = 0 . \tag{2.18}$$

If in addition the motion is assumed to be independent of one of the dependent variables, say y, it reduces to

$$(\partial_t^2 + f^2 - c_0^2\partial_x^2)h = 0 . \tag{2.19}$$

Thus, (2.19) is hyperbolic in t and x, and as detailed in Sect. 6.3, the solution to (2.19) is a gravity wave modified by the Earth's rotation. Furthermore, although the steady-state solution to (2.18) is elliptic, the time marching problem is inherently hyperbolic.

The governing equations describing the time evolution of the circulation of atmospheres and oceans are fundamentally hyperbolic. It is important to keep this in mind when developing numerical methods to solve atmosphere–ocean problems.

The shallow water equations are discussed further in Chap. 6. There they are used to show how the Coriolis term should be treated. In particular, it is this term that makes geophysical fluid dynamics stand out from ordinary fluid dynamics. Conveniently, the shallow water equations are also used as an example problem to show how multiple-variable problems are solved using numerical methods.

2.4 Boundary Conditions

To solve (2.1) and find a solution for Φ, the governing PDE has to be integrated with respect to the dependent variables. Thus, the solution inherently contains integration constants. The number of integration constants is determined by the order of the PDE. For instance, (2.1) has to be integrated twice in x' and y', leading to four integration constants. Another example is provided by integrating the linearized shallow water equations (2.11)–(2.13) or (2.17). The latter equation has to be integrated three times with respect to time t, twice with respect to x, and twice with respect to y, leading to a total of seven integration constants. To assign values to these seven unknown integration constants, seven conditions are needed. Since the equations are only valid within a specified domain, these conditions are usually constructed by knowing something about the solution at the boundary of the domain. The conditions are therefore commonly referred to as *boundary conditions*. Since conditions in time are of particular importance, it is common to refer to them as *initial conditions*.

It is extremely important that the total number of initial and boundary conditions should exactly match the number of integration constants: no more, no less. Mathematically speaking, this is imperative to ensure that the problem is well posed, which in turn ensures that a solution exists and is unique. When solving the equations by numerical means, it is even more important to adhere to this fundamental principle. The reason is that the computer is very good at producing numbers (except when dividing by zero). Thus, even if the number of conditions is overspecified or underspecified, the computer will still produce numbers. These numbers may look realistic, but are nevertheless incorrect. In particular, the solution to any problem is equally dependent on the initial and boundary conditions as on the equations themselves, and this further attests to the importance of matching the number of integration constants and the number of initial and boundary conditions.

For instance, to determine the solution to the elliptic Poisson equation (2.2), four boundary conditions are needed, two in x and two in y. To determine the solution to the diffusion equation (2.5), one initial condition and two boundary conditions are needed. Finally, to determine the solutions to the wave equation, a total of four conditions are needed to determine the integration constants, namely two initial conditions and two spatial (boundary) conditions. As the dimensions of the equation increase, the number of integration constants increases as well, and so also does the number of boundary conditions.

There are essentially two types of spatial boundaries, *natural boundaries* and *open boundaries*. At natural boundaries, the boundary conditions are dictated by the physics of the problem, for instance by the existence of a mountain or a coastline. This is in contrast to open boundaries, which are computational or virtual boundaries. Open boundary conditions are discussed further in Chap. 7.

Regarding conditions to be specified at natural boundaries there are essentially two types:

- Dirichlet conditions, in which case the variable is known at the boundary, and
- Neumann conditions, in which case the derivative normal to the boundary is specified.

Most other boundary conditions are combinations of these.

A classic example of a Dirichlet condition is the condition that prevails at an impermeable wall or a solid boundary. The boundary condition is then that there cannot be any flow through such a boundary. If $\mathbf{n}$ is the unit vector perpendicular to the solid boundary Ω, then the condition of no throughflow is

$$\mathbf{n} \cdot \mathbf{v} = 0 \,, \quad \text{at } \Omega \,. \tag{2.20}$$

A classic example of a Neumann-type condition is the condition of no heat flux or heat exchange across an insulated boundary or surface. To achieve, this it is required that the diffusive heat flux $\mathbf{F} = -\kappa \nabla \theta$ across the boundary Ω must be zero, where θ is the temperature. Thus,

$$\mathbf{n} \cdot \mathbf{F} = 0 , \quad \text{at } \Omega . \tag{2.21}$$

As mentioned, the two conditions may be combined to give other natural boundary conditions. One is the so-called "slip" condition. For instance, consider a flat bottom or surface at $z = 0$ (or $z = -H$) and the requirement

$$\nu \partial_z \mathbf{u} = C_{\mathrm{D}} \mathbf{u} , \quad z = 0 , \tag{2.22}$$

where ν is the vertical eddy viscosity, $\mathbf{u}$ is the horizontal component of the current (or wind), and C_{D} is a drag coefficient (more often than not, the latter is a constant). Note that (2.22) does not specify the gradient or the variable itself. Thus, $\mathbf{u}$ is nonzero if the gradient is nonzero (and vice versa). Hence the name slip condition.

Other common boundary conditions are *cyclic* or *periodic boundary conditions*. A periodic boundary condition is one in which the solution is specified as repeating itself beyond a certain distance. Thus, a periodic boundary condition in x for a given tracer concentration $C(x)$ would be

$$C(x, t) = C(x + L, t) , \tag{2.23}$$

where L is the distance over which the solution repeats itself. An example is the condition that must be applied when considering solutions to atmospheric and oceanic problems contained in a zonal band bounded to the south and north by a zonal wall. In the longitudinal direction, the solutions then "wrap around", that is, repeat themselves every 360°. Another example of such a repeating distance is the wavelength of a monochromatic wave.

2.5 Taylor Series

The basis for all numerical finite difference methods is that all well-behaved functions can be expanded in terms of a Taylor series. A well-behaved function is simply one for which the function itself and all its first- and higher order derivatives exist and are continuous.[3] In addition, a well-behaved function may be represented by an infinite sum of orthogonal functions, such as trigonometric functions (Sects. 2.9 and 2.10).

The Taylor series of any well-behaved function, say $\phi(x, y, z, t)$, expanded in x is defined by

$$\phi|^t_{x+\Delta x,y,z} = \phi|^t_{x,y,z} + \sum_{n=1}^{\infty} \frac{1}{n!} \partial^n_x \phi|^t_{x,y,z} \Delta x^n , \tag{2.24}$$

[3] Note that this definition is slightly different from the one offered in the little known but enlightening book by Lighthill (1970).

where the notation used is such that $\phi|_{x,y,z}^{t}$ is ϕ evaluated at (x, y, z, t), $\phi|_{x+\Delta x,y,z}^{t}$ is ϕ evaluated at $(x + \Delta x, y, z, t)$, and $\partial_x^n \phi|_{x,y,z}^{t}$ is the n th derivative of ϕ evaluated at (x, y, z, t). Similarly, the function ϕ may be expanded in any of the other independent variables, e.g.,

$$\phi|_{x,y,z}^{t+\Delta t} = \phi|_{x,y,z}^{t} + \sum_{n=1}^{\infty} \frac{1}{n!} \partial_t^n \phi|_{x,y,z}^{t} \Delta t^n \, . \tag{2.25}$$

In short, (2.24) provides a way of finding the value of ϕ at the neighboring space point $x + \Delta x$, provided that ϕ and all its first- and higher order derivatives are known at the point x. Likewise, (2.25) provides a way of finding ϕ at the neighboring time $t + \Delta t$.

Consider for instance the well-behaved function $\theta(x, t)$, where x is a spatial coordinate and t is time. Using (2.24),

$$\theta|_{x+\Delta x}^{t} = \theta|_x^t + \partial_x \theta|_x^t \Delta x + \frac{1}{2} \partial_x^2 \theta|_x^t \Delta x^2 + \frac{1}{6} \partial_x^3 \theta|_x^t \Delta x^3 + \mathcal{O}(\Delta x^4) \, . \tag{2.26}$$

Here, following the notation used in (2.24), θ and all its first- and higher order derivatives with respect to x on the right-hand side of (2.26) are evaluated at the point (x, t). The notation $\mathcal{O}(\Delta x^4)$ (read as "order Δx to the four") is used to emphasize that there are more higher order terms, and that the first term neglected is of fourth order in Δx. This procedure may be repeated to find the value of the function $\theta(x, t)$ at the point $x - \Delta x$. To do this, Δx is simply replaced by $-\Delta x$ in (2.26), whence

$$\theta|_{x-\Delta x}^{t} = \theta|_x^t - \partial_x \theta|_x^t \Delta x + \frac{1}{2} \partial_x^2 \theta|_x^t \Delta x^2 - \frac{1}{6} \partial_x^3 \theta|_x^t \Delta x^3 + \mathcal{O}(\Delta x^4) \, . \tag{2.27}$$

The only difference between (2.27) and (2.26) is the alternating sign in front of every other term on the right-hand side of (2.27). Thus, subtracting (2.27) from (2.26),

$$\theta|_{x+\Delta x}^{t} - \theta|_{x-\Delta x}^{t} = 2\partial_x \theta|_x^t \Delta x + \mathcal{O}(\Delta x^3) \, . \tag{2.28}$$

Note that the term $\mathcal{O}(\Delta x^2)$ cancels out, as do all terms of even order in Δx.

As is well known, the Taylor series (2.26)–(2.28) may also be used to define the first-order derivative of θ. For instance, solving (2.26) with respect to $\partial_x \theta|_x^t$ leads to

$$\partial_x \theta|_x^t = \frac{\theta|_{x+\Delta x}^{t} - \theta|_x^t}{\Delta x} + \mathcal{O}(\Delta x) \, . \tag{2.29}$$

Similarly, solving (2.27) with respect to $\partial_x \theta|_x^t$ yields

$$\partial_x \theta|_x^t = \frac{\theta|_x^t - \theta|_{x-\Delta x}^{t}}{\Delta x} + \mathcal{O}(\Delta x) \, . \tag{2.30}$$

Finally, solving (2.28) with respect to $\partial_x \theta|_x^t$ leads to

$$\partial_x \theta|_x^t = \frac{\theta|_{x+\Delta x}^{t} - \theta|_{x-\Delta x}^{t}}{2\Delta x} + \mathcal{O}(\Delta x^2) \, . \tag{2.31}$$

These expressions require knowledge of the function θ at the points $x + \Delta x$ and $x - \Delta x$, as well as at x. In addition, in the limit $\Delta x \to 0$, all the higher order terms in (2.29)–(2.31) tend to zero. These expressions are therefore just three slightly different definitions of the derivative of θ with respect to x in the limit $\Delta x \to 0$.

Note that the choice Δx is arbitrary. An equally valid choice is $2\Delta x$. Replacing Δx by $2\Delta x$ in the Taylor series (2.26) and (2.27) leads to

$$\theta|_{x+2\Delta x}^t = \theta|_x^t + 2\partial_x\theta|_x^t\Delta x + 2\partial_x^2\theta|_x^t\Delta x^2 + \frac{4}{3}\partial_x^3\theta|_x^t\Delta x^3 + \mathcal{O}(\Delta x^4) \tag{2.32}$$

and

$$\theta|_{x-2\Delta x}^t = \theta|_x^t - 2\partial_x\theta|_x^t\Delta x + 2\partial_x^2\theta|_x^t\Delta x^2 - \frac{4}{3}\partial_x^3\theta|_x^t\Delta x^3 + \mathcal{O}(\Delta x^4)\ , \tag{2.33}$$

respectively. Again, subtracting one from the other and solving for $\partial_x\theta|_x^t$ yields

$$\partial_x\theta|_x^t = \frac{\theta|_{x+2\Delta x}^t - \theta|_{x-2\Delta x}^t}{4\Delta x} + \mathcal{O}(\Delta x^2)\ , \tag{2.34}$$

which is an equally valid definition of the derivative of θ with respect to x in the limit $\Delta x \to 0$.

2.6 Finite Difference Approximations

The governing equations for atmospheres and oceans all contain derivatives of various orders of the dependent variables (Chap. 1). When applying the finite difference method, all these derivatives are replaced by so-called *finite difference approximations* (FDAs). The resulting equations are referred to as *finite difference equations* or finite different versions of the equation. To construct the FDAs of the first- and higher order derivatives present in the governing equations, the Taylor series presented in Sect. 2.5 are simply truncated. For instance, to construct a feasible FDA to $\partial_x\theta$, we may use either (2.29), (2.30), or (2.31). However, Δx is kept finite. To construct the FDA, the higher order terms are simply neglected. Thus, by truncating the higher order terms in (2.29)–(2.31), we obtain three viable FDAs to $\partial_x\theta$, viz.,

$$[\partial_x\theta]_x^t = \frac{\theta|_{x+\Delta x}^t - \theta|_x^t}{\Delta x}\ , \tag{2.35}$$

$$[\partial_x\theta]_x^t = \frac{\theta|_x^t - \theta|_{x-\Delta x}^t}{\Delta x}\ , \tag{2.36}$$

and

$$[\partial_x\theta]_x^t = \frac{\theta|_{x+\Delta x}^t - \theta|_{x-\Delta x}^t}{2\Delta x}\ . \tag{2.37}$$

To separate the continuous partial derivative from the FDA, brackets are used. Since (2.37) is centered on the spatial point x, it is referred to as a *centered difference*. In contrast, (2.35) and (2.36) are one-sided, and henceforth referred to as *one-sided differences*. Moreover, since (2.35) steps forward, while (2.36) steps backward, (2.35) is referred to as a *forward one-sided difference*, while (2.36) is referred to as a *backward one-sided difference*. To establish (2.35) and (2.36), the first term neglected is $\mathcal{O}(\Delta x)$, while the first term neglected in (2.37) is $\mathcal{O}(\Delta x^2)$. Thus, a major difference between the centered and the one-sided differences is that the centered one is valid to second order ($\mathscr{O}(\Delta x^2)$), while the two one-sided ones are only valid to first order ($\mathscr{O}(\Delta x)$). Consequently, the centered difference (2.37) is sometimes referred to as a *second-order difference*, while the one-sided differences (2.35) and (2.36) are referred to as *first-order differences*.

Exactly the same calculations based on (2.25) may be performed to derive FDAs to the first and higher order derivatives of θ with respect to time t. For instance, (2.25) leads to

$$\theta|_x^{t+\Delta t} = \theta|_x^t + \partial_t\theta|_x^t \Delta t + \frac{1}{2}\partial_t^2\theta|_x^t \Delta t^2 + \frac{1}{6}\partial_t^3\theta|_x^t \Delta t^3 + \mathscr{O}(\Delta t^4) \tag{2.38}$$

and

$$\theta|_x^{t-\Delta t} = \theta|_x^t - \partial_t\theta|_x^t \Delta t + \frac{1}{2}\partial_t^2\theta|_x^t \Delta t^2 - \frac{1}{6}\partial_t^3\theta|_x^t \Delta t^3 + \mathscr{O}(\Delta t^4)\,. \tag{2.39}$$

By a similar manipulation to the one used to obtain (2.35)–(2.37), we find

$$[\partial_t\theta]_x^t = \frac{\theta|_x^{t+\Delta t} - \theta|_x^t}{\Delta t}\,, \tag{2.40}$$

$$[\partial_t\theta]_x^t = \frac{\theta|_x^t - \theta|_x^{t-\Delta t}}{\Delta t}\,, \tag{2.41}$$

and

$$[\partial_t\theta]_x^t = \frac{\theta|_x^{t+\Delta t} - \theta|_x^{t-\Delta t}}{2\Delta t}\,. \tag{2.42}$$

It is also possible to construct FDAs of the higher order derivatives of θ using Taylor series. For instance, to find a centered FDA to the second-order derivative $\partial_x^2\theta|_x^t$, the two Taylor series (2.26) and (2.27) are first added and the results solved for $\partial_x^2\theta|_x^t$. This leads to

$$\partial_x^2\theta|_x^t = \frac{\theta|_{x+\Delta x}^t - 2\theta|_x^t + \theta|_{x-\Delta x}^t}{\Delta x^2} + \mathscr{O}(\Delta x^2)\,. \tag{2.43}$$

Then by neglecting the higher order terms $\mathscr{O}(\Delta x^2)$ in (2.43), a viable FDA of $\partial_x^2\theta$ is

$$[\partial_x^2\theta]_x^t = \frac{\theta|_{x+\Delta x}^t - 2\theta|_x^t + \theta|_{x-\Delta x}^t}{\Delta x^2}\,. \tag{2.44}$$

Since equal weight is given to the points $x + \Delta x$ and $x - \Delta x$, that is, to the points on either side of x, the FDA is centered. As in (2.37), the neglected terms are $\mathscr{O}(\Delta x^2)$. This turns out to be a general result, so all centered FDAs share the characteristic that the neglected terms are of second order.

As exemplified in (2.44), FDAs to any higher order derivative with respect to t, x, and other spatial independent variables may be formulated by "smart" manipulations of the Taylor series. For instance, to create a centered-in-space FDA for the third-order spatial derivative of $\theta(x, t)$, that is, $\partial_x^3\theta|_x^t$, (2.26) and (2.27) are combined with (2.32) and (2.33) to yield

$$\partial_x^3\theta|_x^t = \frac{\theta|_{x+2\Delta x}^t - 2\theta|_{x+\Delta x}^t + 2\theta|_{x-\Delta x}^t - \theta|_{x-2\Delta x}^t}{2\Delta x^3} + \mathscr{O}(\Delta x^2)\,. \tag{2.45}$$

Neglecting higher order terms, here terms of the $\mathcal{O}(\Delta x^2)$, leads to the FDA

$$[\partial_x^3 \theta]_x^t = \frac{\theta|_{x+2\Delta x}^t - 2\theta|_{x+\Delta x}^t + 2\theta|_{x-\Delta x}^t - \theta|_{x-2\Delta x}^t}{2\Delta x^3} . \tag{2.46}$$

Note that (2.46) is a centered difference, so it should come as no surprise that (2.46) is of second order.

2.7 Truncation Errors

As already mentioned, the main difference between the one-sided and centered differences is the order of the terms neglected when constructing the FDA from Taylor series. Clearly, the neglected terms constitute an error as long as Δx is finite. It is only in the limit $\Delta x \rightarrow 0$ that the FDA becomes an exact replica of the derivative. The size of the error is thus related to the increment Δx. Furthermore, when establishing centered FDAs, the neglected terms were of $\mathcal{O}(\Delta x^2)$, while they were of $\mathcal{O}(\Delta x)$ when constructing the one-sided FDAs. The centered FDAs are therefore more accurate than the one-sided ones. The error made is a direct consequence of the order at which the Taylor series are truncated. Hence, the error is referred to as a *truncation error*. The accuracy of our finite difference version of the governing equations is thus proportional to the lowest order of the truncation errors. For instance, solving a PDE containing both second- and first-order derivatives, the finite difference equations will only be second-order accurate if centered finite difference approximations are employed. So using a one-sided difference for the first-order derivative, even if we apply higher order differences for the higher order derivatives, the finite difference equation will still only be first-order accurate.

As shown in Sect. 10.1, Taylor series may also be used to construct FDAs that are truncated to still higher orders, e.g., to $\mathcal{O}(\Delta x^n)$, where $n \geq 3$. Such FDAs are even more accurate and are therefore referred to as *higher order approximations*. The schemes applying them are consequently referred to as *higher order schemes*. When constructing such approximations, points at distances $2\Delta x$ or more away from the point x have to be included, as also exemplified by (2.46). Although it is desirable that the approximations should be as accurate as possible, we stress that higher order schemes can involve other complications, as we shall see in Sect. 10.1.

Finally, it is good practice to ensure that all the FDAs employed to approximate the various terms in our governing equations have the same truncation errors, but not necessarily to the same order in both time and space. For instance, consider a one-dimensional wave propagating in a direction forming an angle with the x and y directions. The only way to ensure that the numerical solution then has the same accuracy regardless of the propagation direction of the wave is to use FDAs that have the same accuracy along all spatial directions.

2.8 Notation

When solving a PDE using numerical methods, and in particular when using finite difference methods, it is common to define a grid or mesh covering the domain over which the solution is to be found. As an example, consider a two-dimensional spatial problem where we seek a solution to the Laplace equation (2.4) within a quadratic domain. Let the two coordinates x and y start at 0 and end at $x = L$ and $y = L$, respectively. Mathematically, this is expressed by $x \in \langle 0, L \rangle$ and $y \in \langle 0, L \rangle$, where the brackets $\langle\, ,\ \rangle$ indicate that the solution is sought inside the boundaries $x = 0, L$ and $y = 0, L$. This is equivalent to saying that the equations are only valid within the domain. At the boundaries, the solution is assumed known and given by the boundary conditions.

The first step is to cover the domain by a quadratic mesh as shown in Fig. 2.1. To keep track of the grid points in the mesh, these points are counted along the axes. This is done by using dummy indices, for instance, $j = 1, 2, 3, \ldots, J + 1$

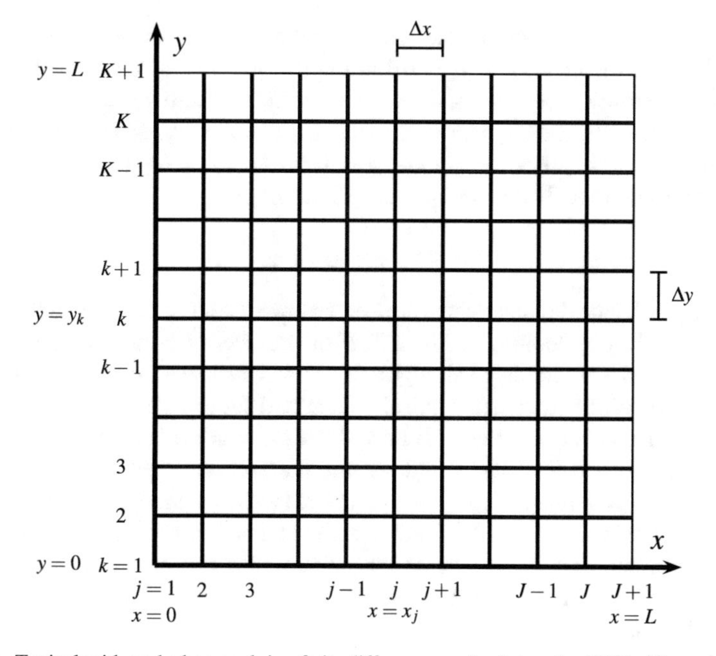

Fig. 2.1 Typical grid used when applying finite difference methods to solve PDEs. The points along the axes in the x, y directions are incremented by Δx, Δy, respectively, so that there are a total of $J + 1$ points along the x-axis and $K + 1$ points along the y-axis, including the start and end points. Grid points are counted using the dummy counters j, k. The total number of grid points is $(J + 1) \times (K + 1)$

along the x-axis, and $k = 1, 2, 3, \ldots, K + 1$ along the y-axis.[4] By this count there are $J + 1$ points along the x-axis and $K + 1$ points along the y-axis, including the start and end points. While $x = 0$ with $j = 1$, $x = L$ is associated with $j = J + 1$. Similarly, $y = 0$ is associated with $k = 1$ and $y = L$ with $k = K + 1$. A random point along the x-axis is denoted by x_j, where the subscript refers to the value of x at the j th point along the x-axis. Similarly, y_k denotes a random point along the y-axis. The coordinates of a random grid point are therefore x_j, y_k, where $j = 1(1)J + 1$ and $k = 1(1)K + 1$. Hereafter, the notation $j = 1(1)J + 1$ will be short for $j = 1, 2, 3, \ldots, J + 1$, implying that j starts at $j = 1$ and increases in unit steps up to and including $J + 1$.

The distances between two adjacent points along the x-axis and the y-axis are denoted by Δx and Δy, respectively. The locations of x_j and y_k along the two axes are therefore

$$x_j = (j - 1)\Delta x\ , \quad j = 1(1)J + 1\ , \tag{2.47}$$

and

$$y_k = (k - 1)\Delta y\ , \quad k = 1(1)K + 1\ , \tag{2.48}$$

respectively. The location of the grid point x_j, y_k is therefore $(j - 1)\Delta x$, $(k - 1)\Delta y$. In particular, $x_1 = y_1 = 0$ and $x_{J+1} = y_{K+1} = L$ constitute the boundaries. If the starting points along the axes are different from zero, say $x = x_0$ and $y = y_0$, respectively, then $x_1 = x_0$, $y_1 = y_0$, $x_{J+1} = x_0 + L$, and $y_{K+1} = y_0 + L$. Under these circumstances, $x_j = x_0 + (j - 1)\Delta x$ and $y_k = y_0 + (k - 1)\Delta y$.

It is important to realize that (2.47) and (2.48) imply

$$\Delta x = \frac{L}{J}\ , \quad \Delta y = \frac{L}{K}\ , \tag{2.49}$$

respectively, whence the increments are known once J and K are chosen for a given L. This is evident by looking at Fig. 2.1. If the number of points along one of the axes is increased, the increment along that axis must decrease for a fixed L. If on the other hand one of the increments must be increased for one reason or another, then the number of points along that axis must decrease in accordance with (2.49). The significance of this is that the increments and the number of points along the axes are inversely proportional. They cannot therefore be chosen independently. Once Δx and Δy are chosen for a fixed L, J and K are calculated from (2.49). Likewise, if J and K are chosen for a fixed L, then (2.49) dictates the size of Δx and Δy.

Hereafter the notation θ_{jk} will be used to denote the value of the variable $\theta(x, y)$ at the grid point x_j, y_k. Thus

$$\theta_{jk} = \theta(x_j, y_k) = \theta\big[(j - 1)\Delta x, (k - 1)\Delta y\big]\ . \tag{2.50}$$

[4]FORTRAN 90/95 and 2003 allows us to use $j = 0$ and $k = 0$ as dummy indices. This leads to $x_j = j\Delta x$, $j = 0(1)J$ and $y_k = k\Delta y$, $k = 1(1)K$, so that $x_0 = 0$ and $y_0 = 0$. Furthermore, $j = J$ is associated with $x = L$ and $k = K$ with $y = L$. Under these circumstances the number of points along the axes are the same and the increments Δx and Δy likewise.

Furthermore,

$$\theta_{j-1k} = \theta(x_{j-1}, y_k) = \theta\big[(j-2)\Delta x, (k-1)\Delta y\big] , \tag{2.51}$$

$$\theta_{jk+1} = \theta(x_j, y_{k+1}) = \theta\big[(j-1)\Delta x, k\Delta y\big] , \tag{2.52}$$

$$\theta_{j+1k+1} = \theta(x_{j+1}, y_{k+1}) = \theta[j\Delta x, k\Delta y] , \tag{2.53}$$

and so on.

Time is a special coordinate in the modeling of atmospheres and oceans. To show its special status, as is common practice, a superscript will be used hereafter for the time counter to discriminate between spatial and temporal variables. Thus, let n be the time counter and Δt the time step or time increment. Then the time elapsed at time n, or after n time steps, is defined by

$$t^n = n\Delta t, \quad n = 0(1)\infty . \tag{2.54}$$

The notation used to denote the variable $\theta(x, t)$ at the space–time grid point x_j, t^n is therefore

$$\theta_j^n = \theta|_{x_j}^{t^n} = \theta(x_j, t^n) = \theta\big[(j-1)\Delta x, n\Delta t\big] . \tag{2.55}$$

The apparent inconsistency in starting the time counter at $n = 0$ and the space counter at $j = 1$ is historical. To save space on the computer, the time levels are never all stored in the computer's random access memory (RAM). How many time levels need to be stored depends on the time stepping scheme employed. If for instance the scheme only needs access to two time levels at any time step, only two time levels are stored at any time. This point is discussed further in Chaps. 4–6, when we present various schemes for the diffusion, advection, and shallow water equations.

If the variable in question is four-dimensional, e.g., $\theta = \theta(x, y, z, t)$, the notation used is

$$\theta_{jkl}^n = \theta(x_j, y_k, z_l, t^n) , \tag{2.56}$$

where $z_l = (l-1)\Delta z$.

With this notation in mind, the Taylor series (2.26) and (2.27) may be rewritten to yield

$$\theta_{j\pm1}^n = \theta_j^n \pm \partial_x\theta|_j^n\Delta x + \frac{1}{2}\partial_x^2\theta|_j^n\Delta x^2 \pm \frac{1}{6}\partial_x^3\theta|_j^n\Delta x^3 + \mathscr{O}(\Delta x^4) , \tag{2.57}$$

while the similar Taylor series in time becomes

$$\theta_j^{n\pm1} = \theta_j^n \pm \partial_t\theta|_j^n\Delta t + \frac{1}{2}\partial_t^2\theta|_j^n\Delta t^2 \pm \frac{1}{6}\partial_t^3\theta|_j^n\Delta t^3 + \mathscr{O}(\Delta t^4) . \tag{2.58}$$

Hence (2.29) and (2.31) can be written

$$\partial_x\theta|_j^n = \frac{\theta_{j+1}^n - \theta_j^n}{\Delta x} + \mathscr{O}(\Delta x) , \tag{2.59}$$

$$\partial_x\theta|_j^n = \frac{\theta_{j+1}^n - \theta_{j-1}^n}{2\Delta x} + \mathscr{O}(\Delta x^2) . \tag{2.60}$$

Regarding the FDAs, the forward-in-time FDA in (2.40) can be written

$$[\partial_t\theta]_j^n = \frac{\theta_j^{n+1} - \theta_j^n}{\Delta t} . \tag{2.61}$$

Similarly, the centered FDAs of the first-order derivative in time and space become

$$[\partial_t \theta]_j^n = \frac{\theta_j^{n+1} - \theta_j^{n-1}}{2\Delta t} , \tag{2.62}$$

$$[\partial_x \theta]_j^n = \frac{\theta_{j+1}^n - \theta_{j-1}^n}{2\Delta x} , \tag{2.63}$$

respectively, while the second-order FDAs to the second-order derivative in time and space are

$$\left[\partial_t^2 \theta\right]_j^n = \frac{\theta_j^{n+1} - 2\theta_j^n + \theta_j^{n-1}}{\Delta t^2} , \tag{2.64}$$

$$\left[\partial_x^2 \theta\right]_j^n = \frac{\theta_{j+1}^n - 2\theta_j^n + \theta_{j-1}^n}{\Delta x^2} . \tag{2.65}$$

So far it has been assumed that the increments Δx, Δy, Δz, and Δt are constant. However, in general they may vary in space, and even in time. If the increments vary in space alone, the grid is referred to as an *unstructured mesh*. If the increments vary in both time and space, the grid is referred to as an *adaptive, unstructured mesh*.

2.9 Orthogonal Functions

When using the finite difference method, only grid-point values of the dependent variables are considered. No assumption is made about how the variables behave between grid points. An alternative approach is to expand the dependent variables in terms of an infinite series of smooth orthogonal functions or eigenfunctions, in which, for numerical reasons, the series is truncated after a finite number of elements. The problem is then reduced to solving a finite set of ordinary differential equations to determine the time dependence of the expansion coefficients. As such the orthogonal functions act as interpolators between the grid points.

As an example, consider the general linear one-dimensional time-dependent problem

$$\partial_t \phi = \mathscr{H}[\phi] , \quad \text{for } x \in \langle -L, L \rangle \text{ and } t > 0 , \tag{2.66}$$

where $\phi = \phi(x, t)$ is a well-behaved function as defined in Sect. 2.5. Furthermore, $\mathscr{H}$ is a linear differential operator in x, e.g., $\mathscr{H} = u_0 \partial_x$, where u_0 is a constant. The computational domain is of length $2L$ in space, and ends at $x = \pm L$, respectively. To solve (2.66), suitable boundary conditions at $x = \pm L$ have to be specified, together with an appropriate initial condition at $t = 0$. Here, it is simply assumed that the initial distribution in ϕ space is known, and that it is cyclic at $x = \pm L$. The cyclic condition is commonly applied in many atmospheric applications, in particular when considering solutions within zonal bands that stretch around the globe at certain latitudes. Under these circumstances, the solution must "wrap" around, given that there are no real boundaries in the latitudinal direction.

Since ϕ is a well-behaved function, it may be expanded in terms of an infinite set of orthogonal functions or eigenfunctions, say $e_n(x)$, $n = 1, 2, 3, \ldots$. The expansion functions $e_n(x)$ are in general complex functions, e.g., $e_n(x) = \mathrm{e}^{\mathrm{i}\alpha_n x}$, where α_n is the wavenumber associated with the n th eigenfunction. Thus,

$$\phi = \sum_{n=-\infty}^{\infty} \varphi_n(t) e_n(x) \ , \tag{2.67}$$

where the $\varphi_n(t)$ are time-dependent expansion coefficients. This is a well-known method, commonly used to separate variables when solving differential equations involving more than one independent variable analytically. Without loss of generality, in addition to being orthogonal, the expansion functions $e_n(x)$ are generally assumed to be orthonormal, i.e.,

$$\int_{-L}^{L} e_n(x) e_m^*(x) \mathrm{d}x = \delta_{mn} \ , \tag{2.68}$$

where δ_{mn} is the Kronecker delta, equal to 1 when $m = n$ and zero otherwise, and $e_n^*(x)$ is the complex conjugate of $e_n(x)$. Consider the situation when the expansion functions $e_n(x)$ are known. Then it is the expansion coefficients $\varphi_n(t)$ that we must determine. Multiplying (2.66) by the complex conjugate e_m^* of the eigenfunction e_m, and then integrating over all possible x-values, leads to

$$\int_{-L}^{L} (\partial_t \phi) \, e_m^*(x) \mathrm{d}x = \int_{-L}^{L} \mathscr{H}[\phi] e_m^*(x) \mathrm{d}x \ , \quad \forall m \ . \tag{2.69}$$

The left-hand side is further developed using (2.67) and (2.68) to obtain

$$\begin{aligned} \int_{-L}^{L} (\partial_t \phi) \, e_m^* \mathrm{d}x &= \int_{-L}^{L} \left[\sum_n (\partial_t \varphi_n) \, e_n \right] e_m^* \mathrm{d}x \\ &= \sum_n \partial_t \varphi_n \int_{-L}^{L} e_n e_m^* \mathrm{d}x = \partial_t \varphi_m \ , \quad \forall m \ . \end{aligned} \tag{2.70}$$

The operator $\mathscr{H}$ only operates on x, so

$$\mathscr{H}[\phi] = \sum_n \varphi_n \mathscr{H}[e_n] \ . \tag{2.71}$$

Finally, using (2.70) and (2.71) leads to

$$\partial_t \varphi_m = \sum_n \varphi_n \int_{-L}^{L} \mathscr{H}[e_n] e_m^* \mathrm{d}x \ , \quad \forall m \ , \tag{2.72}$$

which is a set of m coupled ordinary differential equations for the time rate of change of the expansion coefficients φ_m.

It is now of interest to consider how the choice of eigenfunctions may simplify the problem:

1. If the expansion functions are eigenfunctions of $\mathscr{H}$, then $\mathscr{H}[e_n] = \lambda_n e_n$, where λ_n are the eigenvalues. Insertion into (2.72) and using the orthonormality condition (2.68), (2.72) simplifies to

$$\partial_t \varphi_m = \lambda_m \varphi_m \ , \quad \forall m \ , \tag{2.73}$$

and becomes decoupled.
2. If the original equation is

$$\mathscr{G}[\partial_t \phi] = \mathscr{H}[\phi] \ , \tag{2.74}$$

where $\mathscr{G}$ is a linear operator, then the problem is simplified by using expansion functions that are eigenfunctions of $\mathscr{G}$ with eigenvalues λ_n. This leads to

$$\lambda_m \partial_t \varphi_m = \sum_n \varphi_n \int_{-L}^{L} \mathscr{H}[e_n] e_m^* \mathrm{d}x \ . \tag{2.75}$$

2.10 Fourier Series

A much used orthogonal set of expansion functions is the set of trigonometric functions $\mathrm{e}^{\mathrm{i}\alpha_n x}$, where α_n forms a discrete set of wavenumbers. For instance if the problem to be solved is constrained by cyclic boundary conditions at $x = \pm L$, then the set of wavenumbers is $\alpha_n = n\pi/L$, $n = 0, 1, 2, \ldots$. As in the previous section, the well-behaved function $\phi(x, t)$ may therefore be expanded in terms of trigonometric functions, that is,

$$\phi(x, t) = \sum_{n=-\infty}^{\infty} \varphi_n(t) \mathrm{e}^{\mathrm{i}\alpha_n x} \ . \tag{2.76}$$

The series (2.76) is called a *Fourier series* and the expression

$$\varphi_n(t) \mathrm{e}^{\mathrm{i}\alpha_n x} \tag{2.77}$$

is called a *Fourier component*.

The complex conjugates of the expansion functions are $\mathrm{e}^{-\mathrm{i}\alpha_n x}$, which means that (2.76) can be rewritten in the form

$$\phi(x, t) = \phi_0 + \sum_{n=1}^{\infty} \varphi_n(t) \mathrm{e}^{\mathrm{i}\alpha_n x} \ , \tag{2.78}$$

where it is assumed that $\alpha_0 = 0$. It is important to realize that the subscript n attached to the expansion coefficients implies that they are different for each wavenumber.

Furthermore, the Fourier component is a complex number. Thus, the well-behaved function ϕ is in general a complex function, too. If ϕ is real then it simply equals the real part of the right-hand side of (2.78), that is,

$$\phi = \phi_0 + \sum_{n=1}^{\infty} (a_n \sin \alpha_n x + b_n \cos \alpha_n x) \ , \tag{2.79}$$

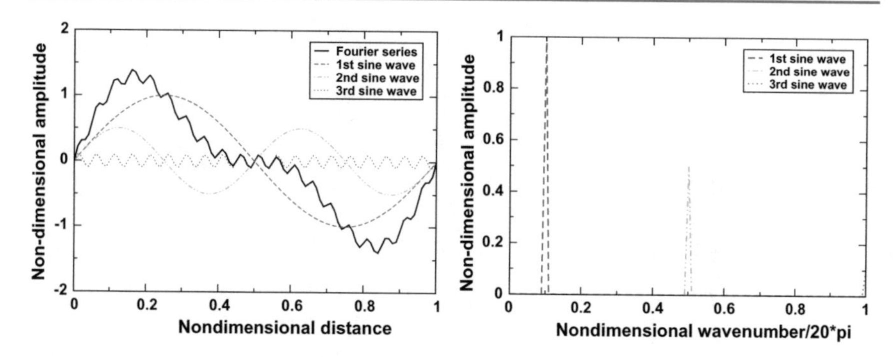

Fig. 2.2 *Left*: A well-behaved function (*solid black line*) whose Fourier series is composed of three sine waves of different wavenumbers and amplitudes. The first sine wave has a dimensionless amplitude/wavenumber of 1.0/2π (*dashed red line*), the second 0.5/4π (*dash-dotted green line*), and the third 0.1/20π (*dotted blue line*). *Right*: Fourier space in which the three waves show up as three spikes at the appropriate wavenumber (dimensionless). The height of the spikes reflects the amplitude (or energy content) of each sine wave (dimensionless)

where ϕ_0 and the coefficients a_n and b_n are all real. Hence, the Fourier series may be interpreted as a decomposition of the well-behaved function into an infinite sum of waves with discrete wavenumbers α_n and amplitudes a_n and b_n. As an example consider a well-behaved function decomposed into a sum of three sine waves of amplitudes $a_1 = a$, $a_2 = 0.5a$, and $a_3 = 0.1a$ and wavelengths $\lambda_1 = 2L$ ($\alpha_1 = 2\pi/L$), $\lambda_2 = L$ ($\alpha_2 = 4\pi/L$), and $\lambda_3 = 0.1L$ ($\alpha_3 = 40\pi/L$). The result is the dimensionless function ϕ/a shown in the left-hand panel of Fig. 2.2. The three waves may also be displayed in the amplitude–wavenumber space, as shown in the right-hand panel of Fig. 2.2. Note that in the latter the individual waves show up as spikes at the appropriate wavenumber.

2.11 Fourier Transforms

If the wavenumber space is continuous rather than discrete, the sum depicted in (2.76) turns into an integral, that is,

$$\phi(x,t) = \frac{1}{2\pi}\int_{-\infty}^{\infty} \varphi(t,\alpha)\mathrm{e}^{\mathrm{i}\alpha x}\mathrm{d}\alpha \ . \tag{2.80}$$

Here $\varphi(t,\alpha)$ is called the *Fourier transform* of $\phi(x,t)$, defined by

$$\varphi(t,\alpha) = \int_{-\infty}^{\infty} \phi(x,t)\mathrm{e}^{-\mathrm{i}\alpha x}\mathrm{d}x \ , \tag{2.81}$$

while (2.80) is called the *inverse Fourier transform*. Since φ is a complex function, the real function ϕ is retrieved by taking the real part of (2.80). Note that the Fourier transform (2.81) is a continuous function of the wavenumber α.

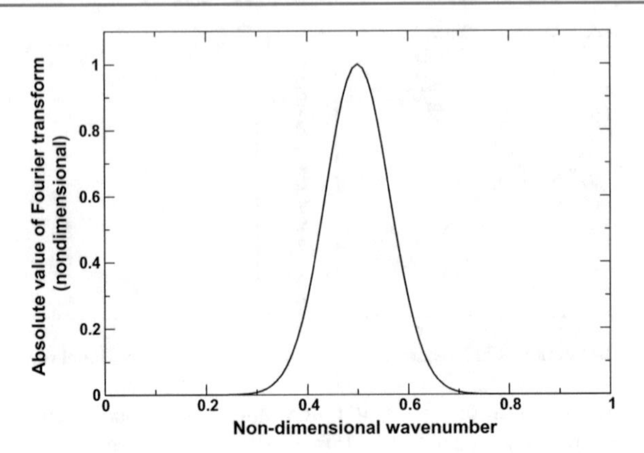

Fig. 2.3 Fourier transform of a Gaussian bell function (which is another Gaussian bell) in the Fourier space, commonly referred to as the Fourier spectrum. Note that the dominant waves are clustered around the (dimensionless) wavenumber $\alpha = 0.5$ corresponding to a dimensionless wavelength of 4π. Since a resolution of at least 4 points per wavelength is necessary to resolve this wave, the nondimensional spatial increment is required to be smaller than $\Delta x = \pi$

Comparing the inverse Fourier transform (2.80) with (2.76) reveals that they are quite similar, except that the "expansion coefficient" φ is a continuous function of the wavenumber space $\alpha \in [-\infty, +\infty]$. Once again, the Fourier transform φ may be plotted in the amplitude–wavenumber space (hereafter called the *Fourier space*), just as the expansion coefficients of (2.76) were plotted in the discrete amplitude–wavenumber space shown in Fig. 2.2. Its distribution in the Fourier space is commonly referred to as the *Fourier spectrum.* The Fourier spectrum is, however, a continuous function of the wavenumber, as exemplified by the distribution shown in Fig. 2.3.

As can be seen from (2.80), the Fourier transform (2.81) is simply the amplitude associated with the wave of wavenumber (or wavelength) α. In a sense, the amplitude reveals how much "energy" is associated with each wavelength. Thus, the Fourier spectrum reveals how much energy is contained at the various wavelengths. The wavelengths associated with the highest amplitudes are also the wavelengths that contain the highest energy content. Knowing the Fourier transform thus reveals information about the wavelengths that dominate the motion.

2.12 Summary and Remarks

Since the equations governing the motion of atmospheres and oceans belong to the class of partial differential equations (PDEs), we discussed certain characteristics of such equations in this chapter (Sects. 2.1–2.3). There are three classes of PDEs, namely, elliptic, parabolic, and hyperbolic. It turns out that the equations governing the motion of atmospheres and oceans are typically hyperbolic. Exceptions are motions dominated by turbulent mixing (diffusive problems), which are parabolic problems, and steady-state problems, which are elliptic. For hyperbolic problems, any property of the dependent variable, e.g., velocity, temperature, or salinity, is propagated in the form of waves of one kind or another, while for parabolic problems the property in question is a diffusive process, e.g., by heat conduction or mixing. While parabolic and hyperbolic problems are time marching problems, elliptic problems do not evolve in time. Thus, any change in the property of a dependent variable anywhere is felt immediately everywhere.

In Sect. 2.4, we emphasized that the solution to any PDE is determined, not only by the PDE itself, but also by the conditions imposed on the boundary of the solution domain. At these boundaries, the governing equations cease to be valid and are replaced by so-called boundary conditions. There are in general two types of boundaries: the natural boundaries where the physics of the problem dictates the condition, and so-called open boundaries required by limitations on computer capacity. At such boundaries, the governing equations are still valid, and it is no straightforward matter to construct a boundary condition that leaves the problem well posed. Open boundary conditions will be discussed further in Chap. 7, where there is more to say about how to avoid specifying boundary conditions that degrade the numerical solution.

Taylor series were introduced in Sect. 2.5. These constitute the basis for all finite difference approximations (FDAs) of the derivatives that enter the equations governing the motion of atmospheres and oceans (Sect. 2.6). The result is what is referred to as a *finite difference version* of the original continuous problem. By using finite differences based on Taylor series, the method of solution is referred to as the *finite difference method*, and the resulting equations as *finite difference equations*. As the name indicates, FDAs are approximations, and not perfect representations of the continuous derivatives. Intrinsic to them are the so-called truncation errors (Sect. 2.7), produced by truncating the Taylor series used to construct the FDA. Moreover, the truncation error differs from one FDA to the next, and depends on the order at which the Taylor series are truncated. For instance, if centered approximations are used, the truncation error is of second order in the space–time increments Δx, Δy, Δz, and Δt, while so-called one-sided approximations are of first order.

Section 2.8 introduced some convenient notation to simplify the presentation in the remaining chapters. For instance any function θ of time t and space x at the grid point x_j, t^n is denoted θ_j^n. Moreover, an FDA of the second-order derivative of $\theta(x,t)$ with respect to x at the grid point x_j, t^n is denoted $[\partial_x^2\theta]_j^n$, while the continuous second-order derivative is denoted $\partial_x^2\theta|_j^n$. As an example, the FDA of $\partial_x^2\theta$ at the grid point x_j, t^n is written

$$[\partial_x^2\theta]_j^n = \frac{\theta_{j-1}^n - 2\theta_j^n + \theta_{j+1}^n}{\Delta x^2} . \tag{2.82}$$

If θ depends on more than one independent space variable, say $\theta = \theta(x, y, t)$, then the FDA of the two-dimensional del operator is written

$$[\nabla_{\mathrm{H}}^2\theta]_j^n = [\partial_x^2\theta]_j^n + [\partial_y^2\theta]_j^n = \frac{\theta_{j-1k}^n - 2\theta_{jk}^n + \theta_{j+1k}^n}{\Delta x^2} + \frac{\theta_{jk-1}^n - 2\theta_{jk}^n + \theta_{jk+1}^n}{\Delta x^2} . \tag{2.83}$$

In this chapter, we also touched upon orthogonal functions, Fourier series, and Fourier transforms (Sects. 2.9–2.11). Why learn about these? The orthogonal functions are the basis for the spectral method presented in Sect. 10.6, while Fourier series and Fourier transforms are essential in constructing the Fourier spectrum, which reveals information about the dominant wavelengths of the problem. Recall that to derive the finite difference equations, finite difference approximations to all derivatives are first established. These finite difference approximations contain space increments and time increments which are finite, and form a four-dimensional grid system with finite distances between the grid points. These distances are equal to the size of the increments, as shown for instance in Fig. 2.1. Moreover, the numerical solution is only known at these grid points, so how does one choose the size of the increments? Here is where knowledge of the Fourier spectrum is useful. If something is known about the Fourier spectrum, for instance, the spectrum of the initial distribution, it informs us about the dominant wavelengths (or wavenumbers). To capture or resolve these wavelengths, the increments must be chosen accordingly. A minimum of four points per wavelength is required, but ideally ten points per wavelength should be chosen. The size of the increments should be chosen in accordance with this rule. We shall return to this point again and again in the remaining chapters.

2.13 Exercises

1. Show that both the Helmholtz and the Laplace equations are elliptic in x and y.
2. Show that the diffusion equation is parabolic in t, x and t, y, but elliptic in x, y.
3. Show by use of Taylor series expansions that a possible centered FDA of $\partial_x^4\theta(x)$ is

$$[\partial_x^4\theta]_j = \frac{\theta_{j+2} - 4\theta_{j+1} + 6\theta_j - 4\theta_{j-1} + \theta_{j-2}}{\Delta x^4} , \tag{2.84}$$

   and that the truncation error is $\mathcal{O}(\Delta x^2)$. Note that we use points that are a distance $2\Delta x$ away from the point x_j itself. This is common when deriving centered FDAs to higher order derivatives [see (2.46)].
4. Assume that $\theta(x, t)$ and all its derivatives tend to zero as $x \to \pm\infty$. Show that under these conditions, the Fourier transforms of $\partial_x\theta(x, t)$ and $\partial_x^2\theta(x, t)$ are

$$\widetilde{\partial_x\theta} = \mathrm{i}\alpha\widetilde{\theta} \quad \text{and} \quad \widetilde{\partial_x^2\theta} = -\alpha^2\widetilde{\theta} , \tag{2.85}$$

   respectively, where the notation $\widetilde{\psi}$ denotes the Fourier transform of ψ.
5. Show, by making use of the results in Exercise 4, that a formal analytic solution to the diffusion equation

$$\partial_t\theta = \kappa\partial_x^2\theta , \tag{2.86}$$

where $\theta = \theta(x, t)$, κ is a constant, and the boundary conditions are

$$\theta = \begin{cases} 0\,, & x \to +\infty, -\infty\,, \\ \theta_0 \mathrm{e}^{-(x/a)^2}\,, & t = 0\,, \end{cases} \tag{2.87}$$

is

$$\theta = \frac{a\theta_0}{2\sqrt{\pi}} \int_{-\infty}^{\infty} \exp\left[-\alpha^2\left(\frac{1}{4}a^2 + \kappa t\right)\right] \mathrm{e}^{\mathrm{i}\alpha x} \mathrm{d}\alpha\,. \tag{2.88}$$

[Hint: Make a Fourier transform of (2.86) to arrive at $\partial_t \widetilde{\theta} = -\kappa\alpha^2\widetilde{\theta}$. Remember to make a Fourier transform of the initial condition too, and use Fourier tables to arrive at the formal solution (2.88).]

2.14 Computer Problems

2.14.1 Truncation Errors in Two Recursion Formulas

All readers are strongly recommended to do this problem. It is simple enough to enable you to refresh your knowledge of Fortran and your Fortran skills without having to write lengthy codes. Moreover, it demonstrates the dramatic consequences of the presence of the insignificant truncation errors that are always present in numerical computations.

Part a

Let

$$\pi = 4\arctan(1)\,, Z_1 = \pi\,, \text{and} S_1 = \pi\,. \tag{2.89}$$

Compute

$$Z_{j+1} = 3.1 Z_j - 2.1 Z_1 \text{ and } S_{j+1} = \left(\frac{9.}{5.}\right) S_j - \left(\frac{4.}{5.}\right) S_1\,, \tag{2.90}$$

for $j = 1(1)N$. Compute also the *relative* error, that is,

$$\varepsilon_j^Z = \frac{Z_{j+1} - Z_j}{Z_j} \text{ and } \varepsilon_j^S = \frac{S_{j+1} - S_j}{S_j}\,, \tag{2.91}$$

for each j. Write π, Z_j, S_j, and the relative error ε_j in percent. The output should be readable and self-explanatory, e.g., should have headings for each column. Do the problem on different platforms available to you (e.g., handhelds, laptops, PCs, and supercomputers). Experiment by using (1) different constants in the recursion formulas, (2) double and single precision, and (3) different numbers of iterations (use at least $N = 1000$). Does this impact the results?

Part b

Show analytically why the recursion formulas for Z_j and S_j should diverge from π. If one or both of the recursion formulas does not diverge numerically, explain why.

Hint: The recursion formulas displayed in (2.90) are of the form $C_{j+1} = aC_j - bC_1$, where a and b are constants. In both cases, we then get $\varepsilon_j = a\varepsilon_{j-1}$, where $a > 1$.

Reference

Lighthill MJ (1970) Fourier analysis and generalised functions. Cambridge University Press, first printed 1958

3 Time Marching Problems

The purpose of this chapter is to present some properties inherent in the governing equations listed in Sect. 1.1. Most problems in the geophysical sciences, including atmospheres, oceans, seas, and lakes, involve solving so-called time marching problems. Typically, the state of the fluid in question is known at one specific time. As postulated by Bjerknes (1904) in his comments on weather forecasting (see quote in the preface), the aim is then to predict the state of the fluid at a later time. This is done by solving the governing equations listed in Sect. 1.1. Such problems are known in mathematics as *initial value problems*.

A closer study of (1.1)–(1.4), and in particular (1.3), reveals that the advection and diffusion processes are of fundamental importance. Thus, advection and diffusion (or turbulent mixing) can be said to be the "heart and soul" of the dynamics of geophysical fluids. Moreover, as can be seen from (1.3) for instance, the two processes often work in concert.

Another prominent feature, evident in the momentum equation (1.1), is the possibility of a stationary solution referred to as the *geostrophic balance* [see (1.39)]. It is this possibility that makes the geophysical fluid dynamics stand out from ordinary fluid dynamics. Any deviation from this balance manifests itself through nonzero acceleration terms, the so-called *ageostrophic terms*. Examples of such terms are the local time rate of change of the velocity, nonlinear terms, etc.

It is thus essential for everyone who aspires to become a meteorologist and/or oceanographer to know how to treat the advection and diffusion processes and the geostrophic balance numerically. In the following chapters, we shall give a detailed account of how to solve the diffusion equation (Chap. 4), the advection equation (Chap. 5), and the shallow water equations (Chap. 6) using numerical finite difference methods. Later we shall try to give some insights into how to solve the combined advection–diffusion problem (Chap. 10). These relatively simple problems will serve to introduce some of the basic concepts needed to solve atmospheric and oceanographic problems using numerical methods. Moreover, and equally important, they will illustrate some of the pitfalls.

L. P. Røed, *Atmospheres and Oceans on Computers*,
Springer Textbooks in Earth Sciences, Geography and Environment,
https://doi.org/10.1007/978-3-319-93864-6_3

Finally, it is important to bear in mind that the properties detailed in this chapter should carry over when replacing the governing equations by their finite difference analogues. These properties should thus be taken into account when constructing the finite difference equations. If not, the solution may simply be false or incorrect. Checking the behavior of the solutions against these fundamental properties is part of what is often referred to as *model verification*. Verification is the first step in a chain of activities commonly referred to as model quality assurance or model evaluation, as presented in Appendix 11 (see also GESAMP 1991; Lynch and Davies 1995; Hackett et al. 1995). When coding errors are found, the process is referred to as *debugging*, which simply means weeding out errors in the program code.

3.1 Advection–Diffusion Equation

Equations like the tracer equation (1.3) balance the time rate of change of a variable in response to advective and diffusive fluxes and source and sink terms. They are therefore commonly referred to as *advection–diffusion equations*. As the name indicates, they combine two very different physical processes. The first is associated with the advection process (Sect. 2.3), while the second is associated with the turbulent mixing or diffusion process (Sect. 2.2).

Neglecting possible tracer sources and sinks ($S_i = 0$) in the tracer equation of a Boussinesq fluid (1.20), it takes the form

$$\partial_t \theta + \nabla \cdot \mathbf{F} = 0 , \tag{3.1}$$

where θ is any dependent variable (or tracer) and $\mathbf{F}$ is a *flux vector* that includes fluxes due to turbulent mixing and advection.[1] For instance, if θ is the potential temperature, (3.1) is the conservation equation for internal energy or heat content without sources and sinks.[2]

To repeat, the flux vector represents two distinct physical processes that transfer properties from one location to the next. One is the advective process transporting or propagating properties from one place to the next via the motion (or waves). The second is the turbulent mixing process, associated with small-scale, inherently chaotic processes, leading to properties being exchanged between two locations without invoking any mean motion. It is therefore common to separate the flux vector $\mathbf{F}$ into two parts:

$$\mathbf{F} = \mathbf{F}_{\mathrm{A}} + \mathbf{F}_{\mathrm{D}} , \tag{3.2}$$

[1] If the dependent variable is a vector, such as the velocity $\mathbf{v}$, the flux becomes a tensor and (3.1) takes the form $\partial_t \mathbf{v} + \nabla \cdot \mathscr{F} = 0$, where $\mathscr{F}$ is the flux tensor.

[2] The total energy of a system consists of the internal energy and the mechanical energy. The internal energy is concerned with the heat content of a fluid and is an important part of its thermodynamics. In contrast, the mechanical energy concerns the motion of the fluid and is part of its dynamics.

where $\mathbf{F}_{\mathrm{A}}$ represents the flux due to the advective process, hence referred to as the *advective flux vector*, and $\mathbf{F}_{\mathrm{D}}$ represents the flux due to turbulent mixing, hence referred to as the *diffusive flux vector*.

Since the advective and diffusive flux vectors represent two very contrasting physical processes, they naturally have very different parameterizations leading to different mathematical formulations. The advective flux vector $\mathbf{F}_{\mathrm{A}}$ depends on the motion alone, and hence the property θ follows the path of the individual fluid parcels. It is therefore commonly parameterized as

$$\mathbf{F}_A = \mathbf{v}\theta \ . \tag{3.3}$$

If the dependent variable is a vector, e.g., the velocity vector $\mathbf{v}$, the flux is a tensor $\mathscr{F}_{\mathrm{A}} = \mathbf{vv}$.

In contrast, the parameterization of the diffusive flux vector is more complex. The reason is that diffusion in a geophysical fluid is caused by turbulent mixing, which in itself is a complicated and somewhat disordered process. Moreover, its impact on the larger scale motion is partly unknown. In essence though, the turbulent mixing acts in many respects to even out disturbances in the tracer fields, thereby making them smoother. So the impact of F_{D} on the larger scale has many characteristics similar to the diffusive processes, and this is the justification behind its name. The most common parameterization of the turbulent mixing of tracers, often referred to as turbulence closure, is therefore in terms of a diffusive process. Accordingly, its mathematical formulation is

$$\mathbf{F}_{\mathrm{D}} = -\mathscr{K} \cdot \nabla\theta \ , \tag{3.4}$$

where

$$\mathscr{K} = \begin{bmatrix} \kappa_{11} & \kappa_{12} & \kappa_{13} \\ \kappa_{21} & \kappa_{22} & \kappa_{23} \\ \kappa_{31} & \kappa_{32} & \kappa_{33} \end{bmatrix} = \kappa_{ij}\mathbf{i}_i\mathbf{i}_j \tag{3.5}$$

is the diffusion tensor (or conductive capacity). Again, if the dependent variable happens to be a vector, for instance, the velocity vector $\mathbf{v}$, the flux becomes a tensor $\mathscr{F}_{\mathrm{D}} = -\mathscr{K} \cdot \nabla\mathbf{v}$. Assuming that all the elements of the diffusion tensor are zero except along the diagonal, and that $\kappa_{11} = \kappa_{22} = \kappa_{33} = \kappa_0$, (3.4) reduces to

$$\mathbf{F}_{\mathrm{D}} = -\kappa_0\nabla\theta \ , \tag{3.6}$$

where κ_0 is a constant. This parameterization is referred to as Fickian diffusion.[3] Equation (3.4) expresses the idea that the absolute value of the diffusive flux vector increases when the difference in the tracer increases. Hence large differences in the tracer θ over small distances will be smoothed out rapidly. Since the elements κ_{ij} depend on the strength of the turbulence, they may vary in time and space according to the local turbulence characteristics. It is nevertheless quite common to use the Fickian diffusion parameterization (3.6) in numerical models.

[3]It takes its name from Dr. Adolf Eugen Fick who formulated the parameterization given in (3.6) in 1855.

3.2 Diffusion

Neglecting the advective flux vector in (3.1), the time rate of change of the tracer is balanced by the diffusive flux alone. Thus (3.1) reduces to

$$\partial_t \theta = -\nabla \cdot \mathbf{F}_\mathrm{D} = \nabla \cdot (\mathscr{K} \cdot \nabla \theta) \,, \tag{3.7}$$

where the last equals sign follows from (3.4). Assuming that (3.6) prevails, (3.7) takes the form

$$\partial_t \theta = \kappa_0 \nabla^2 \theta \,. \tag{3.8}$$

Solving (3.8), which is a parabolic problem (see Sect. 2.2), is referred to as solving the *diffusion problem*.

Recall that turbulent mixing acts to even out small-scale differences in the tracer fields. Thus, parameterizations of the diffusive flux vector should retain this property. A common measure of the small-scale differences, or disturbances, in the tracer field is its variance. It is therefore the sign of the time rate of change of its variance that determines whether a particular parameterization of turbulent mixing acts to smooth the tracer field.

The variance is defined as the square of the deviation from the mean. Let the mean be defined by

$$\overline{\theta} = \int_V \theta \, \mathrm{d}V \,, \tag{3.9}$$

where the integration extends over a constant, but deformable, volume V, bounded by its surface Ω. The latter implies $\partial_t \overline{\theta} = 0$. The deviation is

$$\theta' = \theta - \overline{\theta} \quad \Longrightarrow \quad \overline{\theta'} = 0 \,. \tag{3.10}$$

Thus,

$$\overline{\theta^2} = \overline{\theta}^2 + \overline{\theta'^2} \quad \Longrightarrow \quad \partial_t \overline{\theta^2} = \partial_t \overline{\theta'^2} \,. \tag{3.11}$$

The sign of the time rate of change of the variance $\partial_t \overline{\theta'^2}$ can therefore be studied by investigating the sign of $\partial_t \overline{\theta^2}$.

Multiplying (3.7) by the tracer concentration θ and integrating over the volume V, it takes the form

$$\partial_t \overline{\theta^2} = -2 \int_\Omega \theta \mathbf{F}_\mathrm{D} \cdot \delta \boldsymbol{\sigma} + 2 \overline{\mathbf{F}_\mathrm{D} \cdot \nabla \theta}, \tag{3.12}$$

where use has been made of Gauss' theorem. The vector $\delta \boldsymbol{\sigma}$ is a vector directed along the outward normal to the surface Ω, that is, $\delta \boldsymbol{\sigma} = \mathbf{n} \delta \sigma$, where $\mathbf{n}$ is a unit vector directed along the outward normal to the surface Ω, and $\delta \sigma$ is an infinitesimal surface element on Ω. For the boundary condition at Ω, we assume either a Dirichlet boundary condition, in which case $\theta = 0$ at the surface Ω, or a Neumann condition, in which case $\mathbf{n} \cdot \mathbf{F}_\mathrm{D} = 0$ at Ω. Hence (3.12) reduces to

$$\partial_t \overline{\theta^2} = 2 \overline{\mathbf{F}_\mathrm{D} \cdot \nabla \theta} \,. \tag{3.13}$$

Accordingly,

$$\partial_t \overline{\theta^2} \leq 0, \tag{3.14}$$

if and only if $\overline{\mathbf{F}_\mathrm{D} \cdot \nabla\theta} \leq 0$. Thus, as long as the diffusive flux vector $\mathbf{F}_\mathrm{D}$ is directed opposite to $\nabla\theta$ everywhere, commonly referred to as a down-the-gradient parameterization, it acts to even out small scale noise in the tracer field. For instance, using the Fickian diffusion parameterization (3.6), we obtain

$$\mathbf{F}_\mathrm{D} \cdot \nabla\theta = -\kappa(\nabla\theta)^2 \leq 0 , \tag{3.15}$$

revealing that it is indeed a down-the-gradient parameterization. Moreover, (3.15) shows that small-scale noise is smoothed more rapidly than the longer scales.

Recall that most problems in oceanography and meteorology are nonlinear. While there is no transfer of energy from one wavelength to the next in a linear system, this is not true for nonlinear systems. In such systems, energy input at long wavelengths (small wave numbers) is always transferred to progressively shorter wavelengths (high wave numbers) in the end. This was described elegantly in a rhyme credited to G. I. Taylor[4]: "Big whirls have smaller whirls that feed on their velocity, and little whirls have lesser whirls, and so on to viscosity …in the molecular sense."

When transforming a PDE into a finite difference equation by means of finite difference approximations, the resolved wavelengths are limited by $2\Delta x$, often referred to as the Nyquist wavelength (or frequency in the time domain), Δx being the specified spatial increment. Thus, when solving nonlinear PDEs using finite difference methods, one must mimic this energy cascade across the Nyquist wavelength. Since diffusion has the property of damping small-scale "noise", it is one tool that may prove useful (see Sect. 10.3).

3.3 Advection

Neglecting the diffusive part of the flux vector $\mathbf{F}$ in the advection–diffusion equation (3.1) yields

$$\partial_t\theta = -\nabla \cdot \mathbf{F}_\mathrm{A} = -\nabla \cdot (\mathbf{v}\theta) , \tag{3.16}$$

which is the *advection equation*. Recall that the advection equation is a hyperbolic problem (see Sect. 2.3). Solving it is thus referred to as solving the *advection problem*.

It is interesting to perform an analysis similar to the one performed in the preceding section. As for the diffusion problem, (3.16) is replaced by the boundary condition on the surface Ω of the constant deformable volume V. Let the advective flux be parameterized by the common parameterization $\mathbf{F}_\mathrm{A} = \mathbf{v}\theta$, and let the boundary condition be such that $\mathbf{F}_\mathrm{A} \cdot \delta\boldsymbol{\sigma} = 0$. The latter is achieved by assuming $\mathbf{v} = 0$ (the no-slip condition) or $\mathbf{v} \cdot \delta\boldsymbol{\sigma} = 0$ (no flow across the boundary). Multiplying (3.16) by θ and integrating over the volume V, it takes the form

[4]Geoffrey Ingram Taylor (1886–1975) made fundamental contributions to the study of turbulence, championing the need for a statistical theory and performing the first measurements of the effective diffusivity and viscosity of the atmosphere. He is remembered in the names attributed to several basic fluid flow instabilities (Taylor–Couette, Rayleigh–Taylor, and Saffman–Taylor).

$$\partial_t \overline{\theta^2} = 2\overline{\mathbf{F}_A \cdot \nabla \theta} = \overline{\mathbf{v} \cdot \nabla \theta^2} = -\overline{\theta^2 \nabla \cdot \mathbf{v}} \,. \tag{3.17}$$

Since θ^2 is always positive, the sign of the time rate of change of the variance is determined by the velocity divergence. Thus, if it is positive (negative) the variance decreases (increases). The case $\nabla \cdot \mathbf{v} = 0$ is special. The right-hand side of (3.17) is then zero, and hence any initial disturbances or noise in θ will prevail. Under these circumstances, the total variance is conserved.

The Boussinesq ocean is to a good approximation divergence free due to its incompressibility (see Sect. 1.4, or Gill 1982, p. 85). Thus, in a Boussinesq ocean, the advection process does not lead to any decrease or increase in the property being advected. Hence, any disturbance generated in a limited domain may be advected to other locations undisturbed. This is not true for the atmosphere since the atmosphere is highly compressible. Hence, in limited areas where the individual fluid parcels are drawn apart ($\nabla \cdot \mathbf{v} > 0$), any small-scale disturbances are smoothed. In contrast, any initial small-scale disturbances tend to increase in areas of convergence ($\nabla \cdot \mathbf{v} < 0$).

It is important to bear in mind the properties outlined above when solving the advective problem by numerical means. In particular, when the fluid is divergence free, the total variance should be conserved. This is in contrast to the diffusion problem. Here, any down-the-gradient parameterizations will lead to a decrease in the total tracer variance, a property that should once again be reflected in the finite difference formulation.

3.4 Shallow Water Equations

The third fundamental set of balance equations important in atmosphere and ocean dynamics are the shallow water equations (1.21)–(1.23). Neglecting forcing terms on the right-hand side of (1.23), they take the form

$$\nabla_H \cdot \mathbf{u} + \partial_z w = 0 \,, \tag{3.18}$$

$$\partial_t \mathbf{u} + \nabla_H \cdot (\mathbf{u}\mathbf{u}) + \partial_z (w\mathbf{u}) + f\mathbf{k} \times \mathbf{u} = -g\nabla_H \eta \,. \tag{3.19}$$

Here, use has been made of (1.24) to substitute for the pressure, and $\mathbf{u}$ denotes the horizontal component of the velocity, so that $\mathbf{v} = \mathbf{u} + w\mathbf{k}$.

Once again, to investigate the properties of the time rate of change of the total velocity variance[5] $\overline{\mathbf{u}'^2}$, it assumed that solutions of (3.18) and (3.19) are sought within the constant deformable volume V. Introducing the *total kinetic energy* per unit density

$$E_K = \int_V e_K \mathrm{d}V \,, \tag{3.20}$$

where $e_K = \mathbf{u}^2/2$ is the kinetic energy per unit mass, we obtain

$$2E_K = 2\overline{e_K} = \overline{\mathbf{u}^2} = \overline{\mathbf{u}}^2 + \overline{\mathbf{u}'^2} \,, \tag{3.21}$$

[5]The velocity variance is twice what is commonly referred to as the eddy kinetic energy.

and hence

$$\partial_t E_{\mathrm{K}} = \frac{1}{2}\partial_t \overline{\mathbf{u}'^2} \,. \tag{3.22}$$

To investigate the time rate of change of the velocity variance, it is therefore sufficient to study the time rate of change of the total kinetic energy E_{K}.

Multiplying (3.19) by $\mathbf{u}$, and noting that the Coriolis term vanishes since

$$\mathbf{u}\cdot(\mathbf{k}\times\mathbf{u}) = \mathbf{k}\cdot(\mathbf{u}\times\mathbf{u}) = 0 \,,$$

it takes the form

$$\partial_t e_{\mathrm{K}} + \nabla\cdot(e_{\mathrm{K}}\mathbf{v}) = -g\mathbf{u}\cdot\nabla_{\mathrm{H}}\eta \,, \tag{3.23}$$

where we have used

$$\mathbf{u}\cdot\left[\nabla_{\mathrm{H}}\cdot(\mathbf{u}\mathbf{u}) + \partial_z(w\mathbf{u})\right] = \nabla_{\mathrm{H}}\cdot(e_{\mathrm{K}}\mathbf{u}) + \partial_z(e_{\mathrm{K}}w) = \nabla\cdot(e_{\mathrm{K}}\mathbf{v}) \,, \tag{3.24}$$

and the continuity equation (3.18). Moreover, integrating (3.23) over the constant volume V, using the condition of no flux through the boundary Ω of the volume V, leads to

$$\partial_t E_{\mathrm{K}} = C \,, \tag{3.25}$$

where

$$C = -g\overline{\mathbf{u}\cdot\nabla_{\mathrm{H}}\eta} = -\int_A\left(\int_{-H}^{\eta} g\mathbf{u}\cdot\nabla_{\mathrm{H}}\eta \mathrm{d}z\right)\mathrm{d}A \,. \tag{3.26}$$

Here, as illustrated in Fig. 3.1, A is the projected area of the constant deformable volume V onto a horizontal surface, $H(x, y)$ is the depth of the fluid in its reference state at rest (equilibrium depth), and $\eta(x, y, t)$ is the deviation of the surface away from the level surface of the reference state ($\eta = 0$). Thus, the sign of the time rate of change of the velocity variance depends on the sign of C, a quantity yet to be interpreted.

Since η and $\nabla_{\mathrm{H}}\eta$ are both independent of depth/height, the inner integral of (3.26) may be expanded to yield

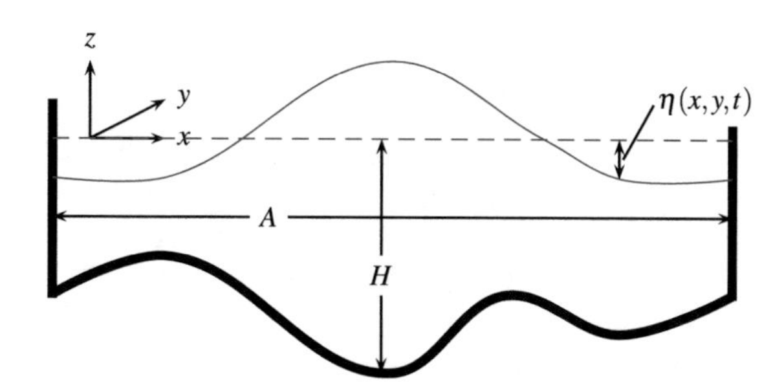

Fig. 3.1 Projected area A of a constant deformable volume in a one-layer model of the atmosphere and/or ocean. The *solid curved blue line* is the deviation of the surface $\eta = \eta(x, y, t)$ away from the depth $H(x, y)$ of the reference state in which $\eta = 0$ (*dashed straight blue line*). The limiting walls are assumed to be vertical down to a level deeper than the maximum surface deviation, so that the projected area of an arbitrary state always equals the projected area of the reference state

$$\int_{-H}^{\eta} g\mathbf{u} \cdot \nabla_{\mathrm{H}} \eta \mathrm{d}z = g\mathbf{U} \cdot \nabla_{\mathrm{H}} \eta = -g\eta \nabla_{\mathrm{H}} \cdot \mathbf{U} + \nabla_{\mathrm{H}} \cdot (g\eta \mathbf{U}) \ , \tag{3.27}$$

where

$$\mathbf{U} = \int_{-H}^{\eta} \mathbf{u} \, \mathrm{d}z \tag{3.28}$$

is the volume transport. Furthermore, by integrating the continuity equation (3.18) from bottom to top, and using the kinematic boundary condition (1.7) and (1.9) (Sect. 1.2), it takes the form

$$\partial_t \eta = -\nabla_{\mathrm{H}} \cdot \mathbf{U} \ . \tag{3.29}$$

Substituting into (3.27) yields

$$\int_{-H}^{\eta} g\mathbf{u} \cdot \nabla_{\mathrm{H}} \eta \mathrm{d}z = \partial_t \left(\frac{1}{2} g\eta^2 \right) + \nabla_{\mathrm{H}} \cdot (g\eta \mathbf{U}) \ . \tag{3.30}$$

Since the second term on the right-hand side of (3.30) vanishes upon integration over the area A, substituting into (3.26) yields

$$C = -\partial_t \left(\int_A \frac{1}{2} g\eta^2 \mathrm{d}A \right) = -\partial_t E_\Phi \ . \tag{3.31}$$

It is now convenient to introduce the *available potential energy* E_Φ per unit density. This is defined as the potential energy of an arbitrary state minus the potential energy of the reference state (see for instance Lorenz 1955; Røed 1997, 1999). Thus,

$$E_\Phi = \int_A \left(\int_{-H}^{\eta} gz \, \mathrm{d}z \right) \mathrm{d}A - \int_{A_0} \left(\int_{-H}^{0} gz \, \mathrm{d}z \right) \mathrm{d}A = \int_A \frac{1}{2} g\eta^2 \mathrm{d}A = -C \ , \tag{3.32}$$

where A is the area of the volume V projected onto a horizontal surface of the random state, and A_0 is the corresponding area for the reference state. To arrive at the second equality, it is assumed that $A_0 = A$, as illustrated in Fig. 3.1. The rationale behind the concept of available potential energy is that the enormous amount of potential energy of the reference state is not available for release into kinetic energy, and is therefore of no interest.

Finally, adding (3.25) and (3.32) yields

$$\partial_t \left(E_{\mathrm{K}} + E_\Phi \right) = 0 \ . \tag{3.33}$$

C is therefore a reversible energy conversion term converting potential energy into kinetic energy or vice versa. Thus, if C is positive, energy is converted to kinetic energy at the expense of the available potential energy, and the velocity variance increases. On the other hand, if C is negative, energy is converted to available potential energy at the expense of kinetic energy, and the velocity variance decreases. Moreover, (3.33) shows that the total mechanical energy $E = E_{\mathrm{K}} + E_\Phi$ is conserved under the assumption of no external forcing. Both E_{K} and E_Φ are positive-definite quantities. Thus, if both E_{K} and E_Φ are zero initially, no motion will ensue. If $E_\Phi > 0$ initially, motion can ensue and E_{K} will increase, as will the velocity variance. To conclude, for any numerical scheme to be trustworthy, it should reflect the fact that the total mechanical energy is conserved, so as not to generate false conversion of energy between the two forms of mechanical energy (see Arakawa and Lamb 1977).

3.5 Summary and Remarks

The important issue discussed in this chapter is that the governing equations and their subsets have certain properties that should be retained when constructing finite difference versions of those equations. Among the important properties are conservation of mass, energy, potential vorticity, and enstrophy (the integral of the square of the vorticity along a bounding surface) in the absence of sources and sinks. These fundamental conservation properties should be taken into account when the governing equations are transformed into their finite difference versions. If not, false (numerical) mass, energy, vorticity, and enstrophy may be produced (or dissipated).

Simpler equations like the diffusion equation, the advection equation, and the shallow water equation have certain properties that should be reflected in their finite difference versions. One is that any initial disturbance in a tracer (or velocity) field should be retained in an incompressible fluid under advection. If not, false decreases or increases in the variance may be produced. Moreover, if the diffusion is not properly parameterized it may cause small disturbances to grow in an unrealistic way, rather than being smoothed in time.

Moreover, since tracers such as temperature, salinity, and humidity have a decisive impact on the dynamics through their influence on the pressure distribution through density, it is important to get the advection (transport) and diffusion by turbulent mixing of these tracers correct to zero order.

Finally, transport of contaminants in atmospheres and oceans is a crucial issue when discussing environmental issues. For instance, emissions of radionuclides in one location are transported via the atmospheric and oceanic circulation to other locations far away from the source region. Other examples are transboundary advection of chemical substances such as sulfur (mainly in the atmosphere) and nutrients (mainly in the ocean). In the ocean, advection processes are also of crucial importance for search and rescue, oil drift, and drifting objects (e.g., fish larvae, rafts, man overboard, shipwrecks, etc.). Transport and spread of the above are governed by an advection–diffusion equation.

References

Arakawa A, Lamb VR (1977) Computational design of the basic dynamical process of the UCLA general circulation model. Methods Comput Phys 17:173–265

Bjerknes V (1904) Das Problem der Vettervorhersage, betrachtet vom Standpunkte der Mechanik und der Physik. Meteor Z 21:1–7

GESAMP (1991) (IMO/FAO/UNESCO/WMO/WHO/IAEA/UN/UNEP Joint Group of Experts on the Scientific Aspects of Marine Pollution), Coastal modelling, Technical report. International Atomic Energy Agency (IAEA), GESAMP Reports and Studies 43, 192 pp

Gill AE (1982) Atmosphere–ocean dynamics. International geophysical series, vol 30. Academic, New York

Hackett B, Røed LP, Gjevik B, Martinsen EA, Eide LI (1995) A review of the metocean modeling project (MOMOP). Part 2: model validation study. In: Lynch DR, Davies AM (eds) Quantitative skill assessment for coastal ocean models. Coastal and estuarine studies, vol 47. American Geophysical Union, Washington, pp 307–327

Lorenz EN (1955) Available potential energy and the maintenance of the general circulation. Tellus 2:157–167

Lynch DR, Davies AM (1995) Quantitative skill assessment for coastal ocean models. Coastal and estuarine studies, vol 47. American Geophysical Union, Washington

Røed LP (1997) Energy diagnostics in a $1\frac{1}{2}$-layer, nonisopycnic model. J Phys Oceanogr 27: 1472–1476

Røed LP (1999) A pointwise energy diagnostic scheme for multilayer, nonisopycnic, primitive equation ocean models. Mon Weather Rev 127:1897–1911

4 Diffusion Problem

In this chapter, we discuss the fundamentals of how to cast a PDE into finite difference form. More specifically, the reader will learn how to discretize the diffusion equation, and learn why some discretizations work and some do not. This will be the opportunity to introduce concepts such as *numerical stability*, *convergence*, and *consistency*. It will be explained how to check whether a discretization is stable and consistent, and the reader will learn about *explicit* and *implicit* schemes, the rudiments of *elliptic solvers*, and the concept of *numerical dissipation* or artificial damping inherent in our discretizations.

4.1 One-Dimensional Diffusion Equation

In its simplest form, the diffusion equation is

$$\partial_t \theta = \kappa \partial_x^2 \theta \ , \tag{4.1}$$

where the tracer θ is any dependent variable (e.g., potential temperature, humidity, speed, salinity, etc.), and κ is the diffusion or mixing coefficient. Thus, the diffusive flux is parameterized as down-the-gradient diffusion. Hereafter, the mixing coefficient is treated as being uniform in time and space (Fickian diffusion), unless specifically stated.

Note that (4.1) is parabolic in nature (see Sect. 2.2). The physical characteristic of the problem is, therefore, to transfer properties from one location to adjacent locations by conduction, without involving any motion. The process acts simply to even out differences. For instance, if the initial distribution is a very narrow bell-type tracer distribution (Fig. 5.7), diffusion acts to transfer the higher values to adjacent locations at the expense of the peak value as time progresses. Thus, the higher values, including the peak value, are diminished, while the values at adjacent locations are increased. If the diffusion process is allowed to go on forever, the tracer value finally becomes uniform within a finite domain. In an infinite domain, the tracer values

L. P. Røed, *Atmospheres and Oceans on Computers*,
Springer Textbooks in Earth Sciences, Geography and Environment,
https://doi.org/10.1007/978-3-319-93864-6_4

will be infinitely small, but will cover the infinite domain. In summary, the diffusion process acts to diminish differences so that the end result is a much smoother field.

An obvious example of a diffusion process in atmospheres, oceans, seas, and lakes is the turbulent mixing of heat. The tracer θ appearing in (4.1) is then the potential temperature. Another classic atmosphere–ocean example is the so-called Ekman problem, in which θ represents the velocity. In the atmosphere, it explains how the wind velocity is reduced in the planetary boundary layer due to friction at the surface. In the ocean, the Ekman problem explains how the momentum due to surface traction is transferred downwards through the water column.

4.2 Finite Difference Form

Consider, for instance, the situation where θ describes the deviation (or anomaly) of the potential temperature away from a given mean temperature profile at zero degrees Celsius. The strength of the turbulent mixing is then measured by the mixing coefficient κ appearing in (4.1). Furthermore, let

$$\theta(x, 0) = f(x)\ , \quad \forall x\ , \quad \text{and}\ \ \theta(0, t) = \theta(L, t) = 0\,^{\circ}\mathrm{C}\ , \quad \forall t\ , \tag{4.2}$$

where $f(x)$ is a known function in the domain $x \in \langle 0, L\rangle$, equal to zero for $x = 0, L$. The aim here is to solve (4.1) numerically and see how the initial anomaly of (4.2) evolves in time between the two positions $x = 0, L$. Taking $\theta = 0\,^{\circ}\mathrm{C}$ at $x = 0$ and $x = L$ to be fixed for all times, the boundary conditions are Dirichlet conditions. In addition, the initial anomaly is assumed to be different from the trivial solution $\theta(x, 0) = 0\,^{\circ}\mathrm{C}$, $\forall x$. Thus, there exists at least one position in space where $f(x) \neq 0\,^{\circ}\mathrm{C}$.

To find a numerical solution to (4.1), we use the notation given in Sect. 2.8. To begin with, the intervals $x \in \langle 0, L\rangle$ and $t \in \langle 0, T\rangle$, where T is some finite time, are divided into J and N increments of size Δx and Δt, respectively. We thus form a grid whose grid points are located at (x_j, t^n), where $x_j = (j-1)\Delta x$ and $t^n = n\Delta t$. The integers j and n are counters counting the number of steps needed to reach the grid point (x_j, t^n). Thus, $j \in [1, J+1]$ and $n \in [0, N]$, where $x_{J+1} = L$ and $t^N = T$ (see Fig. 4.1).

Next, we derive a finite difference approximation to the derivatives $\partial_t\theta$ and $\partial_x^2\theta$ at the grid points. Using a forward-in-time approximation to express $\partial_t\theta|_j^n$, and a centered-in-space approximation to express $\partial_x^2\theta|_j^n$, they take the form

$$[\partial_t\theta]_j^n = \frac{\theta_j^{n+1} - \theta_j^n}{\Delta t}\ , \tag{4.3}$$

$$[\partial_x^2\theta]_j^n = \frac{\theta_{j+1}^n - 2\theta_j^n + \theta_{j-1}^n}{\Delta x^2}\ . \tag{4.4}$$

Substituting (4.3) and (4.4) into (4.1) yields

$$\frac{\theta_j^{n+1} - \theta_j^n}{\Delta t} = \kappa\frac{\theta_{j+1}^n - 2\theta_j^n + \theta_{j-1}^n}{\Delta x^2}\ , \quad j = 2(1)J\ , \quad n = 0(1)N\ , \tag{4.5}$$

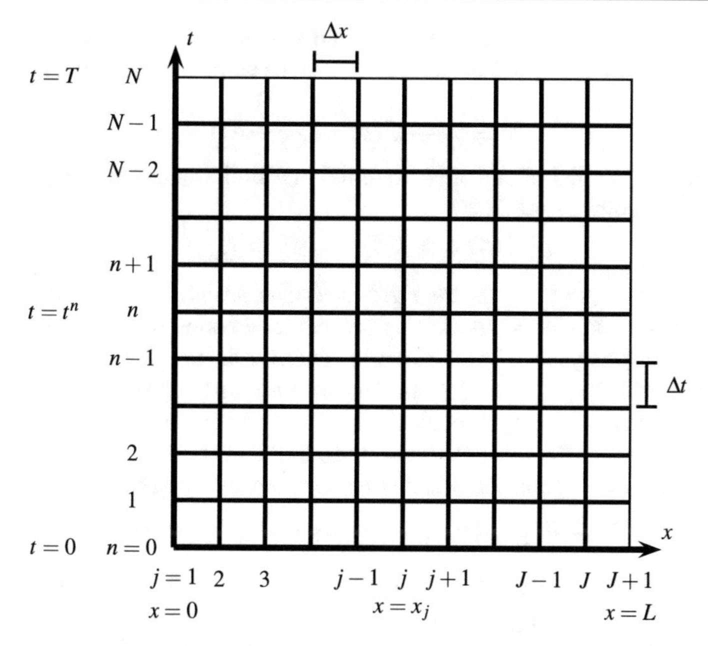

Fig. 4.1 Grid used for numerical solution of (4.1). The grid points in the x, t directions are incremented by Δx, Δt, respectively. There are $J+1$ points along the x-axis and $N+1$ points along the t-axis, counted by the dummy indices j, n. The coordinates of the grid points are $x_j = (j-1)\Delta x$ and $t^n = n\Delta t$, respectively

or

$$\theta_j^{n+1} = \theta_j^n + K\left(\theta_{j+1}^n - 2\theta_j^n + \theta_{j-1}^n\right)\,, \quad j = 2(1)J\,, \quad n = 0(1)N\,, \tag{4.6}$$

where

$$K = \frac{\kappa\Delta t}{\Delta x^2}\,. \tag{4.7}$$

Equations (4.5) and (4.6) are valid for $j = 2(1)J$ and $n = 0(1)N$ alone. At the boundaries $j = 1$ ($x = x_1 = 0$) and $j = J+1$ ($x = x_{J+1} = L$), and for $n = 0$ ($t = t^0 = 0$), the boundary and initial conditions (4.2) hold. In numerical language, they are

$$\theta_j^0 = f_j\,, \quad j = 1(1)J+1\,, \quad \text{and} \quad \theta_1^n = \theta_{J+1}^n = 0\,, \quad n = 0(1)N\,. \tag{4.8}$$

To find θ at the first time level $n = 1$, let $n = 0$ in (4.6):

$$\theta_j^1 = \theta_j^0 + K\left(\theta_{j+1}^0 - 2\theta_j^0 + \theta_{j-1}^0\right)\,, \quad j = 2(1)J\,. \tag{4.9}$$

Using this formula, the value for the first time level $n = 1$ at the grid point $j = 2$ is

$$\theta_2^1 = \theta_2^0 + K\left(\theta_3^0 - 2\theta_2^0 + \theta_1^0\right)\,. \tag{4.10}$$

The values of θ_1^0, θ_2^0, and θ_3^0 on the right-hand side of (4.10) are all known from the boundary and/or initial conditions (4.8). The value at the next grid point and at this time level is

$$\theta_3^1 = \theta_3^0 + K\left(\theta_4^0 - 2\theta_3^0 + \theta_2^0\right) . \tag{4.11}$$

The values at $j = 4, 5, \ldots, J$ may be found in a similar manner. In particular, for the last "wet" point $j = J$

$$\theta_J^1 = \theta_J^0 + K\left(\theta_{J+1}^0 - 2\theta_J^0 + \theta_{J-1}^0\right) , \tag{4.12}$$

where θ_{J-1}^0, θ_J^0, and θ_{J+1}^0 on the right-hand side are all known from the initial and/or boundary conditions (4.8). This procedure provides values for the potential temperature anomaly at all the interior grid points for the time level $n = 1$ (or at time $t = \Delta t$). In addition, θ_1^1 and θ_{J+1}^1 are known from the boundary condition at time level $n = 1$. Hence, looping through all the "wet" grid points $j = 2(1)J$ provides θ at all grid points for the time level $n = 1$. To find θ at the next time level $n = 2$, a similar procedure is followed. Substituting $n = 1$ into (4.6) yields

$$\theta_j^2 = \theta_j^1 + K\left(\theta_{j+1}^1 - 2\theta_j^1 + \theta_{j-1}^1\right) , \quad j = 2(1)J . \tag{4.13}$$

Once again, θ_1^2 and θ_{J+1}^2 are known from the boundary condition, and looping through (4.13) provides θ for all grid points at the time level $n = 2$. Following the same procedure for $n = 3, 4, \ldots, N$ then provides θ at all grid points (j, n), as required.

At the boundaries $j = 1$ and $J + 1$, the variable θ is known from the boundary condition (4.8). This reflects a well-known property of differential equations, whether they are PDEs or ordinary differential equations (ODEs), namely that they are valid only in the interior of a domain. At the boundaries (whether in time or space), the equations are replaced by the boundary condition. This is also true for the finite difference version (4.6) of (4.1).

It should also be observed that since $x_{J+1} = L = J\Delta x$, the values of J, L, and Δx cannot be chosen independently. Once two of them are chosen, the third is given by the formula

$$J = \frac{L}{\Delta x} . \tag{4.14}$$

Likewise,

$$N = \frac{T}{\Delta t} , \tag{4.15}$$

showing that N, Δt, and T depend on each other in a similar manner.

Since the time derivative in (4.1) is of first order, only one boundary condition in time is allowed. This is also true for its finite difference version (4.6). The application of a forward one-sided finite difference approximation in time is, therefore, the obvious choice in order to bring us from the initial time level $n = 0$ to the next time level. The accuracy of this scheme is $\mathcal{O}(\Delta t)$. Since a centered finite difference approximation was applied in space, the spatial accuracy is higher, namely $\mathcal{O}(\Delta x^2)$. A similar accuracy in time would require the application of a centered-in-time scheme for the time rate of change. One such scheme is

$$[\partial_t \theta]_j^n = \frac{\theta_j^{n+1} - \theta_j^{n-1}}{2\Delta t} . \tag{4.16}$$

Substituting (4.16) and (4.4) into (4.1), the new finite difference version takes the form

$$\frac{\theta_j^{n+1} - \theta_j^{n-1}}{2\Delta t} = \kappa \frac{\theta_{j+1}^n - 2\theta_j^n + \theta_{j-1}^n}{\Delta x^2}\ , \quad j = 2(1)J\ , \quad n = 0(1)N\ , \tag{4.17}$$

or

$$\theta_j^{n+1} = \theta_j^{n-1} + 2K\left(\theta_{j+1}^n - 2\theta_j^n + \theta_{j-1}^n\right)\ , \quad j = 2(1)J\ , \quad n = 0(1)N\ . \tag{4.18}$$

To obtain the solution at the first time level $n = 0$, let $n = 0$ in (4.18):

$$\theta_j^1 = \theta_j^{-1} + 2K\left(\theta_{j+1}^0 - 2\theta_j^0 + \theta_{j-1}^0\right)\ , \quad j = 2(1)J\ . \tag{4.19}$$

To find θ_j^1, we thus need to know θ_j^{-1}. This corresponds to knowing the potential temperature anomaly at a time $t < 0$, that is, we must know it at one time level prior to the initial time level. By using the one-sided forward scheme, we avoid this problem, but we thereby lose accuracy. However, as shown in Sects. 4.3–4.8, there are more pressing needs that dissuade us from using a centered-in-time centered-in-space finite difference approximation to solve the diffusion equation numerically.

4.3 Numerical Stability

The scheme (4.18) is said to be *numerically unstable*. This entails that the numerical solution, instead of following the continuous solution, steadily deviates from it, until it explodes. This commonly happens exponentially, just like an analytic instability (think of baroclinic and barotropic instabilities in the atmosphere and ocean). This behavior is, therefore, called a numerical instability, to distinguish it from physical instabilities. Thus, for the numerical solutions to have any legitimacy, they are required to be numerically stable. This is an absolute requirement, and is formulated as follows:

> A numerical scheme is stable if and only if the numerical solution is bounded within any given finite time span.

As a prelude to the stability analysis of the numerical scheme, it is useful to consider the analytic solution to (4.1). Since, as detailed in Sect. 2.10, any well-behaved function may be expanded in a Fourier series, the function $\theta(x, t)$ may be written

$$\theta(x,t) = \sum_{m=-\infty}^{\infty} \Theta_m(\alpha_m, t)\mathrm{e}^{\mathrm{i}\alpha_m x} = \sum_{m=-\infty}^{\infty} \theta_m\ . \tag{4.20}$$

Here, α_m is the wavenumber of the m th Fourier component. Each component

$$\theta_m = \Theta_m \mathrm{e}^{\mathrm{i}\alpha_m x} \tag{4.21}$$

is called a Fourier component, where $\Theta_m(t)$ is the time-dependent amplitude of the m th component. Substituting (4.20) into (4.1) yields

$$\partial_t \Theta_m = -\kappa \alpha_m^2 \Theta_m \ , \tag{4.22}$$

where the summation is dropped. Thus, each Fourier component is analyzed separately. Equation (4.22) is an ordinary differential equation (ODE). Solving it with respect to the amplitude Θ_m yields

$$\Theta_m = \Theta_m^0 \mathrm{e}^{-\kappa \alpha_m^2 t} \ . \tag{4.23}$$

Here, Θ_m^0 is the initial amplitude of mode m, that is, the value of Θ_m at time $t = 0$. To find these initial amplitudes, the initial distribution of θ is expanded in a Fourier series, viz.,

$$\theta(x, 0) = \sum_{m=-\infty}^{\infty} \Theta_m^0 \mathrm{e}^{\mathrm{i}\alpha_m x} \ . \tag{4.24}$$

Thus substituting (4.23) into (4.20) yields

$$\theta(x, t) = \sum_{m=-\infty}^{\infty} \Theta_m^0 \mathrm{e}^{-\kappa \alpha_m^2 t} \mathrm{e}^{\mathrm{i}\alpha_m x} \ , \tag{4.25}$$

which is a formal analytic solution to (4.1). Equations (4.23) and (4.25) reveal that the amplitude of each individual Fourier component decreases exponentially and monotonically as time progresses. Moreover, the shorter waves (high wave numbers) decrease faster than the longer waves (small wave numbers). This is in accord with the conclusion of Sect. 3.2 that diffusion acts to smooth out disturbances. Furthermore, as revealed by (4.25), this smoothing is not the same for all wavelengths. It is rather selective, in the sense that small scale disturbances are smoothed faster, while longer waves are less prone to damping in the same time period. Thus, diffusion acts like a filter, efficiently smoothing small scale noise, if any, without significantly damping larger scale motions.

Clearly, the numerical solution should behave similarly. In particular, the numerical solution is expected to decrease monotonically in time. Hence, if the numerical solution increases in time, it is obviously wrong and possibly unstable. Note that this instability has nothing to do with the accuracy of the chosen scheme. It is the initial truncation error inherent in our scheme that is allowed to grow in an uncontrolled way when the solution is unstable (see computer problem presented in Sect. 2.14.1).

4.4 Von Neumann Stability Analysis

We need a proper mathematical definition of stability in order to analyze whether the chosen scheme is stable. The requirement of numerical stability is commonly formulated by stating that for any finite time T, that is, for $0 < T < \infty$, there must exist a finite number, say B, such that

$$\left|\frac{\Theta_n}{\Theta_0}\right| \leq B \,, \tag{4.26}$$

where Θ_0 is the initial amplitude of the variable θ. For *linear systems*, and to a certain degree also nonlinear systems, the stability of the chosen scheme can be assessed analytically. Such an analysis is always performed *before* implementing the chosen scheme on the computer.

One such method is the so-called *von Neumann method.* To analyze stability, von Neumann suggested a method somewhat similar to solving the equations analytically. The first step is to define a discrete Fourier component similar to the analytic one given in (4.21), viz.,

$$\theta_j^n = \Theta_n \mathrm{e}^{\mathrm{i}\alpha j \Delta x} \,, \tag{4.27}$$

where Θ_n is the discrete amplitude at time level n and α is the wavenumber of that particular discrete mode, and where the subscript m on Θ and α is dropped for clarity. The next step is to define a *growth factor*, viz.,

$$G \equiv \frac{\Theta_{n+1}}{\Theta_n} \quad \Longrightarrow \quad \Theta_{n+1} = G\Theta_n \quad \text{and} \quad \Theta_{n-1} = G^{-1}\Theta_n \,. \tag{4.28}$$

The growth factor G is the amplification of the amplitude Θ as we proceed from one time level to the next. Interestingly, (4.28) is formally similar to (4.26), except that the growth factor G is defined as the ratio between the next and the previous time level, while (4.26) is the ratio between the value at a random time level and the initial value. Letting $n = 0$ in (4.28) yields

$$\Theta_1 = G\Theta_0 \,, \tag{4.29}$$

where Θ_0 is the initial amplitude. By letting $n = 1$ in (4.28) and making use of (4.29), we obtain

$$\Theta_2 = G\Theta_1 = G^2\Theta_0 \,. \tag{4.30}$$

Continuing this procedure by letting $n = 3, 4, \ldots,$ up to a random number $n = m$ gives

$$\Theta_m = G^m\Theta_0 \,. \tag{4.31}$$

Thus, G^m is the ratio between the amplitude at the random time level $n = m$ or random time $t = m\Delta t$ and the initial amplitude. As shown in Part B of the computer problem presented in Sect. 2.14.1, (4.26) is satisfied if

$$|G| \leq 1 \,, \tag{4.32}$$

since under these circumstances Θ^m decreases as the time level or time increases. The criterion (4.32) is called *von Neumann's stability condition.* In fact, (4.32) is a *sufficient condition* for stability in the sense that, if it is satisfied, the scheme is definitely stable. However, it is not automatically a *necessary condition*. The scheme may be stable even when (4.32) is not satisfied.

4.5 Stability of the Discrete Diffusion Equation

The first scheme to analyze is the forward-in-time centered-in-space (FTCS) scheme given by (4.6). Substituting (4.27) into (4.6), it takes the form

$$\Theta_{n+1} = \Theta_n + K\left(\mathrm{e}^{\mathrm{i}\alpha\Delta x} - 2 + \mathrm{e}^{-\mathrm{i}\alpha\Delta x}\right)\Theta_n \,, \tag{4.33}$$

where we have divided through by the common factor $\mathrm{e}^{\mathrm{i}\alpha j\Delta x}$. Since

$$\mathrm{e}^{\mathrm{i}\alpha\Delta x} + \mathrm{e}^{-\mathrm{i}\alpha\Delta x} = 2\cos\alpha\Delta x \,,$$

it follows that

$$\Theta_{n+1} = \left[1 - 2K(1 - \cos\alpha\Delta x)\right]\Theta_n \,, \tag{4.34}$$

and dividing by Θ_n, then provides the growth factor directly as

$$G = 1 - 2K(1 - \cos\alpha\Delta x) \,. \tag{4.35}$$

To satisfy the condition (4.32), i.e., $-1 \le G \le 1$, whence

$$-1 \le 1 - 2K(1 - \cos\alpha\Delta x) \le 1 \,. \tag{4.36}$$

Since $0 \le (1-\cos\alpha\Delta x) \le 2$, the right-hand inequality is satisfied for all wavenumbers α, all time steps Δt, and all space increments Δx. The inequality on the left-hand side, however, is satisfied if and only if

$$K(1 - \cos\alpha\Delta x) \le 1 \,. \tag{4.37}$$

Recall that $0 \le (1-\cos\alpha\Delta x) \le 2$, and hence (4.37) is satisfied for all wavenumbers α if

$$K \le \frac{1}{2} \quad \Longrightarrow \quad \Delta t \le \frac{\Delta x^2}{2\kappa} \,. \tag{4.38}$$

This condition ensures that (4.36) is satisfied. Hence, von Neumann's condition (4.32) is satisfied as well. Furthermore, (4.38) reveals that Δx and Δt cannot be chosen independently. Once Δx is chosen, the time step Δt must be chosen in accordance with (4.38). The forward-in-time centered-in-space scheme (4.6) is, therefore, *conditionally stable* under the condition (4.38).

Equation (4.37) tells us that the waves that first violate the inequality are waves with wavenumbers given by

$$\cos\alpha\Delta x = -1 \,. \tag{4.39}$$

This corresponds to waves that maximize $1-\cos\alpha\Delta x$. The wavenumbers satisfying (4.39) are

$$\alpha_m\Delta x = (2m-1)\pi \,, \quad m = 1, 2, \ldots \,, \tag{4.40}$$

with corresponding wavelengths

$$\lambda_m = \frac{2\pi}{\alpha_m} = \frac{2\Delta x}{2m-1} \,. \tag{4.41}$$

The most dominant of these waves is the wave corresponding to $m = 1$. Thus, the most unstable wave has wavelength

$$\lambda_1 = 2\Delta x \,. \tag{4.42}$$

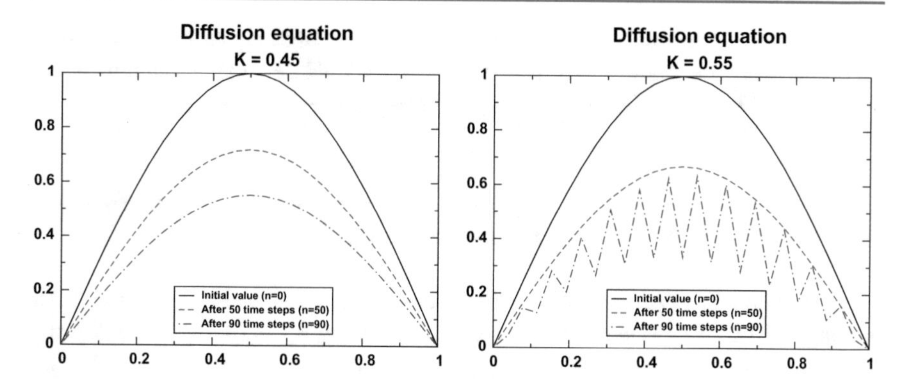

Fig. 4.2 Solutions of the diffusion equation using the FTCS scheme (4.6). Here, $K = \kappa \Delta t / \Delta x^2$ takes the values $K = 0.45$ (*left-hand panel*) and $K = 0.55$ (*right-hand panel*). The dependent variable θ is held fixed at the two boundaries $x = 0, 1$, and the initial condition is $\theta = \sin \pi x$. Solutions are shown for the time levels $n = 0$, $n = 50$, and $n = 90$. The sawtooth pattern in the *right-hand panel* for $n = 90$ is not present in the *left-hand panel*. This indicates that the stability condition (4.38) is violated for $K = 0.55$, but not for $K = 0.45$

This implies that the numerical instability will appear as "$2\Delta x$" noise, or noise at the wavelength $2\Delta x$. This commonly appears as a sawtooth pattern, like the one shown in Fig. 4.2.

In summary, the forward-in-time centered-in-space scheme (4.6) applied to the diffusion equation is a conditionally stable scheme. If the method had returned $|G| > 1$, in which case von Neumann's condition (4.32) is not met, the scheme would have been called an *unconditionally unstable* scheme. If the analysis had returned $|G| \leq 1$ regardless of our choice of Δx and Δt and wavenumbers α, the scheme would have been described as *unconditionally stable*. In the special case $|G| = 1$, the scheme is said to be *neutrally stable*.

When $|G| < 1$, (4.28) implies $|\Theta_{n+1}| < |\Theta_n|$. The amplitude of the next time level is, therefore, smaller than the amplitude of the previous time level. This property is inherent in all schemes for which $|G| < 1$. Since the amplitude decreases as time progresses, it implies that the solution is dissipative. However, the dissipation is artificial, being caused by the applied numerics. Furthermore, the smaller the growth factor the larger the artificial dissipation. This dissipation is, therefore, referred to as *numerical dissipation*. As can be seen from (4.35), the absolute value of the growth factor depends on the size of the spatial and temporal increments. It is, therefore, important to check that the numerical dissipation is as small as possible. This is achieved by choosing the increments so that the absolute value of the growth factor is as close to unity as possible.

Any given problem in oceanography and meteorology may include natural dissipation. Thus, if the scheme exhibits numerical dissipation, it is important to make certain that it is small compared to the natural, physical dissipation. Since neutral schemes (with $|G| = 1$) are energy conserving, a highly desirably property, they are always favored. If this is not possible, it is recommended to choose the time step and the space increments so as to minimize the numerical dissipation. Regarding the FTCS scheme, this implies that the time step Δt should be small enough to satisfy (4.38), but at the same time large enough to make $\Delta t \sim \Delta x^2/2\kappa$.

4.6 Centered-in-Time Centered-in-Space Scheme

Next, the stability of the centered-in-time centered-in-space (CTCS) scheme (4.18) is analyzed using von Neumann's method. Replacing θ_j^n in (4.18) by its discrete Fourier component as defined in (4.27), removing the common factor $\mathrm{e}^{\mathrm{i}\alpha j\Delta x}$, and using the definition of the growth factor (4.28) yields

$$G = G^{-1} - 4K(1 - \cos\alpha\Delta x)\,. \tag{4.43}$$

Multiplying (4.43) by G, and rearranging terms, it takes the form

$$G^2 + 2\lambda G - 1 = 0\,, \tag{4.44}$$

where

$$\lambda = 2K(1 - \cos\alpha\Delta x) \geq 0\,. \tag{4.45}$$

Equation (4.44) has the two solutions

$$G_{1,2} = -\lambda \pm \sqrt{1 + \lambda^2}\,. \tag{4.46}$$

Recall that, in order to be numerically stable, the two solutions must satisfy von Neumann's condition. Since

$$|G_2| = \lambda + \sqrt{1 + \lambda^2} \geq 1\,, \tag{4.47}$$

the CTCS scheme is unconditionally unstable when applied to the diffusion equation. The mantra is:

Never use a centered-in-time centered-in-space scheme for the diffusion problem. It is always unconditionally unstable.

4.7 Necessary Stability Condition

As mentioned above (Sect. 4.4), von Neumann's condition is a sufficient condition. The question arises as to whether it is too strict? Is it also a *necessary condition*?

The original stability requirement is formulated in (4.26). Replacing Θ_n by the expression in (4.31), we get

$$|G|^n \leq B\,. \tag{4.48}$$

Taking the natural logarithm of both sides yields

$$n \ln|G| \leq \ln B \equiv B'\,. \tag{4.49}$$

Even if von Neumann's condition is too strict, $|G|$ cannot be much greater than unity. Thus, writing $|G| = 1+\varepsilon$, it may safely be assumed that ε is a small positive number ($\varepsilon \ll 1$). Hence, $\ln|G| = \ln(1+\varepsilon) \approx \varepsilon$. Furthermore, $t^n = n\Delta t$, which implies that $n = t^n/\Delta t$. Substituting these expressions into (4.49) yields

$$\varepsilon \leq \frac{B'\Delta t}{t^n} = \mathscr{O}(\Delta t) \,. \tag{4.50}$$

Thus, a necessary condition for numerical stability is

$$|G| \leq 1 + \mathscr{O}(\Delta t) \,. \tag{4.51}$$

This proves that von Neumann's condition (4.32) is indeed too strict. However, most physical problems, even those containing instabilities, involve some physical dissipation. Thus, for all practical purposes, von Neumann's sufficient condition for stability ($|G| \leq 1$) may be applied. Likewise, it is good practice to make certain that $|G| \lesssim 1$.

Finally, note that in multivariable and multidimensional problems the growth factor is commonly a tensor or matrix, say $\mathscr{G}$, rather than a scalar as above. The von Neumann sufficient condition then stipulates that the spectral radius should be less than or equal to one. This is equivalent to requiring that the largest eigenvalue of $\mathscr{G}$ is less than or equal to one.

4.8 Explicit and Implicit Schemes

The spatial operators on the right-hand side of the schemes (4.6) and (4.18) are all evaluated at the time level n or earlier ($n-1, n-2, \ldots$). Such schemes are referred to as *explicit schemes*. In contrast, if the spatial operator on the right-hand side includes variables evaluated at the new time level $n+1$, the scheme is referred to as an *implicit scheme*. If all of them are evaluated at time level $n+1$, the scheme is a truly implicit scheme. If only one or a few are evaluated at time level $n+1$, it is referred to as a *semi-implicit scheme*. Likewise, for multivariable problems, e.g., the shallow water equations, where some of the terms are treated explicitly and others implicitly, the scheme is also referred to as a semi-implicit scheme.

As exemplified by the conditionally stable FTCS scheme (4.6), explicit schemes are always simple to solve. Once the unknowns are known for one time level at all grid points, computation of the variables at the next time level is straightforward [see (4.12) in Sect. 4.2]. In contrast, this is not always so for implicit or semi-implicit schemes. This is perhaps best illustrated by turning the explicit FTCS scheme (4.6) into an implicit FTCS scheme, achieved by replacing the centered FDA of $\left[\partial_x^2\theta\right]_j^n$ by one evaluated at time level $n+1$:

$$\left[\partial_x^2\theta\right]_j^{n+1} = \frac{\theta_{j-1}^{n+1} - 2\theta_j^{n+1} + \theta_{j+1}^{n+1}}{\Delta x^2} \,. \tag{4.52}$$

The finite difference version of (4.1) then takes the form

$$\theta_j^{n+1} = \theta_j^n + K\left(\theta_{j-1}^{n+1} - 2\theta_j^{n+1} + \theta_{j+1}^{n+1}\right) \,, \quad j = 2(1)J \,, \quad n = 0(1)N \,, \tag{4.53}$$

which is an implicit FTCS scheme. Rewriting by moving all terms containing evaluation at time level $n+1$ to the left-hand side, we obtain

$$-K\theta_{j-1}^{n+1} + (1+2K)\theta_j^{n+1} - K\theta_{j+1}^{n+1} = \theta_j^n \ , \quad j = 2(1)J \ , \quad n = 0(1)N \ . \tag{4.54}$$

On the left-hand side, θ appears not only at the grid point $(j, n+1)$, but also at the points $(j-1, n+1)$ and $(j+1, n+1)$. Since the value of θ_{j+1}^{n+1} is as yet unknown, it is impossible to simply loop through the j points at this time level. We require a method that solves for θ_{j-1}^{n+1}, θ_j^{n+1}, and θ_{j+1}^{n+1} simultaneously, which is akin to solving an elliptic problem. This illustrates that although the diffusion equation is parabolic, the implicit formulation turns the problem into an elliptic-type problem. To solve it, therefore, requires elliptic solvers (Sect. 4.11).

Turning an explicit scheme into an implicit scheme may affect its numerical stability. Once again, using von Neumann's method to analyze the stability of the scheme (4.53),

$$G = \frac{1}{1+\lambda} \ , \tag{4.55}$$

where λ is as given in (4.45). Since $\lambda \geq 0$, the growth factor $|G| \leq 1$ for all Δt, Δx and wavenumbers α. This implies that the implicit FTCS places absolutely no restriction on the time step Δt or the space increment Δx. It is, therefore, *unconditionally stable*.

As another example, consider an implicit version of the unconditionally unstable explicit CTCS scheme (4.18). As above, it is made implicit by evaluating the FDA of $\partial_x^2\theta$ at time level $n+1$, rather than at time level n. Consequently, (4.52) is used as the FDA for the spatial operator. The finite difference version of (4.1) than takes the form

$$\theta_j^{n+1} = \theta_j^{n-1} + 2K\left(\theta_{j-1}^{n+1} - 2\theta_j^{n+1} + \theta_{j+1}^{n+1}\right) \ , \quad j = 2(1)J, \quad n = 0(1)\ldots \ . \tag{4.56}$$

So once again, by treating the CTCS scheme implicitly, the parabolic diffusion equation is turned into an elliptic-type problem, and an elliptic solver has to be applied to solve it numerically. Furthermore, using von Neumann's method, the growth factor has the two solutions

$$|G_{1,2}| = \pm\frac{1}{\sqrt{1+2\lambda}} \ , \tag{4.57}$$

where λ is as given in (4.45). And once again, $|G| \leq 1$, independently of the choice of increments and for all wavenumbers. The implicit CTCS scheme (4.56) is, therefore, unconditionally *stable*. This is in stark contrast to the explicit CTCS scheme, which was unconditionally *unstable*.

The property of the above schemes of being unconditionally stable turns out to be true for all purely implicit schemes. The implication is that the time step is unconstrained. It can be as long as we like—stability is always guaranteed. For instance, an implicit scheme to forecast the next day's weather could be applied in one sweep using a 24 h time step. For this reason, it might be tempting to use implicit schemes in forecasting. Nevertheless, it is generally wise to shy away from implicit schemes, in particular, under circumstances when diffusion dominates the physics of the problem. As exemplified by (4.55) and (4.57), the growth factor is less than unity, so the scheme exhibits *numerical dissipation*.[1] Moreover, $|G|$ becomes small for long time steps, and consequently, the numerical dissipation becomes large for long time steps. In practice, to control numerical dissipation, the time step Δt has to be chosen small enough to yield a growth factor which is as close to one as possible.

In summary, although implicit schemes are unconditionally stable regardless of how long the time step Δt may be, the need to manage the numerical dissipation restricts the choice of time step. The reader is, therefore, strongly advised *not* to use implicit schemes. Finally, although most semi-implicit schemes turn out to be unconditionally stable too, this is not true for all of them. One example is the Crank–Nicholson scheme (see Sect. 4.10).

4.9 Convergence and Consistency: DuFort–Frankel

The numerical solution of a PDE should mimic the solution of the continuous PDE. This is equivalent to requiring that the numerical solution converge toward the solution of the continuous PDE as the space and time increments Δx, Δy, Δz, and Δt approach zero *independently* of one another. If this is so, the scheme is called a *convergent scheme*. Thus, if it is possible to prove that the scheme is convergent, this ensures that the solution will be close to the true solution for small enough increments.

To prove that a particular numerical scheme is convergent can be a tedious undertaking, especially for complex cases. Here is where the so-called Lax equivalence theorem comes to the rescue (Lax and Richtmyer 1956): "Given a properly posed initial value problem, boundary value problem, and a finite difference approximation that is *consistent* with the PDE, then *stability* is a necessary and sufficient condition for convergence." Thus, if the chosen scheme is stable, it is sufficient to prove that it satisfies the condition of being *consistent* to ensure that it is convergent. A consistent scheme is defined as one in which the finite difference version of the PDE approaches the continuous PDE in the limit as the increments tend to zero independently of one another.[2] If not, it is said to be an *inconsistent scheme*. Note that consistency is a condition on the finite difference version of the PDE, not on the solution. Since consistency is much simpler to analyze, it provides a much simpler way to establish convergence. In summary, requiring the chosen scheme to be stable and consistent is sufficient to ensure that its numerical solution is convergent. Consistency and numerical stability are therefore two fundamental properties that a scheme should possess.

[1] Only wavenumbers for which $\lambda = 0$ have a growth factor $|G| = 1$. However, all wavenumbers are present in any solution. Hence, the growth factor is in general less than unity.

[2] It is not important how fast each of them go to zero.

By definition all numerical schemes in which the FDAs are based on Taylor series expansions satisfy the consistency requirement. Thus, the schemes (4.6) and (4.18) are prime examples of consistent schemes. The scheme (4.6) is also a convergent scheme. In contrast (4.18) is not, since it is unconditionally unstable. However, it is quite easy to construct numerical schemes without using Taylor series expansions. In these cases a consistency analysis is required.

One example of an inconsistent scheme is obtained by replacing the diffusion term in the unconditionally unstable and explicit CTCS scheme (4.18) by a so-called *Dufort–Frankel* FDA. The trick is to replace the term θ_j^n in the centered-in-space FDA of (4.18) by a linearly interpolated expression, namely

$$[\theta]_j^n = \frac{1}{2}\left(\theta_j^{n+1} + \theta_j^{n-1}\right) .$$

The FDA of the diffusion term in (4.18) is then replaced by

$$\left[\partial_x^2\theta\right]_j^n = \left(\theta_{j+1}^n - \theta_j^{n-1} - \theta_j^{n+1} + \theta_{j-1}^n\right) , \tag{4.58}$$

whence

$$\theta_j^{n+1} = \theta_j^{n-1} + 2K\left(\theta_{j+1}^n - \theta_j^{n-1} - \theta_j^{n+1} + \theta_{j-1}^n\right) , \quad j = 2(1)J , \quad n = 0(1)N . \tag{4.59}$$

The appearance of the term θ_j^{n+1} on the right-hand side makes (4.59) implicit. Since all implicit schemes are stable, the introduction of the DuFort–Frankel FDA is expected to make the scheme stable. In contrast to the implicit scheme (4.56), the implicitness in (4.59) is now limited to the single term involving only the space grid point x_j. Moving this term from the right-hand side of (4.59) to the left-hand side and rearranging yields

$$\theta_j^{n+1} = \left[\theta_j^{n-1} + 2K\left(\theta_{j+1}^n - \theta_j^{n-1} + \theta_{j-1}^n\right)\right](1+2K)^{-1} . \tag{4.60}$$

The scheme (4.59) is then turned into an explicit scheme despite the introduction of the implicit term θ_j^{n+1}. This is one reason why it has become so popular, in particular, in oceanography, (e.g., Adamec and O'Brien 1978). A second reason is that, as expected, it is *unconditionally stable*. Applying von Neumann's stability analysis to check stability, we obtain

$$G_{1,2} = \frac{2K\cos\alpha\Delta x \pm \sqrt{1 - 4K^2\sin^2\alpha\Delta x}}{1+2K} . \tag{4.61}$$

Since $|G_{1,2}| < 1$, we do indeed find that (4.60) is unconditionally stable (see Exercise 4).

Since the Dufort–Frankel FDA was constructed by replacing one variable by an interpolated one, it is not based on Taylor series expansions. Therefore, the consistency of the scheme (4.59) has to be analyzed. To do this, we use the Taylor series expansions of Sect. 2.5. Substituting (2.57) and (2.58) into (4.59) yields

$$\partial_t\theta|_j^n - \kappa\partial_x^2\theta|_j^n = -\frac{\Delta t^2}{\Delta x^2}\left[1 + \mathcal{O}(\Delta t^2)\right] + \mathcal{O}(\Delta t^2) + \mathcal{O}(\Delta x^2) . \tag{4.62}$$

If all the terms on the right-hand side of (4.62) go to zero independently of the way the limits $\Delta x \to 0$ and $\Delta t \to 0$ are taken, then the scheme is consistent. Evidently, this is not the case for the first term on the right-hand side. It tends to infinity when Δx tends to zero faster than Δt. This implies that the scheme (4.59) is in general inconsistent.

As mentioned at the end of Sect. 3.2, the diffusion term $\kappa \partial_x^2 \theta$ is often added to the governing equation as a numerical artifact or "trick" to dissipate energy contained on the smaller scales. This small scale "noise" is often created by the presence of nonlinear terms in the equations. Their presence leads to interactions among the various wavelengths, which in turn is responsible for a cascade of energy toward progressively smaller and smaller scales (Phillips 1966). If we do not somehow parameterize this energy cascade at or near the grid resolution, energy tends to accumulate in the $2\Delta x - 4\Delta x$ tail of the energy spectrum. At some point in the integration, this accumulation of energy will lead to a violation of the linear numerical stability criterion and the numerical solution will go unstable (or "blow up").

When the diffusion term is used as an artifact to parameterize the energy cascade, it does not represent any of the physical processes we want to resolve. Rather it is introduced to avoid the model blowing up due to an accumulation of energy near the unresolved scales, by parameterizing a smooth transfer of energy towards the unresolved scales of our grid.[3] Since this parameterization and/or the parameters it contains may change in accordance with the model resolution, it is referred to as *subgrid scale* (SGS) parameterization (e.g., Griffies 2004, Chap. 7, p. 155).

4.10 Crank–Nicholson

Finally, we consider another popular scheme for the diffusion equation, called the *Crank–Nicholson scheme*. Like the Dufort–Frankel FDA, the scheme replaces the FDA of the diffusion term by an FDA called the *Crank–Nicholson* FDA, and once again turns the scheme into an implicit scheme. Its popularity can be put down to two features: it is unconditionally stable, and it is accurate to second order in both time and space.

The starting point is the two schemes

$$\theta_j^{n+1} = \theta_j^n + K\left(\theta_{j-1}^n - 2\theta_j^n + \theta_{j+1}^n\right) \tag{4.63}$$

and

$$\theta_j^{n+1} = \theta_j^n + K\left(\theta_{j-1}^{n+1} - 2\theta_j^{n+1} + \theta_{j+1}^{n+1}\right), \tag{4.64}$$

which are both centered-in-space schemes. While (4.63) is forward-in-time and explicit, (4.64) is backward in time and implicit. Moreover, (4.63) is conditionally stable under the condition $2K \leq 1$, while (4.63) is unconditionally stable. Finally, both are consistent, since they are based on Taylor series expansions.

[3]For a given grid size, $2\Delta x$ is the so-called Nüquist wavelength. Thus, wavelengths smaller than $2\Delta x$ remain unresolved.

Combining (4.63) and (4.64) yields

$$\theta_j^{n+1} = \theta_j^n + K\left[\gamma\left(\theta_{j-1}^{n+1} - 2\theta_j^{n+1} + \theta_{j+1}^{n+1}\right) + (1-\gamma)\left(\theta_{j-1}^n - 2\theta_j^n + \theta_{j+1}^n\right)\right], \tag{4.65}$$

where γ is a number between 0 and 1. If $\gamma = 0$, then (4.65) reduces to the explicit scheme (4.63). If $\gamma = 1$, then (4.65) reduces to the implicit scheme (4.64). If γ is between 0 and 1, the scheme contains both implicit and explicit terms. Hence the scheme (4.65) is semi-implicit and stability is not ensured.

Once again, von Neumann's method is used to analyze the stability of the scheme. The growth factor G is (see Exercise 5)

$$G = \frac{1 - 2K(1-\gamma)(1-\cos\alpha\Delta x)}{1 + 2K\gamma(1-\cos\alpha\Delta x)}. \tag{4.66}$$

Recall that the condition $|G| \leq 1$ or $-1 \leq G \leq 1$ is sufficient for numerical stability. From (4.66), it follows that the growth factor is always less than one, while $G \geq -1$ is satisfied for all wavelengths if and only if

$$2K(1-2\gamma) \leq 1. \tag{4.67}$$

Choosing γ so that $1/2 \leq \gamma \leq 1$, the left-hand side of (4.67) is always negative or zero. Under these circumstances, the scheme is stable regardless of the value chosen for the increments Δx and Δt and the wavenumber α. Thus, the scheme (4.65) is *unconditionally stable*, provided that $1/2 \leq \gamma \leq 1$. This should not come as a surprise, since under these circumstances the weight is on the implicit part.[4] If, however, $0 \leq \gamma < 1/2$, the weight is on the explicit part. In this case, the scheme is *conditionally stable* under the condition (4.67). We note that, for $\gamma = 0$, in which case (4.65) equals the forward-in-time centered-in-space finite difference approximation for the diffusion equation as displayed in (4.6), we do indeed retrieve the condition (4.38) of Sect. 4.4, that is, $2K \leq 1$.

The value $\gamma = 1/2$ is special. It constitutes the lower limit for which the scheme (4.65) is still unconditionally stable ($\gamma \geq 1/2$). Substituting this particular value of γ into (4.65) yields

$$\theta_j^{n+1} = \theta_j^n + \frac{1}{2}K\left(\theta_{j-1}^{n+1} - 2\theta_j^{n+1} + \theta_{j+1}^{n+1} + \theta_{j-1}^n - 2\theta_j^n + \theta_{j+1}^n\right), \tag{4.68}$$

which is the *Crank–Nicholson scheme*.

The scheme is special, not only because it is unconditionally stable, but also in another respect. Despite the fact that we use a forward, one-sided FDA for the time rate of change of $\partial_t\theta$, which normally yields a first-order accuracy in time, the time accuracy is nevertheless of second order. To prove this, we use the Taylor series expansions (2.28) and (2.24) (Sect. 2.5). Substituting these series into the centered differences on the right-hand side of (4.68) leads to

$$\frac{\theta_j^{n+1} - \theta_j^n}{\Delta t} = \frac{1}{2}\kappa\left(\partial_x^2\theta|_j^{n+1} + \partial_x^2\theta|_j^n\right) + \mathcal{O}(\Delta x^2). \tag{4.69}$$

[4] As a corollary, we note that this proves that the truly implicit scheme (4.64), which follows from (4.65) by letting $\gamma = 1$, is indeed unconditionally stable.

Using Taylor series to expand the left-hand side of (4.69), in particular, to replace θ_j^{n+1} and the term $\partial_x^2\theta|_j^{n+1}$, leads to

$$\frac{\theta_j^{n+1}-\theta_j^n}{\Delta t} = \partial_t\theta|_j^n + \frac{1}{2}\partial_t^2\theta|_j^n\Delta t + \mathcal{O}(\Delta t^2) \tag{4.70}$$

and

$$\partial_x^2\theta|_j^{n+1} = \partial_x^2\theta|_j^n + \partial_t(\partial_x^2\theta)|_j^n\Delta t + \mathcal{O}(\Delta t^2)\,. \tag{4.71}$$

Substituting these expressions into (4.69) and rearranging terms, it takes the form

$$\partial_t\theta_j^n = \kappa\partial_x^2\theta|_j^n - \frac{1}{2}\Big[\partial_t^2\theta|_j^n - \kappa\partial_t(\partial_x^2\theta)|_j^n\Big]\Delta t + \mathcal{O}(\Delta t^2) + \mathcal{O}(\Delta x^2)\,. \tag{4.72}$$

Furthermore, applying the continuous diffusion equation (4.1) leads to

$$\partial_t^2\theta|_j^n = \kappa\partial_t(\partial_x^2\theta)|_j^n\,. \tag{4.73}$$

The second term on the right-hand side of (4.72), therefore, vanishes, whence

$$\partial_t\theta|_j^n = \kappa\partial_x^2\theta|_j^n + \mathcal{O}(\Delta t^2) + \mathcal{O}(\Delta x^2)\,, \tag{4.74}$$

showing that the Crank–Nicholson scheme, besides being unconditionally stable, is of second-order accuracy in both time and space.

It nevertheless has one disadvantage compared to the more standard schemes. This may be illustrated by rearranging the terms in (4.68) to yield

$$K\theta_{j-1}^{n+1} - 2(1+K)\theta_j^{n+1} + K\theta_{j+1}^{n+1} = -2\theta_j^n - K\left(\theta_{j-1}^n - 2\theta_j^n + \theta_{j+1}^n\right), \tag{4.75}$$

which requires knowledge of θ_{j-1}^{n+1} and/or θ_{j+1}^{n+1} to solve for θ_j^{n+1}. Consider solving (4.75) by looping through increasing values of j. Then, for arbitrary j, the value of θ_{j+1}^{n+1} is unknown. Hence, using the Crank–Nicholson FDA, the otherwise parabolic diffusion equation is turned into an elliptic problem, just as we observed for the implicit CTCS scheme (4.56). Hence, they both require an elliptic solver for every time step. One such method, called a direct elliptic solver, is outlined in the next section.

4.11 A Direct Elliptic Solver

Many of the model codes used in numerical weather and numerical ocean weather prediction today apply semi-implicit methods, in the sense that some of the terms are treated implicitly. As we observed above for the implicit CTCS and Crank–Nicholson scheme, the consequence of introducing terms that are treated implicitly is that we then have to solve an equation like (4.75) for each time step. We, therefore, need a method whereby problems such as (4.75) can be solved fast and efficiently on a computer. Such methods are commonly referred to as *elliptic solvers*.

The most efficient elliptic solvers are those referred to as direct elliptic solvers. In the infancy of NWP, most elliptic solvers were iterative or indirect elliptic solvers. Even though they may be accelerated, as for instance when applying the iterative elliptic solver known as "successive over-relaxation", they are much slower than the direct methods. To gain an insight into the way direct elliptic solvers work, we now discuss *Gauss elimination*.

Gauss Elimination

The method involves two steps. The first is called a *forward sweep*. The final solution is then found by performing a *backward substitution*. To get started, (4.75) is first rewritten in the more general form

$$a_j^n \theta_{j-1}^{n+1} + b_j^n \theta_j^{n+1} + c_j^n \theta_{j+1}^{n+1} = h_j^n \ , \quad j = 2(1)J \ , \quad n = 0(1)N \ , \tag{4.76}$$

where a_j^n, b_j^n, and c_j^n are the coefficients in (4.75). The subscript j and superscript n attached to these coefficients may be functions of space and time. Likewise, h_j on the right-hand side represents all "forcing" terms, for instance, knowledge of the solution at the previous time step(s). Equation (4.76) is only valid within a finite domain. Thus, at the boundaries associated with the counters 1 and $J+1$, a boundary condition must be specified on θ_1^n and θ_{J+1}^n. For simplicity, it is assumed that these are known at this stage for all time levels n.

Since (4.76) has to be solved for every time step, only one time level is considered. Hence, the superscripts n and $n+1$ are dropped for convenience. Gauss elimination is, therefore, illustrated by solving

$$a_j \theta_{j-1} + b_j \theta_j + c_j \theta_{j+1} = h_j \ , \quad j = 2(1)J \ , \tag{4.77}$$

under the conditions

$$\theta_1 = \hat{\theta}_0 \ \text{ and } \ \theta_{J+1} = \hat{\theta}_L \ , \tag{4.78}$$

where $\hat{\theta}_0$ and $\hat{\theta}_L$ are considered known functions. Observe that (4.77) may be more compactly written as

$$\mathscr{A} \cdot \boldsymbol{\theta} = \mathbf{h}' \ , \tag{4.79}$$

where the tensor $\mathscr{A}$ is the *tridiagonal matrix*

$$\mathscr{A} = \begin{bmatrix} b_2 & c_2 & 0 & \dots & 0 & 0 & 0 \\ a_3 & b_3 & c_3 & \dots & 0 & 0 & 0 \\ 0 & a_4 & b_4 & \dots & 0 & 0 & 0 \\ \vdots & \vdots & \vdots & \ddots & \vdots & \vdots & \vdots \\ \vdots & \vdots & \vdots & \vdots & \ddots & \vdots & \vdots \\ 0 & 0 & 0 & \dots & a_{J-1} & b_{J-1} & c_{J-1} \\ 0 & 0 & 0 & \dots & 0 & a_J & b_J \end{bmatrix} \tag{4.80}$$

and the vectors $\boldsymbol{\theta}$ and $\mathbf{h}'$ are, respectively,

$$\boldsymbol{\theta} = \begin{bmatrix} \theta_2 \\ \theta_3 \\ \vdots \\ \theta_{J-1} \\ \theta_J \end{bmatrix} \tag{4.81}$$

and

$$\mathbf{h}' = \begin{bmatrix} h_2 - a_2\hat{\theta}_0 \\ h_3 \\ \vdots \\ h_J - c_J\hat{\theta}_L \end{bmatrix} . \tag{4.82}$$

Note that the boundary conditions are now contained in the vector $\mathbf{h}'$, and are, therefore, part of the "forcing". This is in line with the mantra that the boundary conditions are as important as the governing equations in determining the solution.

Forward Sweep

The idea is to replace all elements of the matrix $\mathscr{A}$ positioned in the lower left half with zeros. At the same time, it is convenient to normalize the diagonal elements, that is, turn each of them into the value 1. For $j = 2$, (4.79) leads to

$$b_2\theta_2 + c_2\theta_3 = h'_2 . \tag{4.83}$$

Normalizing by dividing through by b_2,

$$\theta_2 + d_2\theta_3 = w_2 , \tag{4.84}$$

where

$$d_2 = \frac{c_2}{b_2} \quad \text{and} \quad w_2 = \frac{h'_2}{b_2} . \tag{4.85}$$

For $j = 3$, (4.79) yields

$$a_3\theta_2 + b_3\theta_3 + c_3\theta_4 = h'_3 . \tag{4.86}$$

Substituting for θ_2 from (4.84) and normalizing gives

$$\theta_3 + d_3\theta_4 = w_3 , \tag{4.87}$$

where

$$d_3 = \frac{c_3}{b_3 - d_2 a_3} \quad \text{and} \quad w_3 = \frac{h'_3 - a_3 w_2}{b_3 - d_2 a_3} . \tag{4.88}$$

Repeating this for $j = 4$ leads to

$$\theta_4 + d_4\theta_5 = w_4 , \tag{4.89}$$

where

$$d_4 = \frac{c_4}{b_4 - d_3 a_4} \quad \text{and} \quad w_4 = \frac{h'_4 - a_4 w_3}{b_4 - d_3 a_4} . \tag{4.90}$$

Repeating this procedure for arbitrary j leads to

$$\theta_j + d_j\theta_{j+1} = w_j , \quad j = 2(1)J - 1 , \tag{4.91}$$

where the coefficients d_j and w_j are defined by the recursion formulas

$$d_j = \begin{cases} \dfrac{c_2}{b_2} , & j = 2 , \\ \dfrac{c_j}{b_j - d_{j-1}a_j} , & j = 3(1)J - 1 , \\ 0 & j = J , \end{cases} \quad w_j = \begin{cases} \dfrac{h'_2}{b_2} , & j = 2 , \\ \dfrac{h'_j - a_j w_{j-1}}{b_j - d_{j-1}a_j} , & j = 3(1)J , \end{cases} \tag{4.92}$$

respectively, where d_J is set to zero. The reason for the latter is that, for $j = J$, (4.79) yields

$$a_J\theta_{J-1} + b_J\theta_J = h'_J \,, \tag{4.93}$$

and, substituting for θ_{J-1} using (4.91) with $j = J - 1$, (4.93) takes the form

$$\theta_J = w_J \,. \tag{4.94}$$

Note that all the coefficients d_j and w_j can be calculated once and for all.

In matrix form, (4.80) now reads

$$\mathscr{A}' \cdot \boldsymbol{\theta} = \mathbf{w} \,, \tag{4.95}$$

where the matrix $\mathscr{A}'$ is

$$\mathscr{A}' = \begin{bmatrix} 1 & d_2 & 0 & \dots & 0 & 0 \\ 0 & 1 & d_3 & \dots & 0 & 0 \\ 0 & 0 & 1 & \dots & 0 & 0 \\ \vdots & \vdots & \vdots & \ddots & \vdots & \vdots \\ 0 & 0 & 0 & \dots & 1 & d_{J-1} \\ 0 & 0 & 0 & \dots & 0 & 1 \end{bmatrix} , \tag{4.96}$$

and the vector $\mathbf{w}$ is

$$\mathbf{w} = \begin{bmatrix} w_2 \\ w_3 \\ \vdots \\ w_{J-1} \\ w_J \end{bmatrix} . \tag{4.97}$$

We, thus, complete the *forward sweep*, in which the equation matrix is normalized and has upper triangular form.

Backward Substitution

First, we note that all the values w_j and d_j are known using the recursion formulas (4.92). Second, (4.94) says that θ_J is simply given by w_J, where the latter is known from (4.92). Solving (4.91) with respect to θ_j, we obtain

$$\theta_j = w_j - d_j\theta_{j+1} \,, \quad \text{for} \quad j = J - 1(-1)2 \,, \tag{4.98}$$

and hence that

$$\theta_{J-1} = w_{J-1} - d_{J-1}\theta_J \,, \quad \theta_{J-2} = w_{J-2} - d_{J-2}\theta_{J-1} \,. \tag{4.99}$$

Continuing this process using (4.98), all the remaining values θ_j may be found by backward substitution.

The Gauss elimination method is very simple to program, and is efficient and fast on the computer. An example of the usefulness of this method, in which the reader is also required to program the method, is presented in the computer problem of Sect. 4.14.3. The reader is encouraged to do this problem, at least to solve the resulting ODE by employing the Gauss elimination method.

4.12 Summary and Remarks

In this chapter, we introduce some of the crucial concepts needed to work with finite difference methods, illustrating with the one-dimensional diffusion equation presented in Sect. 4.1. Section 4.2 begins by detailing the construction of a finite difference equation or scheme to solve the diffusion equation numerically. This is the core of the finite difference method. The method is applicable to any PDE, and the diffusion equations serve merely as an example.

To be relevant, the constructed scheme must be *numerically stable*. This was discussed in Sects. 4.3 and 4.4. The latter outlines von Neumann's stability analysis. Section 4.5 shows how to apply von Neumann's method, and shows that a forward-in-time centered-in-space (FTCS) scheme is conditionally stable. It also explains why a "sawtooth" pattern is the first sign of instability. Section 4.6 then shows that if we replace the FTCS scheme by a centered-in-time centered-in-space (CTCS) scheme, the result will be unconditionally unstable. The von Neumann stability analysis provides only a sufficient condition for stability. Section 4.7 identifies a necessary condition.

Using the finite difference version of the diffusion equation, Sect. 4.8 explains the difference between *explicit* and *implicit* schemes. While the time step in explicit schemes is always constrained by the stability condition, the implicit scheme was shown to be unconditionally stable. However, it is also shown that increasing the time step leads to increasing *numerical dissipation*. In practice, the time step in implicit schemes is, therefore, also constrained.

Section 4.9 introduces the concepts of *consistency* and *convergence* of the finite difference scheme. Convergence is important, since it is only through this requirement that we can be sure that the numerical solution tends to the "true" analytic solution in the limit of infinitely small time steps and space increments. To be convergent a scheme must be stable and consistent. The latter is ensured if the continuous equations are reproduced in the limit of infinitely small time steps and space increments. However, we argue that use of an inconsistent artificial diffusion term is acceptable as a means to control nonlinear instability (Sects. 6.7 and 10.3), as long as the terms that contain the physics are treated consistently.

The Crank–Nicholson scheme presented in Sect. 4.10 is widely used to treat the diffusion term in any PDE. By treating half of the diffusion term implicitly, it becomes second order in time. However, through this introduction, the finite difference equation becomes elliptic, even though the diffusion equation is parabolic, so for each time step one has to solve an elliptic equation. We then require an efficient and fast elliptic solver, and one such solver is presented in Sect. 4.11.

The chapter ends with some exercises and three computer problems. The reader is encouraged to do one of the computer problems in Sects. 4.14.1 or 4.14.2. For readers interested in how a direct elliptic solver works, the computer problem in Sect. 4.14.3 is recommended.

4.13 Exercises

1. Show that the scheme (4.18) is unconditionally unstable.
   Hint: Show that $|G| > 1$, regardless of the choice made for Δt and Δx.
2. Show that if $|G| = 1$, then the chosen scheme has no numerical dissipation.
3. Show that the growth factor associated with the scheme (4.56) is

$$G = \left[1 + \frac{4\kappa\Delta t}{\Delta x^2}(1 - \cos\alpha\Delta x)\right]^{-1/2}, \tag{4.100}$$

   and hence that the scheme is unconditionally stable. Show also that $|G| < 1$ for all wavelengths. Note that $|G|$ decreases as Δt increases.
4. Show that the growth factor for the DuFort–Frankel scheme (4.59) is indeed (4.61).
5. Show that the expression (4.66) is indeed the expression for the growth factor of the scheme (4.65) when using von Neumann's analysis method.

4.14 Computer Problems

4.14.1 Vertical Diffusive Mixing

In the ocean and atmosphere, the vertical (and horizontal) heat exchange is predominantly driven by turbulent mixing. Here, we consider solutions to (4.1) of the form

$$\partial_t \psi = \kappa \partial_z^2 \psi , \tag{4.101}$$

relevant for the atmospheric planetary boundary layer (PBL) and the oceanic mixed layer (OML).

It is assumed that the PBL is contained between two fixed points representing the ground at $z = 0$ and the top of the PBL at $z = H$. Similarly, it is assumed that the OML is contained between the bottom of the OML at $z = -D$ and the ocean surface at $z = 0$. At these levels, (4.101) is replaced by the relevant boundary conditions, as specified below.

In both cases, the aim is to solve (4.101) using numerical methods. The solution is sought in the time span $t \in [0, N\Delta t]$, where t is time, $N = 1201$ is the total number of time steps, and Δt is the time step. Furthermore, let $z_j = (j-1)\Delta z$ be the jth position along the vertical coordinate, where Δz is the space increment.[5] The total number of grid points in each sphere, including the two boundary points, is 27. The height of the PBL is $H = 270\,\text{m}$, while the depth of the OML is $D = 30\,\text{m}$.

[5] This is also commonly referred to as the mesh or grid size.

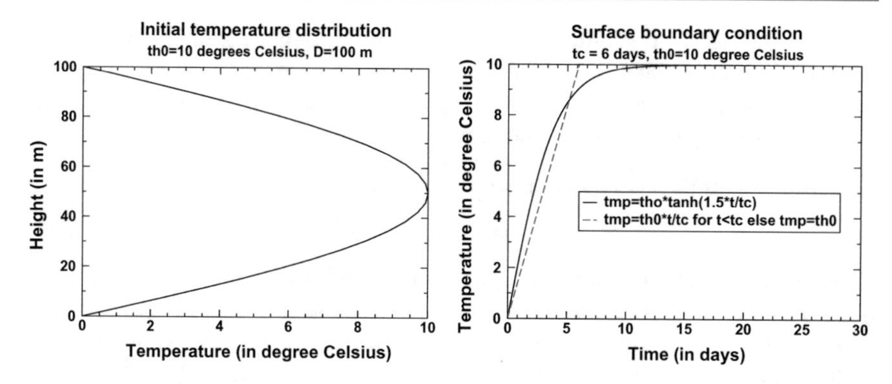

Fig. 4.3 *Left*: Initial temperature distribution according to (4.103). *Right*: Time evolution of the surface temperature according to (4.106) (*red dashed curve*) and (4.107) (*black solid curve*)

A. Develop a forward-in-time centered-in-space scheme (FTCS) to replace (4.101). Show that the derived FTCS scheme is stable under the condition $K \leq 1/2$, where

$$K = \frac{\kappa \Delta t}{\Delta z^2} \,. \tag{4.102}$$

[Hint: Use the von Neumann method.]

B. Next, develop a finite difference version of (4.101) that is centered in both time and space (CTCS), and show that it is unconditionally unstable in a numerical sense.

C. Relatively speaking, oceanic motions are slow compared to atmospheric ones. This is also true regarding mixing. The mixing coefficient in the ocean is a factor 10^{-4} less than in the atmosphere (Gill 1982). Let $K = 0.45$, let the atmospheric mixing coefficient be $\kappa_A = 30\,\text{m}^2\,\text{s}^{-1}$, and let the similar oceanic mixing coefficient be $\kappa_O = \kappa_A \times 10^{-4}$. Compute the time steps Δt needed for the atmospheric and oceanic applications to be stable. These time steps are vastly different. Discuss why and possible consequences.

D. Consider first solutions within the PBL. Let the initial temperature distribution be sinusoidal and given by

$$\psi(z, 0) = \psi_0 \sin \frac{\pi z}{D} \,, \quad z \in [0, H] \,, \tag{4.103}$$

as displayed in the left-hand panel of Fig. 4.3. Here $\psi_0 = 10\,°\text{C}$. Furthermore, let the temperature at ground level ($z = 0$) and at the top of the PBL ($z = D$) be fixed at the freezing point for all times. Thus,

$$\psi(0, t) = \psi(H, t) = 0\,°\text{C} \,, \quad \forall t \,. \tag{4.104}$$

Use the stable FTCS scheme developed in Part A to find the numerical solution to (4.101) for the two cases $K = 0.45$ and $K = 0.55$. Plot the results for $n = 0$, 100, 200, 600, and 1200, in which the height and temperature are made dimensionless by dividing through by H and ψ_0, respectively.

Derive the analytic solution to (4.101) given the initial condition (4.103) and the boundary conditions (4.104). Assess and discuss the solutions by comparing the numerical and analytic solutions. Explain, in particular, why the solution for $K = 0.55$ develops a "sawtooth" pattern.

E. Consider next the OML. Let the initial condition be

$$\psi(z,0) = 0\,^{\circ}\mathrm{C}\,, \quad z \in [-D, 0]\,, \tag{4.105}$$

throughout the water column. Thus, the initial condition is the trivial solution to (4.101). In contrast to the PBL application, the diffusion process is generated by allowing the ocean surface to be heated from above. Specifically, the temperature at the surface $z = 0$ increases from zero to the fixed temperature $\theta_0 = 10\,^{\circ}\mathrm{C}$ after some finite time $t = t_{\mathrm{c}}$. This is achieved by specifying the boundary condition by

$$\psi(0,t) = \psi_0 \begin{cases} t/t_{\mathrm{c}}\,, & 0 < t < t_{\mathrm{c}}\,, \\ 1\,, & t \geq t_{\mathrm{c}}\,, \end{cases} \tag{4.106}$$

or by using a hyperbolic tangent (a well-behaved function), viz.,

$$\psi(0,t) = \psi_0 \tanh \frac{\gamma t}{t_{\mathrm{c}}}\,, \tag{4.107}$$

where $t_{\mathrm{c}} = 6$ days and $\gamma = 1.5$ (see the right-hand panel of Fig. 4.3). At the bottom of the mixed layer $z = -D$, the temperature is fixed at freezing point, i.e.,

$$\psi(-D,t) = 0\,^{\circ}\mathrm{C}\,. \tag{4.108}$$

Once again, we choose the time step so that $K = 0.45$. Plot the results for $n = 0$, $n = 100$, $n = 200$, and $n = 400$. Use either (4.106) or (4.107) as the surface boundary condition.

Assess and discuss the solution, and compare it with the analytic solution given by (4.101).

4.14.2 Vertical Mixing in a Coupled Atmosphere–Ocean Model

Here, we consider temperature evolution in a coupled atmosphere–ocean model. The only active process is heat diffusion by vertical mixing, all other motion and processes being inhibited. Thus, the temperature is governed by a diffusion equation. In its simplest form, neglecting radiation and all other possible sources and sinks, it is

$$\partial_t \theta = \kappa \partial_z \theta\,, \tag{4.109}$$

where θ is the temperature, z the vertical coordinate, and κ the eddy diffusivity or mixing coefficient. The latter is assumed to be constant in time and space, but different for the two spheres (Gill 1982). Their specific values are given in Table 4.1, where κ_{A} is associated with the atmosphere and κ_{O} with the ocean.

The task is to solve (4.109) for the time span $t \in \langle 0, T]$. Let the atmosphere and the ocean span the vertical space $z \in [-D, H]$, where $z \in \langle -D, 0\rangle$ spans the oceanic mixed layer (OML) and $z \in \langle 0, H\rangle$ spans the planetary boundary layer (PBL), so

Table 4.1 Parameter values used in the computer problem of Sect. 4.14.2

Symbol	Description	Value	Unit
κ_A	Atmospheric diffusion/mixing coefficient	30.0	$m^2\,s^{-1}$
κ_O	Oceanic diffusion/mixing coefficient	3.0×10^{-3}	$m^2\,s^{-1}$
H	Height of PBL	270.0	m
D	Depth of OML	30.0	m
θ_T	Temperature at top of PBL	0.0	°C
θ_B	Temperature at bottom of OML	10.0	°C

that $z = 0$ is the location of the interface between them. D and H are considered to be constants and their values are given in Table 4.1.

At the interface, the oceanic heat flux toward the interface from below must be balanced by the atmospheric heat flux away from the interface. Any imbalance would imply that heat accumulates at the interface. Furthermore, the temperature itself has to be continuous at the interface. Thus,

$$\kappa_O \partial_z \theta_O = \kappa_A \partial_z \theta_A \text{ and } \theta_O = \theta_A , \quad \text{at } z = 0 , \tag{4.110}$$

where the subscripts A and O refer to the atmosphere and ocean, respectively. Furthermore, the temperatures at the bottom of the OML (θ_B) and the top of the ABL (θ_T) are fixed, and their values are listed in Table 4.1. Thus, the boundary conditions at the bottom of the OML and top of the PBL are, respectively,

$$\theta_O = \theta_B , \quad z = -D , \tag{4.111}$$

$$\theta_A = \theta_T , \quad z = H . \tag{4.112}$$

Finally, we suppose that the temperature is initially constant and equal to the temperature at the bottom of the OBL everywhere, except at the top of the PBL, where (4.112) holds. Thus,

$$\theta_A(z, 0) = \theta_O(z, 0) = \theta_B , \quad -D \leq z < H , \tag{4.113}$$

$$\theta_A(H, 0) = \theta_T . \tag{4.114}$$

A. Explain why we need exactly the number of boundary and initial conditions specified by (4.110)–(4.114).

B. Show that the solution to (4.109) as $t \to \infty$, and which satisfies the boundary conditions, is

$$\theta = \begin{cases} -\gamma\kappa_A(z + D) + \theta_B , & -D \leq z < 0 , \\ -\gamma\kappa_O(z - H) + \theta_T , & 0 \leq z \leq H , \end{cases} \tag{4.115}$$

where

$$\gamma = \frac{\theta_B - \theta_T}{\kappa_A D + \kappa_O H} . \tag{4.116}$$

C. Develop a forward-in-time and centered-in-space (FTCS) scheme to replace the continuous equations (4.109) for the two spheres. Show that the derived scheme is stable under the conditions $K_A \leq 1/2$ and $K_O \leq 1/2$, where

$$K_A = \frac{\kappa_A \Delta t_A}{\Delta z_A^2} \quad \text{and} \quad K_O = \frac{\kappa_O \Delta t_O}{\Delta z_O^2} \,. \tag{4.117}$$

Here, Δt_A and Δt_O are the time steps of the atmospheric and oceanic parts, respectively, while Δz_A and Δz_O are the respective space increments.

D. Let $j_A = 1(1)J_A + 1$ be the counter in the atmospheric part and $j_O = 1(1)J_O + 1$ the counter in the oceanic part, where $j_A = 1$ is associated with $z = 0$ and $j_O = 1$ with $z = -D$. Thus, the interface is associated with $j_O = J_O + 1$ and $j_A = 1$. Show that under these circumstances $H = J_A \Delta z_A$ and $D = J_O \Delta z_O$, and hence that the ratio between the time steps is

$$\frac{\Delta t_A}{\Delta t_O} = \frac{\kappa_O K_A}{\kappa_A K_O} \left(\frac{\Delta z_A}{\Delta z_O} \right)^2 = \frac{\kappa_O K_A}{\kappa_A K_O} \left(\frac{H J_O}{D J_A} \right)^2 \,. \tag{4.118}$$

E. Assume that the number of grid points in the two spheres is $J_A = J_O = J$, and that $K_A = K_O = K$. Use (4.117) and the numbers given in Table 4.1 to show that under these circumstances $\Delta t_A \ll \Delta t_O$, that is, the time step we have to use for the atmospheric part is much less than the one for the oceanic part. Discuss why there is such a difference in the time steps, and the possible consequences this has in terms of physics and numerics under the condition that $K = 0.45$.

F. Next, assume that $K_O = K_A = K = 0.45$, but that $J_A \neq J_O$. If $\Delta t_A = \Delta t_O$, then $\Delta z_O \neq \Delta z_A$, and if $\Delta z_O = \Delta z_A$, then $\Delta t_O \neq \Delta t_A$. From a numerical point of view it is advantageous to let $\Delta t_A = \Delta t_O$. Explain why.

G. Next let $\Delta t_A = \Delta t_O$ and $J_O = J_A = J$. Show that under these circumstances

$$\frac{K_O}{K_A} = \frac{\kappa_O H^2}{\kappa_A D^2} \,. \tag{4.119}$$

Does K_O satisfy the sufficient condition for stability? Discuss possible implications regarding the growth factor.

H. Use the FTCS scheme of Part A to find the numerical solution to (4.109) that satisfies the initial and boundary conditions. Solve for two cases: one stable case in which $K_A = K_O = 0.45$ and a second, unstable case in which $K_A = K_O = 0.55$. For both, let $\Delta t_A = \Delta t_O = \Delta t$, $J_A = 28$, and $J_O = 301$. The latter implies that $\Delta z_O = 0.1$ m. For the stable case, plot the vertical distribution of the potential temperature at the time levels $n = 0$ (initial distribution), $n = 50$, $n = 500$, $n = 5\,000$, $n = 50\,000$ and $n = 500\,000$. For the unstable case, plot the distribution for the time levels $n = 0$, $n = 10$, $n = 20$, $n = 30$, $n = 40$, and $n = 50$. Compute the time in hours for each case and include a caption (see Fig. 4.4).

In the two cases, the scale along the axes should be such that the temperature range is $\theta \in [\theta_B, \theta_T]$ and the depth/height range $z \in [-D, H]$. The plot should show the

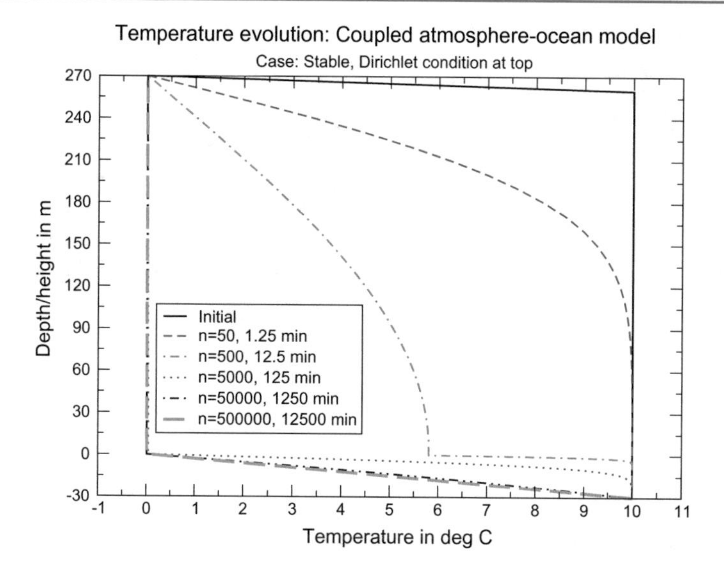

Fig. 4.4 Initial temperature distribution according to (4.113), and time evolution of the vertical temperature profile for the five time levels listed as time progresses

height/depth along the vertical axis and the temperature along the horizontal axis, as in Fig. 4.4, which gives the results for the stable case.

I. Assess and discuss the solution. Compare it with the steady state solution (4.115). Explain in particular why the solution for $K_A = 0.55$ develops a "sawtooth" pattern.

J. What happens if the upper boundary condition is changed to a no-flux condition

$$\partial_z \theta|_{z=H} = 0 \,. \tag{4.120}$$

Solve for this case and plot the vertical potential temperature distribution. Assess the solution by comparing it to the former solution with a fixed temperature condition at the top of the PBL.

4.14.3 Yoshida's Equatorial Jet Current

Following Yoshida (1959), we consider an "infinite" equatorial ocean consisting of two immiscible layers with a density difference $\Delta\rho$ (Fig. 4.5). The density of the lower layer equals the reference density ρ_0. The lower layer is thick with respect to the upper layer. At time $t = 0$, the ocean is at rest and the thickness of the upper layer equals its equilibrium depth H. At this particular time, the ocean is forced into motion by turning on a westerly wind (wind from the west).

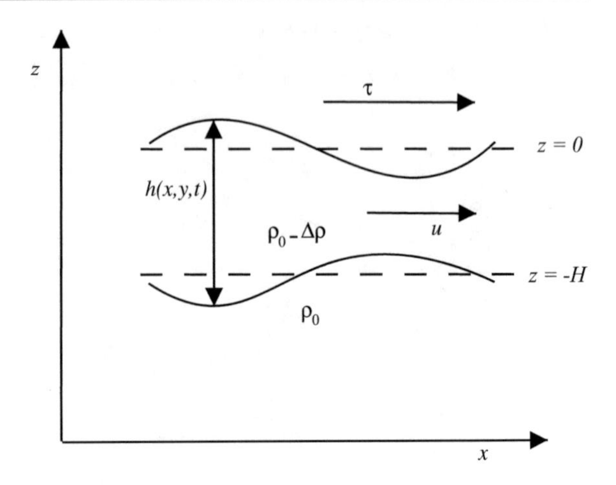

Fig. 4.5 Reduced gravity ocean model consisting of two layers with density difference $\Delta\rho$

The governing equations of such a "reduced gravity" model of the ocean are

$$\partial_t u - \beta y v = \frac{\tau^x}{\rho_0 H} , \tag{4.121}$$

$$\partial_t v + \beta y u = -g' \partial_y h , \tag{4.122}$$

$$\partial_t h + H \partial_y v = 0 . \tag{4.123}$$

Here, $u = u(y, t)$ and $v = v(y, t)$ are, respectively, the east–west and north–south components of the velocity in a Cartesian coordinate system (x, y, z) with x directed eastward along the equator, y directed northwards with $y = 0$ at the equator, and z directed along the negative gravitational direction (see Fig. 4.5). The effect of the Earth's rotation is given by the Coriolis parameter $f = 2\Omega \sin\phi$, where Ω is the rate of rotation and ϕ is the latitude. The westerly wind is given by the wind stress component τ^x, which is fixed in time. Furthermore, is $g' \equiv g(\Delta\rho_0/\rho)$ is the reduced gravity, where g is the gravitational acceleration. The instantaneous thickness of the upper layer is given by $h = h(y, t)$.

At the equator, $f = 0$ and increases with increasing latitude. A simplified parameterization of this effect uses the so-called β-plane approximation, viz.,

$$f = \beta y , \quad \text{where } \beta = \partial_y f|_{y=0} . \tag{4.124}$$

Thus, (4.124) represents the first-order effect of the change in the Earth's rotation rate with latitude.

A. Show that inertial oscillations[6] are eliminated by neglecting $\partial_t v$ in (4.122).

B. Show that the system of equations (4.121)–(4.123) reduces to the ordinary differential equation

$$L^4 \partial_y^2 v - y^2 v = aLy , \tag{4.125}$$

[6]These are oscillations in which the frequency equals the inertial frequency f.

where

$$L = \sqrt{\frac{c}{\beta}}, \quad a = \frac{\tau^x}{\rho_0 \beta L H}, \quad c = \sqrt{g' H}, \tag{4.126}$$

under the condition that the inertial oscillations are eliminated.

C. Two boundary conditions are allowed. Why is this? Let the boundary conditions be $v|_{y=0} = 0$ and $v|_{y\to\infty} = 0$.

D. Make (4.125) dimensionless by letting $y = L\hat{y}$, $(u, v) = a(\hat{u}, \hat{v})$, and $t = (\beta L)^{-1}\hat{t}$. Use a direct elliptic solver, e.g., Gauss elimination, to solve the dimensionless version of (4.125). Let $\Delta y = 0.1$ and plot $\hat{v}$ and $\hat{u}$ at time $\hat{t} = 1$ as a function of $\hat{y}$ from $\hat{y} = 0$ to $\hat{y} = 8$. Note that $v|_{y\to\infty} = 0$, and hence that $\hat{v}$ is different from zero at $\hat{y} = 8$. Explain how to make use of the condition that $v|_{y\to\infty} = 0$.

E. Discuss the numerical solution. Let $\tau^x = 0.1\,\text{Pa}$, $\beta = 2.0 \times 10^{-11}\,\text{m}^{-1}\,\text{s}^{-1}$, $L = 275\,\text{km}$, $\rho = 10^3\,\text{kg}\,\text{m}^{-3}$, and $H = 200\,\text{m}$. What is the maximum current in the equatorial jet for $\hat{t} = 1$?

F. Solve (4.125) analytically. *Hint*: Make a series using Hermitian polynomials (see Abramowitz and Stegun 1965).

References

Abramowitz M, Stegun I (1965) Handbook of mathematical functions, 9th edn. Dover Publications, Inc, New York

Adamec D, O'Brien JJ (1978) The seasonal upwelling in the Gulf of Guinea due to remote forcing. J Phys Oceanogr 8:1050–1060

Gill AE (1982) Atmosphere–ocean dynamics. International geophysical series, vol 30. Academic, New York

Griffies SM (2004) Fundamentals of ocean climate models. Princeton University Press, Princeton. ISBN 0-691-11892-2

Lax PD, Richtmyer RD (1956) Survey of the stability of linear finite difference equations. Commun Pure Appl Math 9:267–293, MR 79,204. https://doi.org/10.1002/cpa.3160090206

Phillips OM (1966) The dynamics of the upper ocean. Cambridge University Press, New York, 269 pp

Yoshida K (1959) A theory of the Cromwell current and of equatorial upwelling. J Oceanogr Soc Jpn 15:154–170

5 Advection Problem

The purpose of this chapter is to study potentially useful schemes and discretizations of the linear advection equation. We discuss various stable and consistent schemes such as the *leap-frog scheme*, the *upstream scheme* (or upwind scheme), the *Lax–Wendroff scheme*, and the *semi-Lagrangian scheme*. The conditions under which they are stable are also discussed, along with ways to avoid the initial problem in centered-in-time schemes. We consider problems like *numerical dispersion*, *numerical diffusion*, and *computational modes*, including ways to minimize their effects, and we discuss *flux corrective schemes*, which improve the ubiquitous numerical diffusion inherent in low-order schemes like the upstream scheme. Finally, we describe the Courant–Friedrich–Levy (CFL) condition and its physical interpretation.

5.1 One-Dimensional Advection Equation

Following Einstein's mantra mentioned in the preface, the advection equation is first reduced to its simplest form, namely the linear one-dimensional advection equation. Recall that the advective flux vector $\mathbf{F}_{\mathrm{A}} = \mathbf{v}\theta$, where $\mathbf{v}$ is the wind in the atmosphere or current in the ocean. Assuming $\mathbf{v} = u\mathbf{i}$ (one-dimensional motion), where u is the speed along the x-axis, the advection equation (3.16) takes the form

$$\partial_t \theta + \partial_x (u\theta) = 0 \,. \tag{5.1}$$

Additionally, assuming that the Boussinesq approximation is valid, in which case $\mathbf{v}$ is non-divergent ($\nabla \cdot \mathbf{v} = 0$), leads to

$$\partial_t \theta + u \partial_x \theta = 0 \,. \tag{5.2}$$

Furthermore, if not explicitly mentioned, it is hereafter assumed that $u = u_0$, where u_0 is uniform in both time and space. Thus, (5.2) takes the form

$$\partial_t \theta + u_0 \partial_x \theta = 0 \,. \tag{5.3}$$

L. P. Røed, *Atmospheres and Oceans on Computers*,
Springer Textbooks in Earth Sciences, Geography and Environment,
https://doi.org/10.1007/978-3-319-93864-6_5

The general analytic solution to (5.3) is (see Exercise 1)

$$\theta = \theta(x - u_0 t) \ . \tag{5.4}$$

To illustrate this, assume that θ is a well-behaved function (see Sect. 2.10). Then it may be represented by the Fourier series

$$\theta(x, t) = \sum_{m=0}^{\infty} \Theta_m(t) \mathrm{e}^{\mathrm{i}\alpha_m x} \ , \tag{5.5}$$

where α_m is the wavenumber of mode m and Θ_m is the associated amplitude of that mode.[1] Substituting (5.5) into (5.3), it takes the form

$$\sum_{m=0}^{\infty} \left(\frac{\mathrm{d}\Theta_m}{\mathrm{d}t} + \mathrm{i}\alpha_m u_0 \Theta_m \right) \mathrm{e}^{\mathrm{i}\alpha_m x} = 0 \ . \tag{5.6}$$

This is true only as long as

$$\frac{\mathrm{d}\Theta_m}{\mathrm{d}t} + \mathrm{i}\alpha_m u_0 \Theta_m = 0 \ , \quad \forall m \ . \tag{5.7}$$

Hence,

$$\Theta_m = \Theta_m^0 \mathrm{e}^{-\mathrm{i}\alpha_m t} \ , \tag{5.8}$$

where Θ_m^0 is the initial value of the amplitude of wavenumber mode m at time $t = 0$. Substituting (5.8) into (5.5) leads to

$$\theta(x, t) = \sum_{n=0}^{\infty} \Theta_m^0 \mathrm{e}^{\mathrm{i}\alpha_m (x - u_0 t)} \ , \tag{5.9}$$

which does indeed show that the solution is a function of $x - u_0 t$ alone. Moreover, (5.9) shows that all the waves propagate with the same speed u_0. An observer traveling with that speed will therefore experience no change in the property θ as time progresses. This is in line with the statement made in Sect. 3.3 that if the motion is non-divergent, there will be no change in the variance.

Consequently, if the initial condition is specified by a single harmonic (monochromatic) wave of wavelength λ and amplitude A, i.e.,

$$\theta(x, 0) = A\mathrm{e}^{\mathrm{i}2\pi x/\lambda} \ , \tag{5.10}$$

the only possible solution is a propagation of that monochromatic wave. Thus, the solution (5.9) takes the form

$$\theta(x, t) = A\mathrm{e}^{\mathrm{i}2\pi(x - u_0 t)/\lambda} \ . \tag{5.11}$$

This particular solution is a wave of wavelength λ propagating with phase speed u_0 in the positive x direction (assuming $u_0 > 0$). This solution is typical of hyperbolic systems (see Sect. 2.3). The solutions (5.4), (5.9), and (5.11) are all such that an observer traveling along with the advection speed will experience no change in the

[1] Solving (5.3) for a limited domain, say $x \in [0, L]$, there is an upper bound to the wavelength with a corresponding lower bound on α_m.

property θ. In contrast, an observer at a fixed position will experience a change in the property θ in accordance with (5.11) as the wave passes by.

The formal solution (5.9) shows that the analytic solution (5.4) consists of waves of various wavelengths (wavenumbers) and amplitudes propagating at the same speed u_0. Since the waves with the highest amplitudes are those containing the most energy, they dominate the solution. These waves will propagate unchanged, and hence correspond to those waves that initially have the greatest amplitude.

5.2 Finite Difference Forms

Here, we consider methods for solving the advection equation (5.3) numerically. This involves discretizing (5.3) by replacing the differential terms in it by appropriate FDAs. Since the solution of the resulting scheme is required to converge to the analytic solution, which is equivalent to requiring that the chosen numerical scheme be numerically stable and consistent, this may not be straightforward. Hence, once the differential terms have been replaced by the chosen FDAs, the resulting scheme has to be analyzed for stability and consistency. Indeed, if the scheme turns out to be unstable or inconsistent, it must be discarded, in which case another scheme must be chosen. Recall that, while stability is checked using the von Neumann method (Sect. 4.4), consistency is analyzed by means of Taylor series (Sect. 4.9).

A scheme that worked well for solving the diffusion equation was a forward-in-time centered-in-space (FTCS) scheme. In accordance with (2.61), a forward-in-time FDA for the time rate of change of θ is

$$[\partial_t \theta]_j^n = \frac{\theta_j^{n+1} - \theta_j^n}{\Delta t} , \tag{5.12}$$

while a centered-in-space FDA for the first-order space derivative of θ, in line with (2.63), is

$$[\partial_x \theta]_j^n = \frac{\theta_{j+1}^n - \theta_{j-1}^n}{2\Delta x} . \tag{5.13}$$

Substituting these FDAs into (5.3) leads to

$$\frac{\theta_j^{n+1} - \theta_j^n}{\Delta t} + u_0 \left(\frac{\theta_{j+1}^n - \theta_{j-1}^n}{2\Delta x} \right) = 0 . \tag{5.14}$$

Solving with respect to θ_j^{n+1}, it takes the form

$$\theta_j^{n+1} = \theta_j^n - u_0 \frac{\Delta t}{2\Delta x} \left(\theta_{j+1}^n - \theta_{j-1}^n \right) . \tag{5.15}$$

The expression (5.15) is commonly referred to as the *Euler scheme*.

Since the Euler scheme (5.15) is based on Taylor series expansions, it satisfies the consistency requirement. To ascertain whether the scheme is convergent, it remains to analyze its stability. Once again, making use of the von Neumann method, the discrete

Fourier component (4.27) is used to substitute for the various terms in (5.15). After some manipulation and dividing through by the common factor $\mathrm{e}^{\mathrm{i}\alpha j \Delta x}$, we obtain

$$\Theta_{n+1} = \Theta_n - \frac{u_0 \Delta t}{2\Delta x}\left(\mathrm{e}^{\mathrm{i}\alpha\Delta x} - \mathrm{e}^{-\mathrm{i}\alpha\Delta x}\right)\Theta_n \,. \tag{5.16}$$

Dividing through by Θ_n and noting that $\mathrm{e}^{\mathrm{i}\alpha\Delta x} - \mathrm{e}^{-\mathrm{i}\alpha\Delta x} = 2\mathrm{i}\sin\alpha\Delta x$, we obtain

$$G = 1 - \mathrm{i}\lambda \tag{5.17}$$

for the growth factor associated with the Euler scheme [see (4.28)], where

$$\lambda = \frac{u_0 \Delta t}{\Delta x}\sin\alpha\Delta x \,. \tag{5.18}$$

Thus, the growth factor is a complex number with real part 1 and imaginary part $-\lambda$. According to von Neumann's method the scheme is stable if the growth factor satisfies $|G| < 1$ for all wave numbers and increments. Using the well-known property that the magnitude of a complex numbers equals the square root of the sum of the squares of the real and imaginary parts,[2] it takes the form

$$|G| = \sqrt{1+\lambda^2} \,. \tag{5.19}$$

Since λ^2 is a positive-definite quantity, the root is always greater than unity, whence $|G| \geq 1$. Only for the special wavenumbers $\sin\alpha\Delta x = 0$ will $|G| = 1$. The Euler scheme is therefore *unconditionally unstable*. To conclude, although the forward-in-time centered-in-space scheme worked for the diffusion problem, it is totally unacceptable for the advection problem. This conclusion may seem somewhat curious, but it should not come as a total surprise. As mentioned in Chap. 2, the advection equation and the diffusion equation represent quite different physics and have quite different characteristics. While the diffusion equation is parabolic, the advection equation is hyperbolic, and there is no reason why a scheme that works well for a parabolic problem should necessarily work well for a hyperbolic problem.

Never use a forward-in-time centered-in-space scheme for the advection problem. It always leads to an unconditionally unstable scheme.

Although the Euler scheme is unacceptable, there are many stable and consistent schemes to choose from (e.g., O'Brien 1986, p. 165ff). The reason is that advection is one of the most prominent processes in the motion of atmospheres and oceans. There is no such thing as a perfect scheme though, so the various schemes put forward all have advantages and disadvantages. A new scheme is often suggested to minimize unwanted properties of another scheme, or for reasons of computer efficiency. Because of the dramatic increase in the power and speed of computers over the years, the requirement of computer efficiency is somewhat less important today. The focus is therefore shifting towards developing schemes that provide better

[2] Let $A = a + ib$ be an imaginary number with real part a and imaginary part b. Then $|A| = \sqrt{AA^*} = \sqrt{a^2+b^2}$, where $A^* = a - ib$ is the complex conjugate of A.

conservation properties or higher order accuracy [schemes of $\mathcal{O}(\Delta t^4)$ and $\mathcal{O}(\Delta x^4)$ or higher, see Sect. 10.1]. Among the former are semi-Lagrangian schemes (Sect. 5.12) and so-called flux corrective schemes (Sect. 5.16).

It is nevertheless constructive to analyze some of the earlier schemes, both for pedagogical reasons and because they often form the basis for many more recently suggested schemes. We thus examine four earlier schemes which are all still in use, namely, the *leap-frog scheme*, the *upstream scheme*, the *diffusive scheme*, and the *Lax–Wendroff scheme*.

5.3 Leap-Frog Scheme

Since the forward-in-time Euler scheme was unconditionally unstable, it is natural to ask whether a centered-in-time scheme would work better. Recall that the centered-in-time and centered-in-space (CTCS) scheme yielded an unconditionally unstable scheme when applied to the diffusion equation.

Once again, using Taylor series expansions to establish the FDAs, the forward-in-time FDA for the time rate of change (5.12) is simply replaced by a centered-in-time FDA, or

$$[\partial_t \theta]_j^n = \frac{\theta_j^{n+1} - \theta_j^{n-1}}{2\Delta t} . \tag{5.20}$$

We keep the centered-in-space FDA (5.13) for $[\partial_x \theta]_j^n$. Replacing the two differential terms in (5.3) by their respective FDAs (5.20) and (5.13), and solving with respect to θ_j^{n+1}, then leads to

$$\theta_j^{n+1} = \theta_j^{n-1} - u_0 \frac{\Delta t}{\Delta x} \left(\theta_{j+1}^n - \theta_{j-1}^n \right) , \tag{5.21}$$

which is referred to as the *leap-frog scheme*. It takes its name from the fact that the scheme includes θ at all the grid points immediately surrounding the point x_j, t^n, but not the grid point x_j, t^n itself, so the point x_j, t^n is in a sense leap-frogged. Since it makes use of centered FDAs in both time and space, the truncation error is $\mathcal{O}(\Delta x^2) + \mathcal{O}(\Delta t^2)$. The scheme is therefore often referred to as a second-order scheme.[3]

5.4 Stability of the Leap-Frog Scheme: The CFL Condition

Since the leap-frog scheme is derived exclusively using Taylor series, it is a consistent scheme. To be convergent, it must also be stable. Once again, to analyze its stability,

[3] In describing this model, people often write "while a second-order scheme is employed for advective terms".

we use the von Neumann method. Each term in (5.21) that contains the dependent variable θ is therefore first replaced by its corresponding discrete Fourier component. Then by dividing through by the common factor $e^{ij\alpha\Delta x}$, it takes the form

$$\Theta_{n+1} = \Theta_{n-1} - u_0 \frac{\Delta t}{\Delta x} (e^{i\alpha\Delta x} - e^{-i\alpha\Delta x})\Theta_n = \Theta_{n-1} - 2iu_0 \frac{\Delta t}{\Delta x} \sin\alpha\Delta x \Theta_n \,. \tag{5.22}$$

The growth factor is found by first dividing through by Θ_n and next making use of (4.28). Multiplying the result by the growth factor then yields

$$G^2 + 2i\lambda G - 1 = 0 \,, \tag{5.23}$$

where, as in (5.18),

$$\lambda = u_0 \frac{\Delta t}{\Delta x} \sin\alpha\Delta x \,. \tag{5.24}$$

The two solutions for the growth factor are

$$G_{1,2} = -i\lambda \pm \sqrt{1-\lambda^2} \,. \tag{5.25}$$

Hence, as expected by inspection of (5.23), the two solutions are complex functions. If $\lambda > 1$, the radical in (5.25) is negative and the solutions are purely imaginary functions. Under these circumstances, $|G_1| \geq 1$ and the scheme is unstable. Hence, to be stable, the radical in (5.25) has to be positive, and this is true if the condition $\lambda^2 \leq 1$ is satisfied. Under these circumstances, the growth factor has a real part equal to $\sqrt{1-\lambda^2}$ and an imaginary part equal to $-\lambda$. Recalling that the magnitude of a complex number is the square root of the complex number itself multiplied by its complex conjugate, we have

$$|G_{1,2}| = \sqrt{G_{1,2}G_{1,2}^*} = \sqrt{1-\lambda^2+\lambda^2} = 1 \,. \tag{5.26}$$

Thus, from the definition (4.28), the growth factor yields $\Theta_{n+1} = \Theta_n$. Hence, the amplitudes of the individual Fourier components neither increase nor decrease as time evolves, implying that there is no artificial or numerical dissipation (damping) involved. This is in line with the property of advection processes outlined in Sect. 3.3, i.e., if $\nabla \cdot \mathbf{v} = 0$, the total variance should be conserved.[4] The leap-frog scheme (5.21) is therefore said to be *neutrally stable* under the condition $\lambda^2 \leq 1$.

The latter condition leads to

$$|\lambda| \leq 1 \,. \tag{5.27}$$

Since $|\sin\alpha\Delta x| \leq 1$, (5.27) is satisfied if

$$|u_0| \frac{\Delta t}{\Delta x} \leq 1 \,. \tag{5.28}$$

The condition (5.28) is thus a sufficient condition for stability of the leap-frog scheme (5.21). It is referred to as the *Courant–Friedrich–Levy condition*, or *CFL condition* for short. The dimensionless number

$$C = |u_0| \frac{\Delta t}{\Delta x} \tag{5.29}$$

[4]The requirement $\partial_x u = 0$ is satisfied here since $u = u_0 = \text{const}$.

is the so-called *Courant number*. The CFL condition is therefore often written as $C \leq 1$. Since Δx is commonly prescribed by the need to resolve the spatial structure, or typical wavelengths of the physical problem, the CFL condition is a rigorous upper bound on the time step Δt, i.e.,

$$\Delta t \leq \frac{\Delta x}{|u_0|} \, . \tag{5.30}$$

Therefore, the greater the advection speed and/or the smaller the grid size, the smaller the time step.

In atmospheres and oceans, the dominant wavelength (or length scale) to resolve is associated with the first baroclinic Rossby radius, say L_{R}. At northerly latitudes, say ~60°N, L_{R} is ~1000 km in the atmosphere and two orders of magnitude smaller in the ocean ($L_{\mathrm{R}} \sim 10\,\mathrm{km}$). In the atmosphere, the Rossby radius is therefore well resolved using a grid size $\Delta x \sim 100\,\mathrm{km}$. In the ocean, the similar grid size has to be two orders of magnitude smaller, namely, $\Delta x \sim 1\,\mathrm{km}$. However, the typical advection speed in the atmosphere ($\sim 10\,\mathrm{m\,s^{-1}}$) is much faster than in the ocean ($\sim 0.1\,\mathrm{m\,s^{-1}}$), so the time step Δt satisfying the CFL condition is about 1 day for both of them. As explained in Chap. 6, in most real applications, the stability is not determined by the advective part of the motion, but rather by the propagation speed of baroclinic waves. As discussed in Chap. 6, this has a serious impact on the time step.

The leap-frog scheme is consistent and stable if $C \leq 1$. Hence, it satisfies the Lax theorem and is convergent. Applying the leap-frog scheme to solve the advection equation (5.3) numerically therefore ensures that its solution approaches the continuous solution as Δx and Δt tend to zero independently. The CTCS leap-frog scheme was employed early on to solve the advective part in models of the atmosphere and ocean. It is in fact still fairly popular and is widely used as one of several options in many models today. One reason for this is that the scheme is neutrally stable, so has no numerical or artificial dissipation (damping), a highly desirable property. A second reason is that the scheme is of second-order accuracy. In addition, the scheme is explicit and thus easy to implement. A final reason is that it works fast and efficiently on most computers.

However, the leap-frog scheme also has some disadvantages. For instance, it contains *numerical dispersion*, sometimes leading to negative tracer concentrations. This is treated in detail in Sect. 5.5. Furthermore, the scheme contains *unphysical modes* (Sect. 5.7). Finally, as with all centered-in-time schemes, we must deal with the initial value problem discussed in Sect. 4.6.

5.5 Numerical Dispersion

The concept of dispersion is well known from other branches of physics and geophysics. In particular, it is a common phenomenon in wave dynamics. Most of us have experienced this by throwing stones in still water. After the initial splash, we observe circular waves propagating away from the original splash point in such a way that longer waves run ahead of progressively shorter waves. The reason for this is that the phase speed, say c, depends on the wavenumber (or wavelength), i.e., waves of different wavelengths propagate at different speeds. Regarding gravity waves in deep water, the longer the wavelength, the faster the phase speed, whence the longer waves will run ahead of progressively shorter waves. The same is also true for other types of waves, such as planetary Rossby waves.

Mathematically, this is expressed through the dispersion relation $\omega = \omega(\alpha)$, where ω is the frequency and α the wavenumber. Recall that the phase speed is defined by

$$c = \frac{\omega}{\alpha} . \tag{5.31}$$

The phase speed is therefore a constant if the frequency is a linear function of α. Then $\partial_\alpha c = 0$, implying that all the waves propagate at the same speed, independently of their wavelength, and the solution is said to be *nondispersive*. In general, ω is a nonlinear function of the wavenumber α, in which case $\partial_\alpha c \neq 0$. Waves of different wavelengths then travel with different speeds and the solution is said to be *dispersive*. The energy contained in the wave propagates with the group velocity, which is defined by

$$c_g = \partial_\alpha \omega = \alpha \partial_\alpha c + c . \tag{5.32}$$

If the wave is nondispersive, $c_g = c$ and the energy then propagates at the same speed as the wave itself. If the wave is dispersive, the phase speed and the group velocity both depend on the wavelength. Moreover, if $\partial_\alpha c < 0$, which is the case for gravity waves, the group velocity is smaller than the phase speed. Under these circumstances, the gravity waves tend to travel faster than the energy is propagated.[5]

Let us assume that the solution to the advection equation is a single wave with phase speed c, i.e.,

$$\theta = \theta_0 \mathrm{e}^{\mathrm{i}\alpha(x-ct)} . \tag{5.33}$$

Substituting into (5.3), it follows that the phase speed is $c = u_0$. Under these circumstances, all the waves propagate with the same constant advection velocity u_0. Hence, the solution to the advection equation is nondispersive. We may wonder whether this is true for the numerical solution as well. To investigate this, we carry out a similar analysis, substituting a discrete version of (5.33) into the finite difference version of the advection equation, i.e.,

$$\theta_j^n = \Theta_0 \mathrm{e}^{\mathrm{i}\alpha(j\Delta x - c^* n\Delta t)} , \tag{5.34}$$

where c^* is the wave speed of the numerical solution. If the analysis results in $\partial_\alpha c^* \neq 0$, the scheme is said to exhibit *numerical dispersion*.

As an example, consider the leap-frog scheme (5.21). To start the analysis, the discrete Fourier component (5.34) is first substituted into (5.21). This leads immediately to

$$\sin(\alpha c^* \Delta t) = u_0 \frac{\Delta t}{\Delta x} \sin \alpha \Delta x , \tag{5.35}$$

[5]For capillary waves, $\partial_\alpha c > 0$, implying that a capillary wave travels more slowly than its energy.

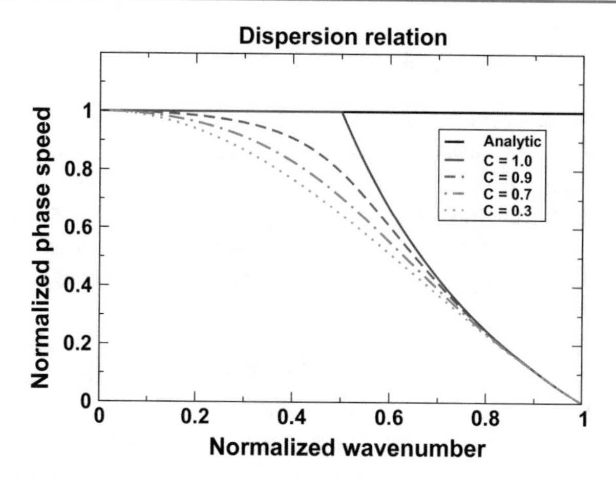

Fig. 5.1 Numerical dispersion for the leap-frog scheme. *Curves* depict the numerical phase speed c^* as a function of the wavenumber α, using (5.36) for various values of the Courant number $C = |u_0|\Delta t/\Delta x$. The *vertical axis* shows the phase speed normalized by the advection speed u_0 and the *horizontal axis* shows the wavenumber normalized by the reciprocal of the space increment Δx or $\pi/\Delta x$. The analytic dispersion curve is just a straight line corresponding to the phase speed $c^* = u_0$, i.e., $c^*/u_0 = 1$. As the wavenumber increases (the wavelength decreases), the numerical phase speed deviates more and more from the correct analytic phase speed, and more so for small Courant numbers. For wavenumbers which give $\alpha\Delta x/\pi > 0.5$, i.e., for waves of wavelength $\lambda < 4\Delta x$, the slope of the curves indicates that the numerical group velocity is negative. Thus, for waves of wavelength shorter than $4\Delta x$, the energy propagates in the opposite direction to the waves

or

$$c^* = \frac{1}{\alpha\Delta t}\arcsin\left(u_0\frac{\Delta t}{\Delta x}\sin\alpha\Delta x\right) . \tag{5.36}$$

Obviously, $\partial_\alpha c^* \neq 0$, so the leap-frog scheme is dispersive. Since it is associated with our numerical solver, this is artificial dispersion, hence referred to as *numerical dispersion*. If Δx and Δt go to zero while the stability condition $|u_0|\Delta t \leq \Delta x$ is respected, then $c^* \to u_0$. This complies with the earlier finding that the leap-frog scheme is convergent. Thus, the dispersion diminishes as the numerical solution converges towards the analytic solution.

Figure 5.1 shows the normalized numerical phase speed as a function of the normalized wavenumber for various Courant numbers. The figure clearly shows the dispersive nature of the leap-frog scheme. Closer inspection reveals that the dispersion gets more pronounced at smaller values of the Courant number (Grotjhan and O'Brien 1976). Moreover, from (5.32), it follows that c_g is zero when

$$\alpha\partial_\alpha c = -c . \tag{5.37}$$

Applying this to the numerical phase speed c^* given by (5.36), we obtain $c_g^* = 0$ for wavenumbers satisfying $\cos\alpha\Delta x = 0$, that is, for wavenumbers

$$\alpha_m = \frac{1}{2}(2m - 1)\pi \ , \quad m = 1, 2, \dots \ . \tag{5.38}$$

The longest wave for which $c_g^* = 0$ is the one with $m = 1$, which is equivalent to $\lambda = 2\pi/\alpha_1 = 4\Delta x$. As can be seen from Fig. 5.1, this corresponds to the normalized wavenumber being equal to 0.5. For higher wavenumbers, that is, waves whose wavelengths are shorter than $4\Delta x$, the group velocity actually becomes negative. Thus, if the wave is poorly resolved, the leap-frog scheme will actually propagate energy in the opposite direction to the wave itself. This is clearly unphysical and must be avoided.

To conclude, it is therefore extremely important that the length scales that dominate the advected property should be well resolved. Let the dominant wavenumber be α. Then looking at Fig. 5.1, Δx must be chosen so that $\alpha\Delta x < 0.3\pi$ for Courant numbers close to one, and even less for smaller Courant numbers. In Sect. 10.6, we touch upon a method for avoiding the dispersion inherent in the leap-frog scheme, viz., the spectral method. However, this method is only applicable to global models. For limited area models, no such remedy is available.

5.6 Initial Value Problem in CTCS Schemes

Another problem is associated with the leap-frog scheme being a CTCS scheme. As mentioned earlier (Sect. 4.2), a CTCS scheme requires more than one initial condition, while the advection equation only allows the specification of one. This is referred to as the initial value problem. When applied to the diffusion equation, the CTCS scheme turned out to be unconditionally unstable, and hence the initial value problem did not arise. In contrast, the CTCS scheme turns out to be an excellent scheme when applied to the advection equation, whereas the FTCS scheme resulted in an unconditionally unstable scheme (Sect. 5.2). Thus, the initial value problem has to be dealt with. We have to decide how to start the time marching procedure when using centered-in-time schemes, and in particular the leap-frog scheme. The answer is to use the Euler scheme (5.15), or any other forward-in-time scheme, just for the first time step, despite the fact that it is unconditionally unstable.

Consequently, to compute the variable θ at time level $n = 1$, we use the scheme

$$\theta_j^1 = \theta_j^0 - u_0 \frac{\Delta t}{2\Delta x}\left(\theta_{j+1}^0 - \theta_{j-1}^0\right) . \tag{5.39}$$

To compute the variable at time level $n = 2$ and onwards, we return to the leap-frog scheme (5.21). Even though the Euler scheme (5.15) is unconditionally unstable, it does not spoil the solution when applied for just one time step. It may even be used from time to time as a way of avoiding the computational or unphysical mode inherent in the leap-frog scheme (Sect. 5.7).

5.7 Computational Modes and Unphysical Solutions

Even though the leap-frog scheme is excellent in many respects, it is not perfect. For instance, as discussed in Sect. 5.5, it contains numerical dispersion. Although this property has a physical interpretation, as its name suggests, it is nonetheless an artifact of the scheme, i.e., it is actually unphysical.

Here, we discuss the fact that the leap-frog scheme contains yet another unphysical property, namely, the so-called *computational modes*, which lead to unphysical solutions. To show this, an analytic solution to the leap-frog scheme (5.21) is compared to the solution of its continuous counterpart (5.3). The starting point is an initial condition in the form of a single harmonic wave with wavenumber α_0 and amplitude Θ_0, like the one in (5.10). The true or analytic solution to the continuous advection equation is then

$$\theta = \Theta_0 e^{i\alpha_0(x-u_0t)} , \tag{5.40}$$

a single monochromatic wave with wavenumber α_0 and phase speed u_0 in the *positive* x direction. Recall that any well-behaved function can be written in terms of an infinite sum of waves (Sect. 4.3). Thus, if we find a solution for one monochromatic wave, the solution for an arbitrary initial condition is found by summing over all possible wavenumbers (summing in wavenumber space).

To solve (5.21) analytically, an initial condition is needed. As for the continuous equation, a monochromatic wave of wavenumber α_0 and amplitude A is chosen. In its discrete form, it is

$$\theta_j^0 = A e^{ij\alpha_0\Delta x} , \quad \forall j . \tag{5.41}$$

In general the analytic solution to (5.21) is

$$\theta_j^n = \Theta_n e^{ij\alpha\Delta x} = G\Theta_{n-1} e^{ij\alpha\Delta x} = G^n \Theta_0 e^{ij\alpha\Delta x} , \tag{5.42}$$

where we have used the definition of the growth factor $G = \Theta_{n+1}/\Theta_n$ to derive the second equality. The last equality follows by induction. Furthermore, the stability analysis of the leap-frog scheme (Sect. 4.4) shows that the growth factor for this scheme has two solutions. Thus, the full analytic solution to the numerical leap-frog scheme applied to the advection equation is

$$\theta_j^n = (G_1^n \Theta_1 + G_2^n \Theta_2) e^{ij\alpha\Delta x} , \tag{5.43}$$

where Θ_1 and Θ_2 are two as yet unknown constants. The two possible growth factors are given by (5.25), and may be rewritten to yield

$$G_1 = P^* \quad \text{and} \quad G_2 = -P , \tag{5.44}$$

where

$$P = \sqrt{1-\lambda^2} + i\lambda \quad \text{and} \quad P^* = \sqrt{1-\lambda^2} - i\lambda \tag{5.45}$$

are true complex conjugate numbers, and λ is given by (5.24).[6] Recall that any complex number $B = a + ib$ may be written $B = |B|e^{i\phi}$, where $|B| = \sqrt{a^2+b^2}$ and $\phi = \arcsin(b/|B|)$. Here $|P| = |P^*| = 1$, and hence,

[6]The CFL condition for stability is assumed to be satisfied, so $\sqrt{1-\lambda^2}$ is a real number.

$$P^* = \mathrm{e}^{\phi_1} , \quad \text{where } \phi_1 = \arcsin(-\lambda) = -\arcsin\lambda , \tag{5.46}$$

$$P = \mathrm{e}^{\phi_2} , \quad \text{where } \phi_2 = \arcsin\lambda . \tag{5.47}$$

Equation (5.36) implies that

$$\arcsin\lambda = \arcsin\left(u_0 \frac{\Delta t}{\Delta x} \sin\alpha\Delta x\right) = \alpha c^* \Delta t . \tag{5.48}$$

Thus (5.44) takes the form

$$G_1 = \mathrm{e}^{-\mathrm{i}\alpha c^* \Delta t} \quad \text{and} \quad G_2 = -\mathrm{e}^{\mathrm{i}\alpha c^* \Delta t} . \tag{5.49}$$

Substituting these solutions into (5.42) yields

$$\theta_j^n = \Theta_1 \mathrm{e}^{\mathrm{i}\alpha(j\Delta x - c^* n\Delta t)} + (-1)^n \Theta_2 \mathrm{e}^{\mathrm{i}\alpha(j\Delta x + c^* n\Delta t)} . \tag{5.50}$$

Invoking the one and only initial condition (5.41), we obtain

$$\theta_j^n = (A - \Theta_2)\mathrm{e}^{\mathrm{i}\alpha_0(j\Delta x - c^* n\Delta t)} + (-1)^n \Theta_2 \mathrm{e}^{\mathrm{i}\alpha_0(j\Delta x + c^* n\Delta t)} . \tag{5.51}$$

Equation (5.51) indicates that the analytic solution to the finite difference version of the advection equation, applying the leap-frog scheme, leads to a solution containing two waves. One propagates in the positive x direction and the other in the negative x direction. They both have phase speed c^*, but they have different amplitudes. In contrast, the true solution (5.11) to the advection equation contains only one wave that propagates in the *positive* x direction with phase speed u_0. This suggests that the wave propagating in the negative x direction is unphysical. Moreover, it has an amplitude alternating between $\pm\Theta_2$, depending on whether it is an even or odd time step. This reinforces the idea that it might be a false wave mode. However, it remains to find the unknown amplitude Θ_2 before we can reach a firm conclusion.

The problem of finding the second integration constant Θ_2 in (5.51) is associated with the initial boundary problem discussed in Sect. 5.6 above. One remedy suggested there was to apply an Euler step as the first step. This corresponds to specifying a second initial condition, since using the Euler scheme is just one remedy among several to start the time marching problem without having to assign a value to θ at the time step prior to the initial time. Recall that the Euler step is

$$\theta_j^1 = \theta_j^0 - u_0 \frac{\Delta t}{2\Delta x}\left(\theta_{j+1}^0 - \theta_{j-1}^0\right) . \tag{5.52}$$

Hence, letting $n = 1$ in (5.51) and substituting the result into (5.52) leads to

$$(A - \Theta_2)\mathrm{e}^{\mathrm{i}\alpha_0(j\Delta x - c^* \Delta t)} - \Theta_2 \mathrm{e}^{\mathrm{i}\alpha_0(j\Delta x + c^* \Delta t)} = \theta_j^0 - u_0 \frac{\Delta t}{2\Delta x}\left(\theta_{j+1}^0 - \theta_{j-1}^0\right) . \tag{5.53}$$

Next, replace the terms containing θ_j^0 and $\theta_{j\pm1}^0$ on the right-hand side of (5.53) using the discrete version of the initial condition (5.10). Then, rearranging terms and noting that, according to (5.35),

$$u_0 \frac{\Delta t}{\Delta x} \sin(\alpha_0 \Delta x) = \sin(\alpha_0 c^* \Delta t) ,$$

Equation (5.53) takes the form

$$\Theta_2 = -A\frac{1-\cos(\alpha_0 c^* \Delta t)}{2\cos(\alpha_0 c^* \Delta t)} . \tag{5.54}$$

Finally, substituting this result into (5.51) leads to

$$\theta_j^n = A\frac{1+\cos(\alpha_0 c^* \Delta t)}{2\cos(\alpha_0 c^* \Delta t)} e^{i\alpha_0 (j\Delta x - c^* n\Delta t)} + (-1)^{n+1} A\frac{1-\cos(\alpha_0 c^* \Delta t)}{2\cos(\alpha_0 c \Delta t)} e^{i\alpha_0 (j\Delta x + c^* n\Delta t)} , \tag{5.55}$$

which is the complete analytic finite difference solution of the advection equation when we apply the leap-frog scheme.

Recall that, as long as the stability criterion (5.28) is satisfied, then $c^* \to u_0$ when Δt and Δx go to zero independently. Interestingly, in this case,

$$\frac{1+\cos(\alpha_0 c^* \Delta t)}{2\cos(\alpha_0 c^* \Delta t)} \longrightarrow 1 \quad \text{and} \quad \frac{1-\cos(\alpha_0 c^* \Delta t)}{2\cos(\alpha_0 c \Delta t)} \longrightarrow 0 . \tag{5.56}$$

Thus, in the limit, the first term on the right-hand side of (5.55) approaches the true solution, while the second term vanishes. As a result, the second term is definitely an unphysical or computational mode, while the first term represents the physical mode.

In summary, using the leap-frog scheme to solve the advection equation numerically leads to a solution in the form of two waves propagating in opposite directions with a phase speed given by (5.36). One of the two corresponds to the physical mode in that it approaches the true solution in the limit as Δt and Δx go to zero independently of each other. The second mode, which approaches zero in the limit $\Delta t \to 0$ and $\Delta x \to 0$, is therefore unphysical. It is commonly referred to as the *computational mode*. The occurrence of two waves stems from the fact that the leap-frog scheme is second order. As such it requires two initial conditions, one of which is superfluous. This shows why it is important to keep the number of boundary and initial conditions in line with the number of integration constants in the underlying continuous PDE. The false mode in the leap-frog scheme is commonly referred to in the literature as the *leap-frog splitting mode* (e.g., Haltiner and Williams 1980; Griffies 2004).

Nevertheless, the leap-frog splitting mode has to be dealt with, and in such a way that the correct solution is not deteriorated. If not dealt with properly, it creates $2\Delta t$ noise in the solution, since the computational mode alternates between positive and negative values. Moreover, in extreme cases, it may generate numerical instabilities.

5.8 Getting Rid of the Computational Mode: Asselin Filter

The simplest way to get rid of the computational mode is to replace the leap-frog scheme with an Euler scheme at odd times, just as we did at the beginning to avoid the initial value problem. Although the Euler scheme is unconditionally unstable, it may still be applied from time to time for single time steps without destroying the stability of the leap-frog scheme or any other stable scheme.

Another method, originally suggested by Robert (1966) and further developed by Asselin (1972), is to apply a time filtering technique. Since the computational mode alternates between positive and negative values from time step to time step, it is obvious that a time filter will do the trick. In general a time filter, invoking only the neighboring time levels, takes the form

$$\overline{\theta}(x,t) = \gamma\theta(x,t+\Delta t) + (1-2\gamma)\theta(x,t) + \gamma\theta(x,t-\Delta t)\ , \tag{5.57}$$

where $\overline{\theta}(x,t)$ is the filtered function and γ is a weighting parameter. Using the notation introduced in Sect. 2.8, it takes the form

$$\overline{\theta}_j^n = \theta_j^n + \gamma\left(\theta_j^{n+1} - 2\theta_j^n + \theta_j^{n-1}\right) . \tag{5.58}$$

If $\gamma = 1/4$, the filter is the standard 1-2-1 filter,

$$\overline{\theta}_j^n = \frac{1}{4}\left(\theta_j^{n+1} + 2\theta_j^n + \theta_j^{n-1}\right) , \tag{5.59}$$

which gives twice the weight to the mid-time level n. The right-hand side of (5.58) is recognizable as the centered-in-space FDA of $\partial_x^2\theta$, as displayed by (4.4). Thus, (5.58) may be written

$$\overline{\theta}_j^n = \theta_j^n + \gamma\Delta x^2\left[\partial_x^2\theta\right]_j^n , \tag{5.60}$$

indicating that the filter acts by adding a diffusion term to the scheme, with a diffusion coefficient $\gamma\Delta x^2$.

To investigate the properties of the filter, we introduce the so-called *response function* defined by

$$R_\gamma = \frac{\overline{\theta}_j^n}{\theta_j^n} . \tag{5.61}$$

Once again, it is enough to analyze the response function for just one period or frequency ω. Representing the function θ by its Fourier component in time, or

$$\theta_j^n = \hat{\theta}_j e^{i\omega n\Delta t} , \tag{5.62}$$

(5.61) takes the form

$$R_\gamma = 1 - 2\gamma + 2\gamma\cos\omega\Delta t . \tag{5.63}$$

The response function associated with the 1-2-1 filter ($\gamma = 1/4$) is then

$$R_{1/4} = \frac{1}{2}(1 + \cos\omega\Delta t) . \tag{5.64}$$

For frequencies satisfying $\omega = (2m-1)\pi$, $m = 1, 2, \ldots$, in which case we have $1 + \cos\omega\Delta t = 0$, the relation (5.64) returns $R_{1/4} = 0$. This corresponds to periods $T_m = 2\Delta t/(2m-1)$. For $m = 1$ this period equals $T_1 = 2\Delta t$, which is exactly the shortest period resolved by the grid (the Nüquist frequency). Since the computational mode inherent in the leap-frog scheme alternates between negative and positive values from time step to time step, its dominating period is exactly $2\Delta t$. Thus, the standard 1-2-1 filter is a perfect filter to get rid of the computational mode inherent in the leap-frog scheme. For waves of period $4\Delta t$, (5.64) returns $R_{1/4} = 1/2$. Hence, noise in the high-frequency band (or low period band) near the Nüquist frequency is also damped to some extent by the 1-2-1 filter, while the effect on longer periods is minimal. It was because of these advantages of the 1-2-1 filter that Robert (1966) and Asselin (1972) suggested using it to filter out the computational mode inherent in the leap-frog scheme.

On the computer this is performed as follows. Assume that the filtered solution $\overline{\theta}_j^{n-1}$ has been determined for time level $n-1$. Replacing θ_j^{n-1} in the leap-frog scheme (5.21) by the filtered value at time level $n-1$, it takes the form

$$\theta_j^{n+1} = \overline{\theta}_j^{n-1} - u_0 \frac{\Delta t}{\Delta x}\left(\theta_{j+1}^n - \theta_{j-1}^n\right) . \tag{5.65}$$

Next, the values at time level n are filtered by using the filter (5.58) or the 1-2-1 filter (5.59) to yield

$$\overline{\theta}_j^n = \theta_j^n + \gamma\left(\theta_j^{n+1} - 2\theta_j^n + \overline{\theta}_j^{n-1}\right) . \tag{5.66}$$

We can then move on to the next time level by reusing (5.65).

As mentioned, applying a time filter like the Asselin filter produces numerical diffusion. Hence it impacts the numerical stability.[7] It can be shown that, while the numerical diffusion increases with increasing values of the weighting parameter γ, the critical value for stability decreases. This implies that the stability criterion becomes stricter and that the time step Δt has to be diminished. Thus, employing the simple 1-2-1 filter to kill the computational mode requires a decrease in the time step, and computation of the advective part becomes less efficient. For this reason, it is common to apply a lower value than 0.25 for the weighting function, say $\gamma = 0.08$. Even a weak Asselin filter, however, eventually modifies the longer wave periods by diffusion. Hence, the Asselin filter must be applied with some care, and not necessarily after each time step.

5.9 Upstream Scheme

Although the leap-frog scheme has many advantages, it is not perfect. One of the least desirable properties is that it is inherently dispersive (Sect. 5.5). The solution therefore exhibits negative tracer concentrations (see Fig. 5.9 in the computer problem of Sect. 5.19.1), even though tracer concentrations are always positive-definite quantities.

A scheme that conserves the positive-definite nature of concentrations is the so-called *upstream scheme* (ocean) or *upwind scheme* (atmosphere). (Henceforth, it will be referred to in both cases as the upstream scheme.) For this reason, it was quite popular early on, and it is still quite common, in particular as one of the options for the advection of tracers in numerical models. It reads

[7] Since we use (5.65) and (5.66) on the computer, the diffusion term is not as straightforward as alluded to on the basis of (5.58).

$$\theta_j^{n+1} = \theta_j^n - |u_0| \frac{\Delta t}{\Delta x} \begin{cases} \theta_j^n - \theta_{j-1}^n \,, \; u_0 \geq 0 \,, \\ \theta_j^n - \theta_{j+1}^n \,, \; u_0 < 0 \,. \end{cases} \tag{5.67}$$

For later convenience, (5.67) may be rewritten to take the form

$$\theta_j^{n+1} = (1 - C)\theta_j^n + C \begin{cases} \theta_{j-1}^n \,, \; u_0 \geq 0 \,, \\ \theta_{j+1}^n \,, \; u_0 < 0 \,, \end{cases} \tag{5.68}$$

where C is the Courant number, as defined by (5.29).

As indicated by its name, the scheme makes use of information exclusively from points upstream in space to calculate the value at the new time level. If $u_0 \geq 0$, then information from the point x_{j-1} is used, and if $u_0 < 0$, information from the point x_{j+1} is used. In addition, the FDA used to replace the time rate of change is forward in time, and the FDA that replaces the first-order derivative in space is one-sided. The resulting finite difference scheme is therefore of first-order accuracy in both time and space [truncation error of $\mathcal{O}(\Delta t)$ and $\mathcal{O}(\Delta x)$]. This implies that only two time levels are needed at any time when looping through the grid points. In summary, the upstream scheme is a first-order, two time level scheme. Note that the leap-frog scheme was a second-order, three time level scheme.

Since the FDAs that enter (5.67) are based on Taylor series expansions, the scheme is consistent. Thus, to be convergent, it must also be stable. As it turns out, the upstream scheme is conditionally stable under the CFL condition (5.28), that is, as long as the Courant number $C \leq 1$ (Exercise 2). Rearranging the CFL condition, it takes the form

$$|u_0|\Delta t \leq \Delta x \,, \tag{5.69}$$

which implies that the distance $|u_0|\Delta t$ must be smaller than the grid size Δx. The implications of this statement will be discussed further when we consider the physical interpretation of the CFL condition (Sect. 5.13).

In summary, the upstream scheme conserves the positive-definite nature of the tracer concentration, is convergent, and, being a two time level scheme, works fast and efficiently on the computer. Nevertheless, it has one major drawback. It contains artificial diffusion, or so-called *numerical diffusion* (Sect. 5.15). Depending on the choices made for the time step and the space increment, the numerical diffusion may be large and sometimes larger than the actual physical diffusion of the original problem. The result is that it tends to even out gradients artificially as time goes by (see, for example, Fig. 5.7). Thus, areas where large gradients appear, e.g., frontal areas, are especially prone to numerical diffusion. Fronts are quickly diffused by the scheme, inhibiting baroclinic instability processes, which are important ingredients in the formation of lows in the atmosphere and eddies in the ocean. In addition, the upstream scheme has truncation errors that are first order in time and space. This is one order of magnitude less than, for instance, the leap-frog scheme. Because of these disadvantages, the present author does not recommend the use of the upstream scheme. One reason why the upstream scheme is presented in this book is that, in modern numerical models, it is often used as the basis for more complicated schemes, e.g., flux corrective schemes (Sect. 5.16). Another reason is that so-called third- or higher order upstream schemes (Sect. 10.1) are still popular in certain models, e.g., ROMS (Haidvogel et al. 2008).

5.10 Diffusive Scheme

As shown in Sect. 5.2, the FTCS scheme applied to the advection equation leads to an unconditionally unstable scheme. In an attempt to avoid this numerical instability, but retain a forward-in-time scheme, it was suggested early on to replace θ_j^n in (5.15) by a linear interpolation using the adjacent grid points, that is, replace it by $(\theta_{j+1}^n + \theta_{j-1}^n)/2$. This leads to the so-called *diffusive* scheme

$$\theta_j^{n+1} = \frac{1}{2}(\theta_{j+1}^n + \theta_{j-1}^n) - u_0 \frac{\Delta t}{2\Delta x}\left(\theta_{j+1}^n - \theta_{j-1}^n\right) . \tag{5.70}$$

To be useful, the scheme must be stable and consistent.

This is not obvious in the present case, since the scheme is no longer based on Taylor series. To analyze its consistency, $-\theta_j^n$ is first added to both sides of the equals sign in (5.70). It then takes the form

$$\theta_j^{n+1} - \theta_j^n = \frac{1}{2}\left(\theta_{j+1}^n - 2\theta_j^n + \theta_{j-1}^n\right) - u_0 \frac{\Delta t}{2\Delta x}\left(\theta_{j+1}^n - \theta_{j-1}^n\right) . \tag{5.71}$$

Using Taylor series, the terms on the left-hand side may be replaced by

$$\theta_j^{n+1} - \theta_j^n = \partial_t \theta|_j^n \Delta t + \frac{1}{2}\partial_t^2 \theta|_j^n \Delta t^2 + \mathcal{O}(\Delta t^3) , \tag{5.72}$$

the first term on the right-hand side by

$$\frac{1}{2}(\theta_{j+1}^n - 2\theta_j^n + \theta_{j-1}^n) = \frac{1}{2}\partial_x^2 \theta|_j^n \Delta x^2 + \mathcal{O}(\Delta x^4) , \tag{5.73}$$

and the last term on the right-hand side by

$$u_0 \frac{\Delta t}{2\Delta x}\left(\theta_{j+1}^n - \theta_{j-1}^n\right) = u_0 \Delta t \left[\partial_x \theta|_j^n + \frac{1}{6}\partial_x^3 \theta|_j^n \Delta x^2 + \mathcal{O}(\Delta x^4)\right] . \tag{5.74}$$

Substituting (5.72)–(5.74) into (5.71), and rearranging terms leads to

$$\partial_t \theta|_j^n + u_0 \partial_x \theta|_j^n = \frac{1}{2}\frac{\Delta x^2}{\Delta t}(1 - C^2)\partial_x^2 \theta|_j^n + \mathcal{O}(\Delta x^2) + \mathcal{O}(\Delta t) , \tag{5.75}$$

where C is the Courant number as defined in (5.29). The scheme is therefore inconsistent, since the first term on the right-hand side of (5.75) goes to infinity when $\Delta t \to 0$ independently of Δx. Nevertheless, the first term on the right-hand side of (5.75) acts as a diffusion term with numerical diffusion coefficient

$$\kappa^* = \frac{1}{2}\frac{\Delta x^2}{\Delta t}(1 - C^2) . \tag{5.76}$$

The finite difference equation (5.70), in which finite space and time increments are used, therefore contains a diffusive term not present in the continuous advection equation. This is why the numerical scheme (5.70) is commonly referred to as the diffusive scheme. Notice that $\kappa^* = 0$ when $C = 1$. Thus, the artificial diffusion vanishes when the so-called *grid velocity* $\Delta x/\Delta t$ exactly matches the advective velocity $|u_0|$.

What about its stability? Once again, using the von Neumann method, the growth factor takes the form

$$G = \sqrt{1 - (1 - C^2)\sin^2 \alpha\Delta x}\ , \tag{5.77}$$

where C is the Courant number. Since $\sin^2 \alpha\Delta x$ is positive definite, it follows that $|G| \leq 1$ if and only if $C \leq 1$. The scheme is therefore conditionally stable under the condition that the Courant number is less than or equal to one. Notice that this condition is exactly the same as the one derived for the leap-frog and upstream schemes.

5.11 Lax–Wendroff Scheme

To avoid, or at least lessen, the impact of the first-order numerical diffusion inherent in the diffusive scheme, and to make it consistent and increase its accuracy, Richtmyer and Morton (1967) advocated the use of a scheme based on the work of Lax and Wendroff (1960), now commonly referred to as the Lax–Wendroff scheme. It is a two step scheme which combines the diffusive scheme and the leap-frog scheme by first performing a diffusive step (also known as the predictor step) and then adding a leap-frog step (also known as the corrective step). First, the solutions are derived at the mid-time level $t^{n+1/2}$ and at the mid-increments in space $x_{j+1/2}$ (the crosses marked in the dashed grid of Fig. 5.2), using the diffusive scheme. Next the leap-frog scheme is used to calculate the value of the variables at the new time level t^{n+1} and at the regular grid points (the circled points in the solid grid in Fig. 5.2), based on the solution at the time levels n and $n + 1/2$.

Thus, to proceed to time level $n + 1/2$ (the predictor step), the diffusive FTCS scheme is used. This leads to

$$\theta_{j+1/2}^{n+1/2} = \frac{1}{2}\left(\theta_{j+1}^n + \theta_j^n\right) - u_0\frac{\Delta t}{2\Delta x}\left(\theta_{j+1}^n - \theta_j^n\right)\ . \tag{5.78}$$

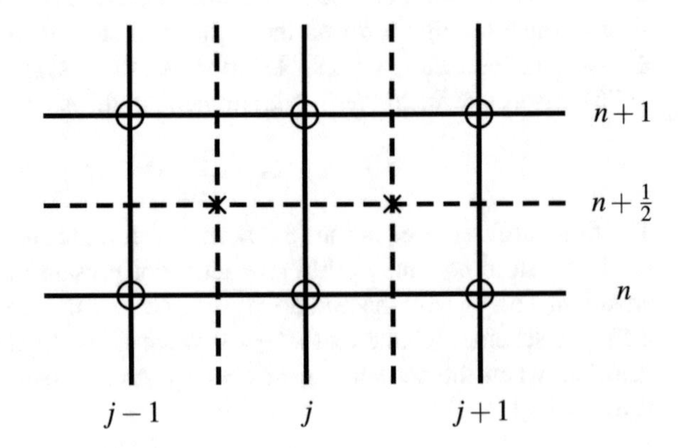

Fig. 5.2 Grid layout for the Lax–Wendroff scheme. *Solid lines* denote the grid through the circled points x_j, t^n. *Dashed lines* denote the grid through the points $x_{j+1/2}, t^{n+1/2}$ marked by crosses

Even though the forward-in-time centered-in-space scheme is unstable, the trick of replacing θ_j^n by half the sum of its nearest space neighbors makes the scheme stable under the condition that the Courant number is less than or equal to one (Sect. 5.10).

The second step applies the leap-frog scheme (5.21) to find the solution at time level $n+1$ at the point j, using the values found at the mid-time level. Thus,

$$\theta_j^{n+1} = \theta_j^n - \mathrm{sgn}\,(u_0)C\left(\theta_{j+1/2}^{n+1/2} - \theta_{j-1/2}^{n+1/2}\right), \tag{5.79}$$

where C is the Courant number, and the function sgn returns the sign of its argument, e.g., $u_0 = \mathrm{sgn}(u_0)|u_0|$. Eliminating the dependence on $n \pm 1/2$ and $j \pm 1/2$ by substituting (5.78) into (5.79), we obtain

$$\theta_j^{n+1} = \theta_j^n - \frac{1}{2}\,\mathrm{sgn}(u_0)C\left(\theta_{j+1}^n - \theta_{j-1}^n\right) + \frac{1}{2}C^2\left(\theta_{j+1}^n - 2\theta_j^n + \theta_{j-1}^n\right), \tag{5.80}$$

which is the Lax–Wendroff scheme.

As shown by (5.80), the Lax–Wendroff scheme is an explicit scheme. Hence, it is expected to be conditionally stable. To analyze its stability, the von Neumann method is used. We first replace the variables by their discrete Fourier components, then find the growth factor G. After some straightforward manipulations, this leads to

$$G = (1 - C^2 + C^2\cos\alpha\Delta x) - \mathrm{i}\,\mathrm{sgn}(u_0)C\sin\alpha\Delta x\,. \tag{5.81}$$

Thus, the growth factor is a complex function and its magnitude therefore takes the form

$$|G| = \sqrt{\left(1 - C^2 + C^2\cos\alpha\Delta x\right)^2 + C^2\sin^2\alpha\Delta x}\,. \tag{5.82}$$

After some straightforward manipulations, we find

$$|G| = \sqrt{1 - (1-C)(1+C)C^2(1-\cos\alpha\Delta x)^2}\,. \tag{5.83}$$

The sign of the last term in the radical is determined by the factor $1-C$, the remaining factors being positive-definite quantities. Consequently, as long as $1-C$ is positive, $|G| \le 1$. Therefore, like the leap-frog, the upstream, and the diffusive schemes, the Lax–Wendroff scheme is conditionally stable if the Courant number is less than one. Moreover, no computational mode is present, in contrast to the leap-frog scheme (Sect. 5.7), since the Lax–Wendroff is forward in time. Furthermore, it lacks the numerical diffusion inherent in the upstream and diffusive schemes (Sect. 5.15). Finally, the initial value problem is absent, since Lax–Wendroff is a two time level scheme. However, as shown in Fig. 5.3, the scheme is numerically dispersive and contains some numerical dissipation, since the growth factor is less than one except for $C = 1$. The latter is minimized by choosing $C \lesssim 1$, which is equivalent to choosing $\Delta x \lesssim u_0\Delta t$. It may also be minimized by choosing C close to zero, that is, $\Delta t \ll \Delta x/u_0$ (not to be recommended).

To make the scheme convergent, it remains to analyze whether it is consistent. This is not obvious, since the diffusive scheme, which was used as the first step, was shown to be inconsistent. Using Taylor series to expand the individual terms appearing in (5.80), we find

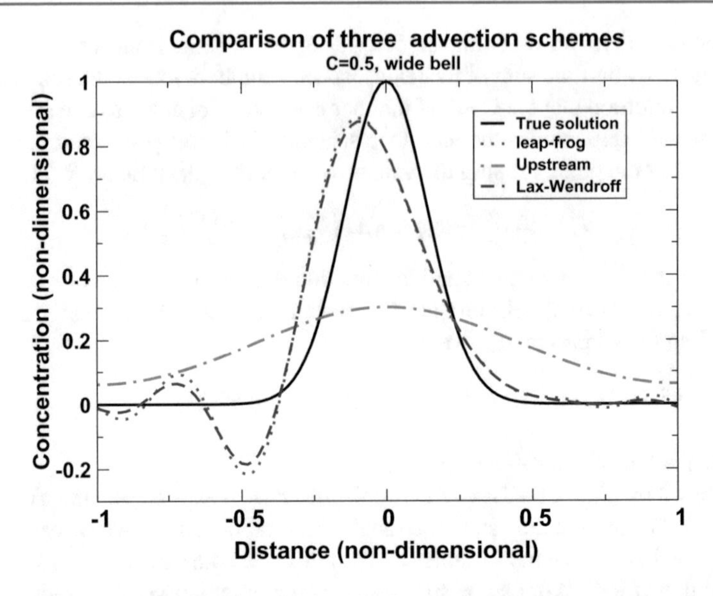

Fig. 5.3 Comparison of the leap-frog (*red dotted line*), upstream/upwind (*green dash-dotted line*), and Lax–Wendroff (*blue dashed line*) schemes applied to the advection equation (5.3). The Courant number is $C = 0.5$. A cyclic boundary condition is used to mimic an infinite domain. The solutions shown are after 10 cycles. The true solution is an initial bell function (*solid black line*). Notice that the leap-frog and the Lax–Wendroff schemes both display numerical dispersion, and that the upstream scheme is highly diffusive

$$\partial_t \theta|_j^n + u_0 \partial_x \theta|_j^n = -\frac{1}{2}\left(\partial_t^2 \theta|_j^n - u_0 \partial_x^2 \theta|_j^n\right) \Delta t + \mathcal{O}(\Delta t^2) + \mathcal{O}(\Delta x^2) + \mathcal{O}(\Delta x^2 \Delta t) \,. \tag{5.84}$$

Since (5.3) implies that $\partial_t^2 \theta = -u_0 \partial_x (\partial_t \theta) = u_0^2 \partial_x^2 \theta$, the first term on the right-hand side vanishes, which finally leads to

$$\partial_t \theta|_j^n + u_0 \partial_x \theta|_j^n = \mathcal{O}(\Delta t^2) + \mathcal{O}(\Delta x^2) + \mathcal{O}(\Delta x^2 \Delta t) \,. \tag{5.85}$$

Since (5.3) is recovered when $\Delta x \to 0$ and $\Delta t \to 0$ independently, the scheme is indeed consistent. Furthermore, the higher order terms neglected are $\mathcal{O}(\Delta x^2)$ and $\mathcal{O}(\Delta t^2)$, and the scheme is therefore second-order accurate even though it is forward in time. Moreover, since the first term on the right-hand side of (5.84) vanishes, the diffusive term inherent in the diffusive scheme does not carry over to the Lax–Wendroff scheme.

5.12 Semi-Lagrangian Method

Some years ago, several attempts were made to construct stable time integration schemes allowing longer time steps than those schemes limited by the CFL condition treated here, e.g., the leap-frog, upstream, and Lax–Wendroff schemes. To this end, Robert (1981) proposed a technique, referred to as the *semi-Lagrangian method*, to treat the advective part of the equations governing the motion of the atmosphere. As an introduction, it is useful to apply the technique to the one-dimensional advection equation (5.3). Later it will also be applied to solve the shallow water equations (Sect. 6.3), and to understand why the upstream and leap-frog schemes are unstable when the CFL condition is violated (Sect. 5.13).

The scheme evolved from earlier analytic methods developed to solve the dam-breaking problem (e.g., Stoker 1957, p. 513), a highly nonlinear problem. At that time, it was referred to as the method of characteristics. In the 1960s, the method was developed into a numerical scheme, (e.g., Lister 1966), but since the work of Robert (1981), it has been commonly referred to as the semi-Lagrangian technique.

The starting point is to define a special differential operator $\mathrm{D}^*/\mathrm{d}t$ by

$$\frac{\mathrm{D}^*}{\mathrm{d}t} \equiv \partial_t + \frac{\mathrm{D}^* x}{\mathrm{d}t}\partial_x \, , \tag{5.86}$$

where $\mathrm{D}^* x/\mathrm{d}t$ is a velocity along the x-axis, yet to be defined. Defining it by

$$\frac{\mathrm{D}^* x}{\mathrm{d}t} = u_0 \, , \tag{5.87}$$

the advection equation (5.3) may be written

$$\frac{\mathrm{D}^* \theta}{\mathrm{d}t} = 0 \, , \quad \text{along} \quad \frac{\mathrm{D}^* x}{\mathrm{d}t} = u_0 \, . \tag{5.88}$$

Since u_0 is constant, integration of (5.87) yields straight lines[8] in the x, t plane with slopes $1/u_0$ (Fig. 5.4). These lines are referred to as the *characteristics*, while Eq. (5.87) defining them is referred to as the *characteristic equation*. Moreover, since the solutions to (5.3) are compatible with solutions to (5.88), the latter is referred to as the *compatibility equation*. Thus, either of them may be used to arrive at the solution.

For example, assume that the tracer θ is known at time $t = 0$ from the initial condition. The task is to find the solution θ_Q at a particular point Q with coordinates (x_Q, t_Q) in the x, t plane (Fig. 5.4), using (5.88). According to (5.87), there is one characteristic passing through Q that crosses the time level $t = 0$ at the point P. The compatibility equation (5.88) then says that θ is conserved (does not change) along this characteristic. Thus, θ_Q is simply found by equating it to θ_P, the latter being known from the initial condition. This procedure may be repeated for a randomly chosen point in the x, t plane, until the whole x, t space is covered. Notice that, for times longer than $t > t_c = L/u_0$, all information about the initial distribution of θ is lost. Thus, for $t > t_c$ the solution in the domain $0 < x < L$ is determined by the boundary condition at $x = 0$ (Fig. 5.4).

Since (5.3) only contains the first derivative with respect to x, just one condition in x is allowed. The boundary at $x = L$ (Fig. 5.4) is therefore open in the sense that there is no boundary condition that replaces the differential equation there.

[8] If u_0 is a function of space and time, then the characteristics are curved lines with $\mathrm{D}^* x/\mathrm{d}t$ defining the slope at any point in the x, t plane.

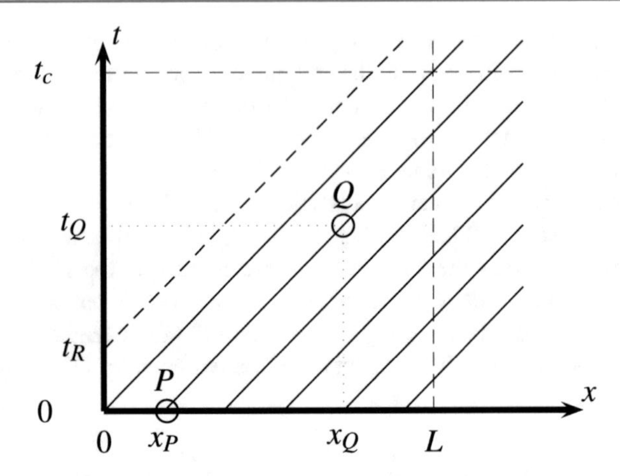

Fig. 5.4 Sketch of the characteristics in the x, t plane. For $u = u_0 = \text{const.} > 0$, the characteristics are straight lines sloping to the right in the x, t plane, as given by (5.87). Following the characteristic back in time through the point marked Q, it hits the horizontal axis ($t = 0$) at the point marked P. Since the tracer θ is conserved along the characteristic, the value of the tracer at the point (x_Q, t_Q), say θ_Q, equals the initial tracer value at the point $(x_Q, 0)$, say θ_P. Since the latter is known from the initial condition, it leads to $\theta_Q = \theta_P$. Notice that the end of the computational domain is at $x = L$. For $x > L$ all information about the initial condition is therefore lost for times $t > t_c$. At these times, the characteristics, e.g., the dashed characteristic, hit the vertical line at $x = 0$, and the solution at a random point $x, t > t_c$ is determined by the boundary condition rather than the initial condition

Indeed the advection equation is also valid at the artificial boundary $x = L$. In principle, therefore, the physical space continues to infinity. Thus, the boundary $x = L$ is a numerical boundary due to the fact that any computer, however large, is limited in its capacity. This problem is particularly relevant for oceanographic models, since oceanic spatial scales are small compared to the corresponding scales in the atmosphere. Conditions constraining the solutions at open boundaries are investigated in some detail in Chap. 7.

The above procedure may be used to solve (5.88) numerically (Lister 1966; Robert 1981; Røed and O'Brien 1983). To this end, we construct FDAs to replace (5.87) and (5.88). The starting point is to overlay the computational domain in the x, t space by a grid, as shown in Fig. 5.5. A straightforward finite difference approximation to (5.88) is

$$\theta_j^{n+1} = \theta_Q^n , \tag{5.89}$$

where θ_Q^n is the tracer value at the point Q in space where the characteristic through the grid point (x_j, t^{n+1}) crosses the time level n (Fig. 5.5). This point is denoted by x_Q. To find its position along the x-axis, a forward-in-time finite difference approximation to (5.87) leads to

$$\frac{x_j - x_Q}{\Delta t} = u_0 , \quad \text{or} \quad x_Q = x_j - u_0 \Delta t . \tag{5.90}$$

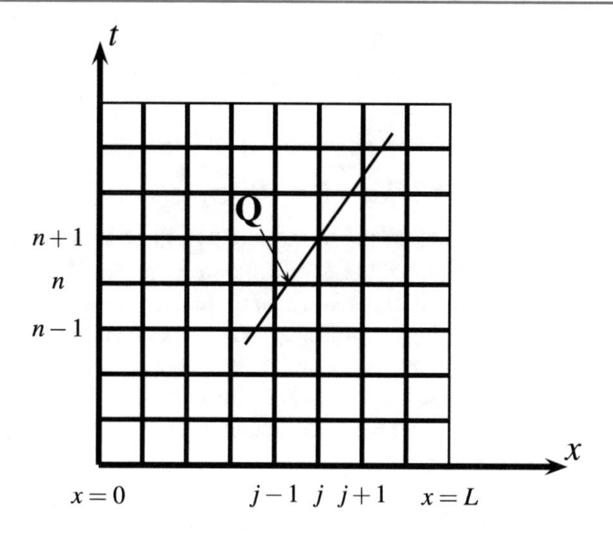

Fig. 5.5 Sketch of the method of characteristics. The distances between the grid points are Δt in the vertical and Δx in the horizontal. The *sloping solid line* is the characteristic through the point $j, n+1$. It is derived from (5.87) and the slope is given by $1/u_0$ ($u_0 > 0$). The point labeled Q is therefore a distance $u_0\Delta t$ to the left of x_j. As long as $u_0\Delta t < \Delta x$, the point Q is located between x_{j-1} and x_j. If, however, $u_0\Delta t > \Delta x$, then the point Q is located to the left of x_{j-1}

It is clear from (5.90) that it may be used to determine the location of x_Q, since both u_0 and x_j are known. Note that, as long as $u_0\Delta t < \Delta x$, x_Q is located between x_j and x_{j-1}. Since θ_j^n is known for all grid points, θ_Q^n may be found by interpolating linearly between the adjacent grid points. Thus, as long as $x_{j-1} \le x_Q \le x_j$, application of a two-point Newton interpolation leads to

$$\theta_Q^n = \theta_{j-1}^n + \frac{\theta_j^n - \theta_{j-1}^n}{\Delta x}(x_Q - x_{j-1}) \,. \tag{5.91}$$

Substituting x_Q from (5.90) into (5.91) then yields

$$\theta_Q^n = (1-C)\theta_j^n + C\theta_{j-1}^n \,, \tag{5.92}$$

where

$$C = |u_0|\frac{\Delta t}{\Delta x} \tag{5.93}$$

is the Courant number. Since, according to (5.88), θ is conserved along the characteristic this leads finally to

$$\theta_j^{n+1} = (1-C)\theta_j^n + C\theta_{j-1}^n \,, \quad C \le 1 \,, \tag{5.94}$$

which is commonly referred to as the *semi-Lagrangian scheme*.

Notice that (5.94) is only valid for $C \le 1$. However, this is a self-inflicted restriction. The advantage of the semi-Lagrangian scheme is that, if $1 < C < 2$ ($\Delta x < u_0\Delta t < 2\Delta x$), in which case the position of x_Q falls between x_{j-2} and

x_{j-1}, then θ_Q^n is approximated by linearly interpolating using θ_{j-1}^n and θ_{j-2}^n. Under these circumstances, the semi-Lagrangian scheme is

$$\begin{aligned}\theta_j^{n+1} &= \theta_{j-2}^n + \frac{\theta_{j-1}^n - \theta_{j-2}^n}{\Delta x}(x_Q - x_{j-2}) \\ &= (2 - C)\theta_{j-1}^n + (C - 1)\theta_{j-2}^n\,, \quad 1 < C \leq 2\,. \end{aligned} \tag{5.95}$$

Thus, as long as we keep track of the position of x_Q, there is no restriction on the time step Δt. The scheme is therefore *unconditionally stable*. This is in contrast to the other schemes treated in this chapter, where we had to impose the CFL condition to ensure stability, whence a longer time step may be used without losing accuracy.

If the constant speed u_0 is replaced by a speed varying in time and space, say $u(x, t)$, then the characteristics are no longer straight lines. Under these circumstances, the position x_Q can be found, for instance, by using a higher order finite difference approximation, say,

$$\frac{x_j - x_Q}{\Delta t} = \frac{1}{2}\left(u_j^n + u_j^{n+1}\right)\,, \quad \text{or} \quad x_Q = x_j - \frac{1}{2}\left(u_j^n + u_j^{n+1}\right)\Delta t\,. \tag{5.96}$$

Again, θ_j^{n+1} may be found by performing a two-point Newton interpolation. Assuming that $x_{j-1} \leq x_Q \leq x_j$ then leads to

$$\theta_j^{n+1} = (1 - C_j^n)\theta_j^n + C_j^n\theta_{j-1}^n\,, \tag{5.97}$$

where

$$C_j^n = \frac{1}{2}\left(u_j^n + u_j^{n+1}\right)\frac{\Delta t}{\Delta x} \tag{5.98}$$

is a local Courant number.

In many models, the semi-Lagrangian technique is a bit different from the above. Instead of finding θ_j^{n+1} by going back in time along the characteristics, one may find the new value by going forward in time along the characteristics. This is in line with what was originally suggested by Robert (1981). Thus, following the characteristic emanating from the grid point (x_j, t^n) forward in time, it crosses the new time level at the point Q with coordinates (x_Q, t^{n+1}), where $x_Q = x_j + u_0\Delta t$ (Fig. 5.6). Thus, using a simple forward-in-time FDA of the compatibility equation (5.88), the tracer value at the point Q is

$$\theta_Q^{n+1} = \theta_j^n\,, \tag{5.99}$$

where θ_Q^{n+1} denotes the value of the tracer at the point Q.

Similarly, following a characteristic emanating from the grid point (x_{j-1}, t^n) forward in time, it crosses the new time level at the point P with coordinates (x_P, t^{n+1}), and where $x_P = x_{j-1} + u_0\Delta t$. Using the compatibility equation leads to

$$\theta_P^{n+1} = \theta_{j-1}^n\,, \tag{5.100}$$

where θ_P^{n+1} denotes the value of the tracer at the point P. Note that, if $u_0 > 0$ and $C \leq 1$, then x_Q satisfies $x_j < x_Q < x_{j+1}$, while x_P satisfies $x_{j-1} < x_P < x_j$. Nevertheless, $C \leq 1$ is not a requirement. Assuming, nonetheless, that this is the

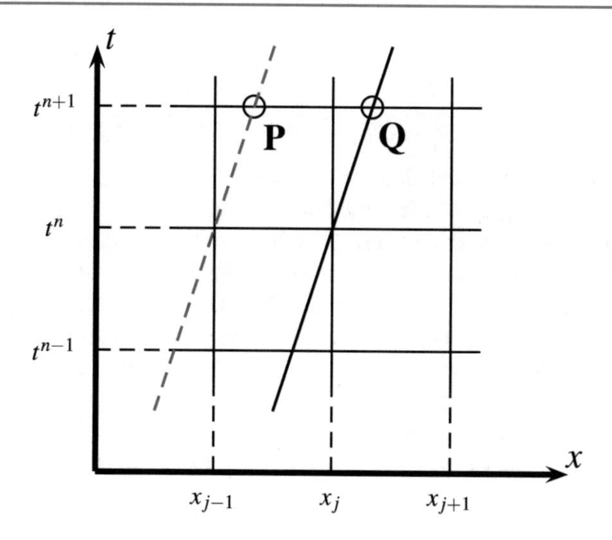

Fig. 5.6 Semi-Lagrangian method. The distance between the grid points is Δt in the vertical and Δx in the horizontal. The characteristic through the grid point (x_j, t^n) (*solid black line*) crosses the time level $n+1$ at the point marked P with coordinates (x_P, t^{n+1}), while the characteristic through the grid point (x_{j-1}, t^n) (*dashed red line*) crosses the time level $n+1$ at the point Q with coordinates (x_Q, t^{n+1}). They are both derived from (5.87) and the slope is given by $1/u_0$ ($u_0 > 0$). Notice that, as long as the Courant number C satisfies $C \leq 1$, then $x_{j-1} < x_P < x_j$ and $x_j < x_Q < x_{j+1}$. If $C > 1$ the points P and Q will be to the right of x_j and x_{j+1}, respectively

case, the tracer value at the new time level at a grid point, say (x_j, t^{n+1}), may again be found by interpolation. This is called *regridding*, and a linear two-point Newton interpolation leads to

$$\theta_j^{n+1} = \theta_P^{n+1} + \frac{\theta_Q^{n+1} - \theta_P^{n+1}}{x_Q - x_P}(x_j - x_P) \,. \tag{5.101}$$

Using (5.99) and (5.100) to substitute for θ_P^{n+1} and θ_Q^{n+1}, and substituting for x_P and x_Q, (5.101) takes the form

$$\theta_j^{n+1} = (1 - C)\theta_j^n + C\theta_{j-1}^n \,, \quad C \leq 1 \,. \tag{5.102}$$

Not surprisingly, (5.102) is the same as (5.94). The restriction $C \leq 1$ is superfluous. As when following the characteristics back in time, we must keep track of where the characteristics cross the new time level when following them forward in time. Thus, if $C > 1$, the points P and Q cross to the right of x_j and x_{j+1}, respectively, and (5.102) must be replaced by (5.95). Furthermore, if the advection velocity is a function of time and space, say $u = u(x, t)$, then the characteristics are no longer parallel straight lines. Nonetheless, if Δt is small enough, they may still be approximated by two nonparallel straight lines with slopes $1/u_{j-1}^n$ and $1/u_j^n$, respectively. Once again, assuming that the points P and Q are located so that $x_j < x_Q < x_{j+1}$ and $x_{j-1} < x_P < x_j$, respectively, then using (5.101), (5.102) takes the form

$$\theta_j^{n+1} = (1 - C^{*n}_j)\theta_j^n + C^{*n}_j\theta_{j-1}^n \,, \tag{5.103}$$

with

$$C^{*n}_j = \frac{C^n_j}{1 + C^n_j - C^n_{j-1}} = \frac{|u^n_j|}{|u^n_j| - |u^n_{j-1}| + \Delta x/\Delta t}\,, \tag{5.104}$$

where C^{*n}_j is a modified local Courant number.

To conclude, as long as we keep track of the points where the characteristics cross the respective time levels, there is no restriction on the time step. The latter implies that the semi-Lagrangian scheme is unconditionally stable, whether following the characteristics backward or forward in time. This is also true if the advection velocity is a function of time and space, but then the time step must be small enough to allow approximation of the characteristics by straight lines. Finally, it is worth mentioning that the semi-Lagrangian scheme is applicable to much more complex problems and systems than the simple advection equation treated here (e.g., Lister 1966; Robert 1981; Røed and O'Brien 1983). These matters will be discussed further in Sect. 6.4.

5.13 Physical Interpretation of the CFL Condition

Figure 5.5 may also be used to understand the physics behind the CFL criterion in the context of the above schemes, and in particular the upstream scheme. Recall that, when $u_0 \geq 0$, we have $u_0 = |u_0|$. Under these circumstances, the upstream scheme (5.68) simplifies to

$$\theta^{n+1}_j = (1 - C)\theta^n_j + C\theta^n_{j-1}\,, \tag{5.105}$$

where $C = |u_0|\Delta t/\Delta x$ is the Courant number.[9] Thus, the upstream scheme may be interpreted as a simple weighting scheme. It then uses the values θ^n_j and θ^n_{j-1} at the previous time step to compute θ at the new time step $n+1$, with the Courant number as weight.

As shown in Fig. 5.5, the characteristic through the grid point $x_j, t^n + 1$ only crosses the time level n to the right of x_{j-1} if $|u_0|\Delta t < \Delta x$. Under these circumstances, the semi-Lagrangian scheme (5.94) matches the upstream scheme (5.105) exactly. If, however, $|u_0|\Delta t > \Delta x$, then the semi-Lagrangian scheme changes to (5.95), since the characteristics then cross the previous time level n to the left of x_{j-1}, while the upstream scheme continues to use (5.105) to compute θ at the new time level. The latter is obviously wrong, since use of (5.105) then leads to an error. If this is allowed to continue for time step after time step, the error accumulates and eventually gives rise to a numerical instability.

As mentioned, the speed defined by Δx and Δt, that is, $\Delta x/\Delta t$, is often referred to as the signal speed of the grid or simply the grid speed. The CFL criterion (5.28) may therefore be interpreted as a condition which constrains the grid speed to be greater than the advection speed u_0. The advection speed must therefore be small enough to allow the area of dependence to lie between x_{j-1} and x_{j+1} at time level n.

[9] If $u_0 < 0$, then $u_0 = -|u_0|$ and $x_Q > x_j$, and the upstream scheme is $\theta^{n+1}_j = (1-C)\theta^n_j + C\theta^n_{j+1}$.

5.14 Implicit Scheme

As for the diffusion equation an implicit scheme for the advection equation may also be constructed. This is easily realized by using a backward-in-time centered-in-space scheme. Thus,

$$\theta_j^{n+1} = \theta_j^n - u_0 \frac{\Delta t}{2\Delta x}\left(\theta_{j+1}^{n+1} - \theta_{j-1}^{n+1}\right), \quad j = 2(1)J-1, \quad n = 0(1), \ldots . \tag{5.106}$$

This scheme is accurate to first order in time and second order in space [$\mathscr{O}(\Delta t)$ and $\mathscr{O}(\Delta x^2)$], Since it is implicit, it requires the use of an elliptic solver for each time step. In this regard, the direct elliptic solver outlined in Sect. 4.11 is a good choice.

To investigate its stability, the von Neumann method is used. Thus,

$$\Theta_{n+1} = \Theta_n - \mathrm{i}C \sin\alpha\Delta x \Theta_{n+1}, \tag{5.107}$$

which leads to a complex growth factor

$$G = \frac{1}{1 + \mathrm{i}C \sin\alpha\Delta x}. \tag{5.108}$$

Thus, its magnitude is

$$|G| = \frac{1}{\sqrt{1 + C^2 \sin^2\alpha\Delta x}}, \tag{5.109}$$

which satisfies $|G| \leq 1$ for all finite time steps. As expected, the implicit scheme is therefore unconditionally stable, and hence, like the semi-Lagrangian scheme, avoids the restrictive CFL condition. Note, however, that if Δt becomes too large, the growth factor will be small, implying that the scheme will contain a large numerical dissipation. So, just as for any other implicit scheme, the choice of time step is restricted by the need to control numerical dissipation.

Finally, note that the time step, just like the grid size, must be sufficient to resolve the typical periods of the physical problem. Commonly, in atmospheric and oceanic applications, the typical period is much longer than the Nüquist frequency $2\Delta t$. Therefore, in most cases, the CFL condition puts a much more stringent restriction on Δt than the requirement of resolving the typical periods of the physical problem. Thus, for most meteorological and oceanographic problems, the resolution requirement is on the grid size.

5.15 Numerical Diffusion

Although the leap-frog scheme is neutrally stable, we have just shown that it has at least one major disadvantage: it is dispersive. In particular, as can be seen from Fig. 5.1, this is true when the resolution is poor, that is, in areas where Δx is inadequate to resolve the dominant wavelength, i.e., where $\alpha\Delta x$ is not necessarily small. Furthermore, the impact of the dispersion increases with decreasing Courant number. In addition, as shown in Sect. 5.7, the leap-frog scheme contains an unwanted computational mode giving rise to an unphysical solution.

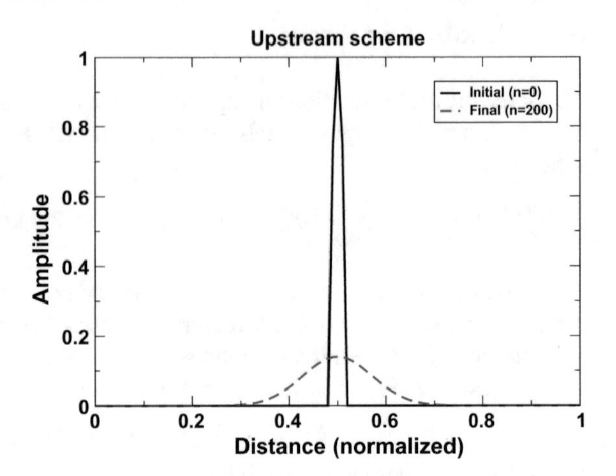

Fig. 5.7 Example of the diffusion inherent in the upstream scheme. The *solid curve* shows the initial distribution at time level $n = 0$, while the *dashed red curve* shows the distribution at time level $n = 200$ (one cycle). The Courant number is $C = 0.5$. Cyclic boundary conditions are imposed at the boundaries of the computational domain

As a result, the upstream scheme was for a long time the favored advection scheme. But unfortunately, the upstream scheme is far from perfect. It contains what is referred to as *numerical diffusion*. The name derives from the fact that the error made by truncating the Taylor series to first order to arrive at the finite difference equation (5.71) acts similarly to physical diffusion, that is, the inherent truncation error tends to diminish differences in the tracer distribution. This is exemplified in Fig. 5.7, where an initial narrow, peak-like tracer distribution spreads out while being advected. In contrast, the analytic solution that the numerical solution tries to mimic is one in which the initial tracer distribution is advected without change. We stress that this does not imply that any tracer content is lost. The numerical diffusion process, just like its physical counterpart, conserves the total tracer content. What happens is that the numerical diffusion smooths out any differences in the initial tracer concentration. Thus, it redistributes the initial tracer distribution while conserving the initial total tracer content. This is evident in Fig. 5.7. Comparing the area under the dashed curve and the area under the solid curve, we find that they are actually the same. Note once again that this redistribution is artificial and arises because we use the upstream scheme to solve the advection equation (5.3).

To analyze the origin of the numerical diffusion in the upstream scheme, let us reconsider (5.67). We first rewrite it in terms of an advective flux defined by

$$F_j^n = \frac{1}{2}\Big[(u_0 + |u_0|)\theta_j^n + (u_0 - |u_0|)\theta_{j+1}^n\Big]\frac{\Delta t}{\Delta x}\,. \tag{5.110}$$

We note that, since $u_0 = \text{sgn}(u_0)|u_0|$, the last term on the right-hand side of (5.110) is zero when $u_0 \geq 0$ and the first term is zero when $u_0 < 0$. Under these circumstances, $F_j^n = C\theta_j^n$ if $u_0 \geq 0$ and $F_j^n = -C\theta_{j+1}^n$ if $u_0 < 0$. Thus (5.67) is written

$$\theta_j^{n+1} = \theta_j^n - (F_j^n - F_{j-1}^n) , \tag{5.111}$$

which is valid regardless of the sign of u_0. If we substitute each of the terms in (5.111) by its associated Taylor series expansion, that is,

$$\begin{aligned} \theta_j^{n+1} &= \theta_j^n + \partial_t \theta|_j^n \Delta t + \tfrac{1}{2}\partial_t^2 \theta|_j^n \Delta t^2 + \mathcal{O}(\Delta t^3) , \\ \theta_{j\pm 1}^n &= \theta_j^n \pm \partial_x \theta|_j^n \Delta x + \tfrac{1}{2}\partial_x^2 \theta|_j^n \Delta x^2 + \mathcal{O}(\Delta x^3) , \end{aligned} \tag{5.112}$$

we get

$$\partial_t \theta|_j^n + \frac{1}{2}\partial_t^2 \theta|_j^n \Delta t + \mathcal{O}(\Delta t^2) = -u_0 \partial_x \theta|_j^n + \frac{1}{2}|u_0| \partial_x^2 \theta|_j^n \Delta x + \mathcal{O}(\Delta x^2) . \tag{5.113}$$

We note that by differentiating (5.3), we get $\partial_t^2 \theta|_j^n = u_0^2 \partial_x^2 \theta|_j^n$, and hence by rearranging terms that

$$\partial_t \theta|_j^n + u_0 \partial_x \theta|_j^n = \frac{1}{2}|u_0|(\Delta x - |u_0|\Delta t)\partial_x^2 \theta|_j^n + \mathcal{O}(\Delta x^2) + \mathcal{O}(\Delta t^2) . \tag{5.114}$$

Defining

$$\kappa^* = \frac{1}{2}|u_0|(\Delta x - |u_0|\Delta t) , \tag{5.115}$$

Equation (5.114) may be written

$$\partial_t \theta|_j^n = -u_0 \partial_x \theta|_j^n + \kappa^* \partial_x^2 \theta|_j^n + \mathcal{O}(\Delta x^2) + \mathcal{O}(\Delta t^2) . \tag{5.116}$$

Thus to second order in time and space, we solve the equation

$$\partial_t \theta + u_0 \partial_x \theta = \kappa^* \partial_x^2 \theta . \tag{5.117}$$

We recognize (5.117) as an advection–diffusion equation (Chap. 3) with a diffusion coefficient κ^* given by (5.115). The terms of order $\mathcal{O}(\Delta x)$ and $\mathcal{O}(\Delta t)$ which we neglected when employing the upstream scheme therefore give rise to diffusion. This diffusion is unphysical, that is, it is an artifact that appears due to the numerical method used. It is therefore referred to as *numerical diffusion*. The strength of the numerical diffusion is determined by the diffusion coefficient defined in (5.115). We note that the diffusion is insignificant if the Courant number is close to or actually equals unity. This corresponds to the upper limit of the CFL criterion (5.28) for stability, and is associated with a nearly neutrally stable scheme. The diffusion term also goes to zero when Δx and Δt go to zero, hence showing that the upstream scheme is consistent.

5.16 Flux Corrective Schemes

In contrast to the second-order leap-frog scheme, the first-order upstream scheme has the advantage of being positive definite. If the distribution of, say, $\theta(x, t)$ at some arbitrary time t is such that $\theta \geq 0$ for all x, then we also have $\theta \geq 0$ for all later times $t = t + n\Delta t$, $n = 1, 2, \ldots$. Another important property is that the positions of the extrema are correctly propagated without dispersion. This is exemplified by

Fig. 5.7. These are valuable properties, well worth retaining. The question arises as to whether it is possible to construct a scheme that holds on to these properties and which avoids, or at least minimizes, the numerical diffusion inherent in the upstream scheme.

There are several schemes that offer an answer. One is the family of so-called total variation diminishing (TVD) schemes, originally introduced by Harten (1997). Another is the family of *flux corrective schemes*. One of the first schemes in the latter family was the one introduced by Zalesak (1979). Here we present another, the multidimensional positive-definite advection transport algorithm (MPDATA). This was first suggested by Smolarkiewicz (1983, 1984), and is nicely reviewed by (Smolarkiewicz and Margolin 1998). The key feature of MPDATA is precisely that it corrects the numerical diffusion inherent in the upstream scheme.

The starting point is the one-dimensional advection equation (3.16). Written in flux form, this becomes

$$\partial_t \theta + \partial_x F_{\mathrm{A}} = 0 \,, \tag{5.118}$$

where $F_{\mathrm{A}} = u\theta$. Here, the advective velocity u is considered to be a function of time and space (non-Boussinesq fluid). Using the upstream scheme results in what is effectively an advection–diffusion equation of the form

$$\partial_t \theta + \partial_x (F_{\mathrm{A}} + F_{\mathrm{D}}^*) = 0 \,, \tag{5.119}$$

where $F_{\mathrm{D}}^* = -\kappa^* \partial_x \theta$ is a diffusive flux, and the diffusion coefficient is

$$\kappa^* = \frac{1}{2}|u|(\Delta x - |u|\Delta t) \,. \tag{5.120}$$

Since the upstream scheme (5.67) is consistent, the diffusion term F_{D}^* is artificial. To avoid this unwanted numerical diffusion, Smolarkiewicz (1983) suggested replacing (5.118) by

$$\partial_t \theta + \partial_x (F_{\mathrm{A}} + F_{\mathrm{A}}^*) = 0 \,, \quad \text{or} \quad \partial_t \theta + \partial_x \left[(u + u^*)\theta \right] \,. \tag{5.121}$$

Here $F_{\mathrm{A}}^* = u^* \theta$ is an artificial *anti-diffusive flux*, where u^* is called the *anti-diffusion velocity*. It is introduced to oppose the diffusive flux F_{D}^* inherent in the upstream scheme. The latter is defined so as to oppose and match the diffusive flux in strength ($F_{\mathrm{A}}^* = -F_{\mathrm{D}}^*$). This leads to

$$u^* = \frac{\kappa^* \partial_x \theta}{\theta} \,. \tag{5.122}$$

The strength of the anti-diffusive flux is therefore determined by the magnitude of $\partial_x \theta$, while its sign is determined by the slope of θ. In regions where the slope is positive ($\partial_x \theta > 0$), the anti-diffusive flux is directed along the positive x-axis, and works to steepen the slope there. In regions where the slope is negative ($\partial_x \theta < 0$), the advective flux is negative, so the slope is steepened there as well. Moreover, at extrema, where $\partial_x \theta = 0$, the anti-diffusive flux is zero ($u^* = 0$). The propagation of extrema is therefore unaffected by adding the anti-diffusive flux, so they are correctly advected.

This may be illustrated by studying the evolution of the initial tracer distribution shown in Fig. 5.7. This shows the initial distribution as well as the solution after

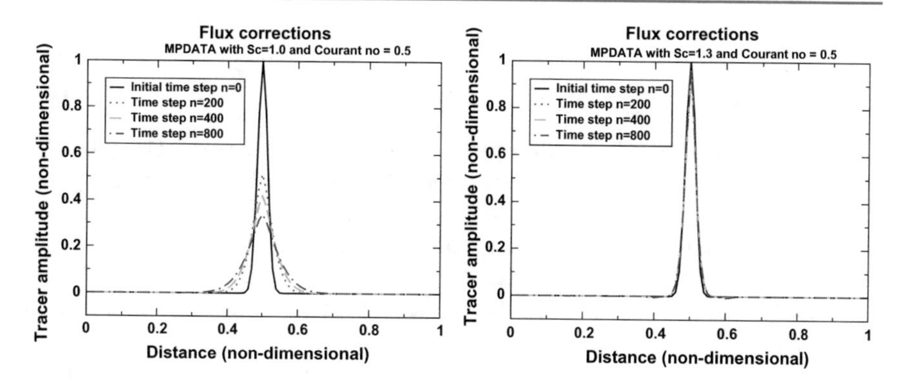

Fig. 5.8 Solutions to the advection equation using the MPDATA scheme suggested by Smolarkiewicz (1983). *Left*: Scaling factor 1.0 (no scaling). *Right*: Scaling factor 1.3. The Courant number is 0.5 in both cases. *Solid black lines* show the initial value (time step $n = 0$), and *red dotted lines* show the solution after 200 time steps (one cycle). *Green dashed lines* are after 400 time steps (two cycles) and *blue dash-dotted lines* are after 800 time steps (four cycles). Periodic boundary conditions are used (see the computer problem in Sect. 5.19.2)

200 time steps (one cycle) using only the upstream scheme (5.67) (no anti-diffusive flux added). The effect of using the MPDATA is shown in Fig. 5.8 (left-hand panel). Observe that adding the anti-diffusive flux F_{A}^* does not affect the propagation of the maximum value. Moreover, to the left and right of the maximum, the addition of the anti-diffusive flux tends to steepen the slopes relaxed by the numerical diffusion. Furthermore, the solution is most affected where $\partial_x\theta$ is steepest, that is, where $\partial_x^2\theta$ changes sign. The solution invoking MPDATA is clearly closer to the correct analytic solution. Nevertheless, the maximum value is diminished, and the slopes are not completely returned to their original steepness. This suggests that, even though the addition of the anti-diffusive flux helps, there is still some numerical diffusion left. As suggested by Smolarkiewicz (1983), this may for instance be remedied by multiplying the anti-diffusive velocity by a scaling factor. The effect of the latter is shown in the right-hand panel of Fig. 5.8.

The numerical implementation suggested by Smolarkiewicz (1983) is equally simple. He suggested making the correction using the so-called *predictor–corrector method*. The numerical calculation is carried out in two steps. In the first step, the *predictor step*, a low-order advection algorithm is used to produce a prediction θ^* without adding the anti-diffusive flux term. The numerical diffusion is then hidden in the $\mathscr{O}(\Delta t)$ and $\mathscr{O}(\Delta x)$ terms. Then using, for instance, the upstream scheme for this purpose, we obtain

$$\theta_j^* = \theta_j^n - \left(F_{\mathrm{A}}|_j^n - F_{\mathrm{A}}|_{j-1}^n \right) , \tag{5.123}$$

where[10]

$$[F_A]_j^n = \frac{1}{2}\Big[(u_j^n + |u_j^n|)\theta_j^n + (u_{j+1}^n - |u_{j+1}^n|)\theta_{j+1}^n\Big]\frac{\Delta t}{\Delta x} . \tag{5.124}$$

This first step retains all the advantages of the upstream scheme. The predictor solution θ_j^* is, however, "infected" by numerical diffusion, where the strength of the diffusion is given by κ^* (5.120). Since κ^* is proportional to $(1 - C)$, the prediction θ_j^* for Courant numbers less than one will appear to be smoother than its analytic or continuous counterpart, and dramatically so for small Courant numbers. This is particularly evident in areas where the initial distribution features steep gradients (Fig. 5.7).

In the second step, the *corrector step*, the advection equation (5.121) is solved *neglecting* the original advection term, that is,

$$\partial_t \theta + \partial_x F_A^* = 0 . \tag{5.125}$$

Applying the same low-order upstream scheme based on the predictor then leads to a corrected solution θ_j^{n+1} of the form

$$\theta_j^{n+1} = \theta_j^* - \left(F_A^*|_j - F_A^*|_{j-1}\right) , \tag{5.126}$$

where

$$F_A^*|_j = \frac{1}{2}\Big[(u_j^* + |u_j^*|)\theta_j^* + (u_{j+1}^* - |u_{j+1}^*|)\theta_{j+1}^*\Big]\frac{\Delta t}{\Delta x} . \tag{5.127}$$

To ensure that the anti-diffusive velocity $u_j^* = 0$ at the extrema, a centered scheme is used to compute it. Referring to (5.122), the FDA for u^* then takes the form

$$u_j^* = \frac{1}{4\Delta x}|u_j^n|\left(\Delta x - |u_j^n|\Delta t\right)\frac{\theta_{j+1}^* - \theta_{j-1}^*}{\theta_j^* + \varepsilon} . \tag{5.128}$$

As suggested by Smolarkiewicz (1983), a small number ε is added to the denominator to ensure that u_j^* goes to zero when θ_j^* and $\theta_{j+1}^* - \theta_{j-1}^*$ go to zero simultaneously. Making use of (5.128) ensures that the position of the maximum is unchanged during the corrector step. As an example, consider the red dashed curve in Fig. 5.7, which is the result of the predictive step. Consequently, since the corrections are proportional to the gradient of the predictive step solution, the largest corrections are made along the two flanks. Since the area under the curve is conserved when using the upstream scheme, the maximum increases during the corrector step (Fig. 5.8). The solution therefore retains all the advantages of the upstream scheme and appears to avoid the artificial smoothing of the steep gradients when applying the upstream scheme alone. Moreover, we observe that the corrector step makes the solution correct to second order [$\mathcal{O}(\Delta x^2)$ and $\mathcal{O}(\Delta t^2)$]. Hence, MPDATA is a second-order scheme which in theory compensates exactly for the artificial diffusive flux inherent in the low-order upstream scheme.

[10]The FDA for the advective flux F_A presented in (5.124) is only valid for a non-staggered grid. In his original paper (Smolarkiewicz 1983), he used a staggered grid (see Sect. 6.5), which results in a slightly different FDA.

Since a low-order upstream scheme was chosen to correct the fluxes, the numerical implementation of MPDATA also contains some artificial diffusion of higher order. This artificial diffusion may in turn be further corrected by running a second corrector step using the corrected solution as input, which in turn contains even higher order diffusion which may be corrected by a third corrector step, and so on. The number of iterative steps is determined by the user, and may be chosen according to the accuracy required. A simpler and cheaper method (in terms of computer time) involves slightly overestimating the anti-diffusive velocity by multiplying (5.128) by a scaling factor greater than unity, a method already suggested by Smolarkiewicz (1983). The anti-diffusive velocity is then redefined by

$$u_j^* = \frac{1}{4} S_c (1 - C_j^n) |u_j^n| \frac{\theta_{j+1}^* - \theta_{j-1}^*}{\theta_j^* + \varepsilon} , \tag{5.129}$$

where $S_c \geq 1$ is the scaling factor. As an example, Fig. 5.8 shows the solution to (5.118), using a scaling factor $S_c = 1.3$ (right-hand panel).

5.17 Summary and Remarks

This chapter discusses ways to discretize the advection equation. Focus is placed on a few specific schemes, rather than going through all the schemes suggested over the years (for a review of schemes applicable to the advection equation, see O'Brien 1986).

Section 5.1 gave a brief introduction to the one-dimensional advection equation. Section 5.2 then discussed the forward-in-time centered-in-space (FTCS) scheme that worked well for the diffusion equation. Interestingly, the scheme turned out to be unconditionally unstable. It was therefore abandoned, and in Sect. 5.3, the conditionally stable centered-in-time centered-in-space (CTCS) scheme, or leap-frog scheme, was introduced, and shown to be conditionally stable under the Courant–Friedrich–Levy (CFL) condition (Sect. 5.4). In contrast, application of this scheme to solve the diffusion equation led to an unconditionally unstable scheme (Sect. 4.6). It was concluded that this difference in stability properties was due to a difference in the physics, the advection equation being a hyperbolic problem and the diffusion equation a parabolic problem. The *CFL condition* states that the leap-frog scheme is stable if $C \leq 1$, where $C = |u_0|\Delta x/\Delta t$ is the *Courant number*, u_0 is the advection velocity, and Δx and Δt are the time step and space increment, respectively. The CFL condition is a fundamental criterion to be satisfied in numerical modeling of geophysical fluid dynamics. Essentially, it requires the grid speed $\Delta x/\Delta t$ to be greater than the advective velocity. This is equivalent to saying that a particle must travel a distance less than the space increment Δx during the time interval Δt, otherwise the information will have passed the next grid point.

After presenting the details of the leap-frog scheme and its stability, important concepts such as *numerical dispersion* (Sect. 5.5), the *initial value problem* (Sect. 5.6), and *computational modes* (Sect. 5.7) were considered in some detail. Even though the leap-frog scheme contains all these unwanted properties, it is nevertheless a popular scheme, widely used in numerical modeling of geophysical fluid dynamics. The reasons for this are that it is neutrally stable (Sect. 5.4), involving no numerical dissipation, its computational modes are easily controlled by applying the Asselin filter outlined in Sect. 5.8, the initial problem can be handled using a Euler scheme, and the numerical dispersion is handled by choosing the grid size to be small enough for the dominant length scales of the problem to be well resolved. Nonetheless, if the numerical dispersion is not properly attended to, it can cause serious problems, such as loss of energy to waves at high wave numbers creating negative tracer values, slowing down the advection of the dominant signal, and reducing peak values.

We then considered three popular forward-in-time, conditionally stable schemes. These were the upstream (Sect. 5.9), diffusive (Sect. 5.10) and Lax–Wendroff (Sect. 5.11) schemes. The main take-home lesson was that all were stable under the CFL condition, but they also exhibit some disadvantages. For instance, as shown in Sect. 5.15, the upstream scheme contains *numerical diffusion*. Since this diffusion tends to even out the fronts (large gradients), it inhibits processes which would in reality lead to physical instabilities, which would in turn create lows and highs in the atmosphere and eddies and jets in the ocean. On the other hand, the upstream scheme conserves the positive-definite nature of tracers, which the leap-frog scheme and Lax–Wendroff schemes do not. The upstream scheme, like the leap-frog scheme, is popular and widely used in many modern codes, at least as one of several options to choose from. The reason for this is that the numerical diffusion may be kept at bay by choosing a Courant number close to unity. It is also possible to use a higher order upstream scheme to minimize numerical diffusion, as will be detailed in Sect. 10.1.

The diffusive scheme turned out to be inconsistent and therefore non-convergent. However, it is the first step in constructing the convergent Lax–Wendroff scheme. The latter has many of the same properties as the leap-frog scheme, but is forward in time and thus contains no computational modes. Moreover, it is accurate to second order in both time and space, even though it is forward in time. It therefore avoids the initial value problem, has the same accuracy as the second-order leap-frog scheme, and has no computational modes. For these reasons, the Lax–Wendroff scheme is also fairly popular, notably in fluid mechanics.

We next introduced more recent schemes such as the semi-Lagrangian scheme and flux corrective schemes. The semi-Lagrangian scheme, treated in Sect. 5.12, is fairly popular in many current NWP and NOWP models, the main reason being that it is unconditionally stable. However, it uses interpolation to compute the variables at the new time level, a process called regridding, which leads to some overheads when compared to the other schemes. It is also applicable to more complex problems, and we shall return to it when discussing the shallow water equations numerically (Chap. 6). The semi-Lagrangian scheme exploits the existence of so-called characteristics, a property of hyperbolic systems. These characteristics carry information about the past and the future. By constructing them, we can establish where information is coming from or where it is going. Moreover, in the absence of any external forcing, the tracer value is conserved along characteristics. The value at one time level is therefore found by following the characteristic forward or backward in time from one grid point. However, a characteristic that starts at a grid point at one time level does not necessarily hit a grid point at the next time level (going forward in time) or at the previous time level (going backward in time). Regridding by linear or higher order interpolation, which adds overheads to the computation, is therefore essential.

Regarding flux corrective schemes, there are several to choose from. For instance, the so-called TVD (total variation diminishing) schemes are widely used, in particular in fluid mechanics. In Sect. 5.16, the focus was on the MPDATA (multidimensional positive-definite advection transport algorithm) scheme, which is widely used within the geophysical fluid dynamics community. Essentially, MPDATA uses a low-order scheme, e.g., the upstream scheme, as a predictor step. Thus, in the predictor step, gradients are relaxed by the numerical diffusion inherent in the low-order scheme. In a second corrector step, these relaxed gradients are re-steepened by introducing an artificial anti-diffusive flux that opposes the numerical diffusive flux almost exactly. Moreover, since the low-order scheme is used in the predictor as well as the corrector step, it conserves the positive-definite nature of tracers, e.g., salinity, and/or humidity, and advects all extrema to their exact position.

As in the previous chapter, two computer problems are suggested below (see Sects. 5.19.1 and 5.19.2). Of these, the reader is particularly encouraged to do the one in Sect. 5.19.1, which discusses and visualizes many of the properties of the various schemes presented in this chapter, and will help the reader to gain insight into how these properties manifest themselves in the numerical results.

5.18 Exercises

1. Show that the true solution to (5.3) is indeed (5.4).
   Hint: Use Fourier series.
2. Show that the CFL criterion for the leap-frog, the diffusive, the upstream, and the Lax–Wendroff schemes is always $C \leq 1$, where $C = |u_0|\Delta t/\Delta x$ is the Courant number.
   Hint: Express the growth function as $G = \sqrt{1-(1-C)f}$, where $f = f(C, \alpha, \Delta x)$.
3. Assume that the Courant number equals one. Show that, under these circumstances, the upstream scheme has no truncation errors. Use the semi-Lagrangian method to illustrate why this is the case.
4. Show that (5.50) is a solution to (5.21). Moreover, show that (5.55) follows from (5.50) if the initial distribution is given by (5.52), and (5.15) is used to find θ_j^{-1}.
5. Show that (5.103) follows from (5.101) when the advection velocity is assumed to be nonconstant.

5.19 Computer Problems

5.19.1 Advection in Atmospheres and Oceans

We consider here numerical solutions of the one-dimensional advection equation

$$\partial_t \psi + u_0 \partial_x \psi = 0 , \tag{5.130}$$

with appropriate boundary and initial conditions. Here, u_0 is the advective speed along the x-axis, assumed to be uniform in time and space. To this end, three schemes will be employed: the *leap-frog*, the *upstream*, and the *Lax–Wendroff* schemes.

Part 1

A. Show that the *leap-frog*, *upstream*, and *Lax–Wendroff* schemes are are all numerically stable under the CFL condition

$$C = |u_0| \frac{\Delta t}{\Delta x} \leq 1 , \tag{5.131}$$

where C is the Courant number, Δx is the space increment, and Δt is the time step.
B. Show that a forward-in-time centered-in-space (FTCS) finite difference approximation applied to (5.130) results in an unconditionally unstable scheme.
C. Show that the upstream (or upwind) scheme includes inherent numerical diffusion with diffusion coefficient

$$\frac{1}{2} |u_0| \Delta x (1 - C) , \tag{5.132}$$

where C is the Courant number defined by (5.131).
D. Let

$$\psi(x, 0) = \psi_0 \mathrm{e}^{-(x/\sigma)^2} , \quad \forall x \in [-\infty, \infty] , \tag{5.133}$$

be the initial distribution of the tracer ψ, where ψ_0 is the maximum initial tracer concentration, and σ is a measure of the width of the Gaussian bell. Show that the analytic solution to (5.130) is

$$\psi(x, t) = \psi_0 \exp \left[- \left(\frac{x - u_0 t}{\sigma} \right)^2 \right] . \tag{5.134}$$

Part 2

Here the aim is to solve (5.130) numerically for $x \in [-L, L]$. Since the governing equation (5.130) is valid for $x \in [-\infty, \infty]$, it is also valid at the two boundaries $x = -L$ and $x = L$. Hence, these boundaries are artificial, and commonly referred to as "open" boundaries (see Chap. 7). Furthermore, the analytic solution is such that the peak of the bell is positioned at $x = 2mL$ at the particular times $t_c^m = 2mL/|u_0|$, where $m = 0, 1, 2, \ldots$. Mapping the analytic solution for $x \in [(2n-1)L, (2n+1)L]$, $n = 1, 2, \ldots$ onto $x \in [-L, L]$ means that the solutions all lie on top of each other at these times. To mimic this, we require

$$\psi(x, t) = \psi(x + 2L, t) , \tag{5.135}$$

which is equivalent to imposing cyclic boundary conditions at the open boundaries $x = -L, L$. Thus, whatever leaves the interior domain at $x = L$ should be present at the left-hand boundary $x = -L$ at the same time level. This is accomplished by replacing ψ_{-1}^n by ψ_J^n, and ψ_{J+2}^n by ψ_2^n whenever they appear, when looping through all the spatial grid points x_j, where $j = 1(1)J + 1$.

E. Let $x_j = -L + (j-1)\Delta x$ for $j = 1(1)J + 1$. Show that

$$J = \frac{2L}{\Delta x} \, . \tag{5.136}$$

F. Show that the numerical analogue of the cyclic boundary condition (5.135) is

$$\psi_{J+j}^n = \psi_j^n \, , \quad \text{for} \quad j = 1(1), \ldots \, , \tag{5.137}$$

and that the numerical analogue of the initial condition (5.133) is

$$\psi_j^0 = \psi_0 \exp\left\{-\left[\frac{-L + (j-1)\Delta x}{\sigma}\right]^2\right\} \, , \quad \text{for} \quad j = 1(1)J + 1 \, . \tag{5.138}$$

G. Let the grid size $\Delta x = \sigma/10$, $L = 50\,\text{km}$, and $\sigma = 2L/10$. The latter corresponds to a broad initial Gaussian bell, which is then properly resolved using the specified Δx. Solve (5.130) using the leap-frog, upwind/upstream, and Lax–Wendroff schemes, subject to the conditions (5.137) and (5.138). Stop the computations after 10 cycles, that is, when the peak of the bell has traversed ten times the distance $2L$. Do one experiment with the Courant number $C = 0.5$ and another with $C = 1$. Feel free also to experiment with other Courant numbers $1/2 < C < 1$ and other space increments Δx. Plot the solution after 1/2, 1, 2, 5, and 10 cycles, together with the initial tracer distribution for each of the two values of the Courant number. Lump the plots for the concentration together into one plot for each scheme, that is, six curves in each plot. Plot the dimensionless values of the tracer $\psi' = \psi/\psi_0$ and use $x' = x/L$ as the dimensionless distance, as exemplified in Fig. 5.9.

H. Repeat the above with $\sigma = 2L/1000$, that is, for a narrow initial bell. In particular, use the same grid size. This implies that the narrow bell is poorly resolved.

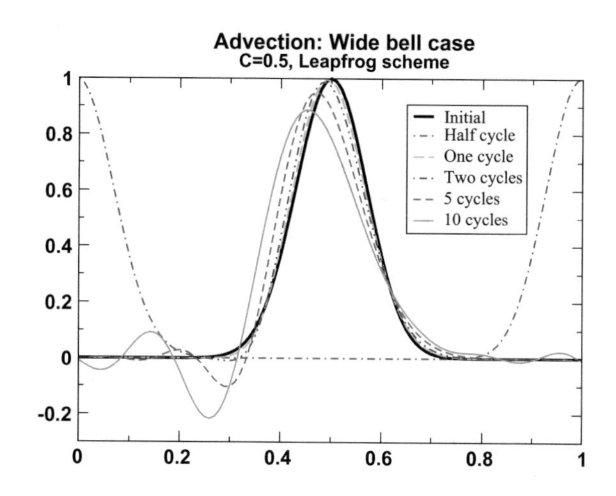

Fig. 5.9 Solution for the wide bell case using the leap-frog scheme with Courant number $C = 0.5$

I. Discuss the solutions of the narrow and wide bells based on these plots. What characterizes the solutions as they evolve in time? Which of the solutions are diffusive and which are dispersive? What are the characteristics of these processes?
J. Use the semi-Lagrangian method to explain why the upstream (upwind) scheme is perfect when $C = 1$. Show also that, under these circumstances, all higher order numerical "diffusion" terms for the upwind scheme are zero.

5.19.2 Flux Corrective Methods

Here, we consider numerical solutions to the advection equation (5.130) in which the advection speed is not necessarily a constant. Writing this equation in flux form, we have

$$\partial_t \theta + \partial_x (u\theta) = 0 \, , \tag{5.139}$$

where θ is the tracer concentration (or fraction). To solve (5.139), consider a non-staggered grid, as shown in Fig. 5.10, with a grid distance of Δx. We also consider the scheme

$$\theta_j^{n+1} = \theta_j^n - \left(F_j^n - F_{j-1}^n \right) \, , \tag{5.140}$$

where

$$F_j^n = \frac{1}{2}\Big[\left(u_j^n + |u_j^n| \right) \theta_j^n + \left(u_{j+1}^n - |u_{j+1}^n| \right) \theta_{j+1}^n \Big] \frac{\Delta t}{\Delta x} \, . \tag{5.141}$$

Part 1
A. Show that (5.140) is a first-order scheme in time and space.
Hint: Do the analysis for $u_j^n \geq 0$ and $u_j^n < 0$ separately.
B. Show that the scheme (5.140) is stable under the condition

$$\max_{j,n} \left(\frac{u_j^n \Delta t}{\Delta x} \right) \leq 1 \, , \tag{5.142}$$

where the velocity is assumed to be slowly varying in space ($u_{j+1}^n \approx u_j^n$).

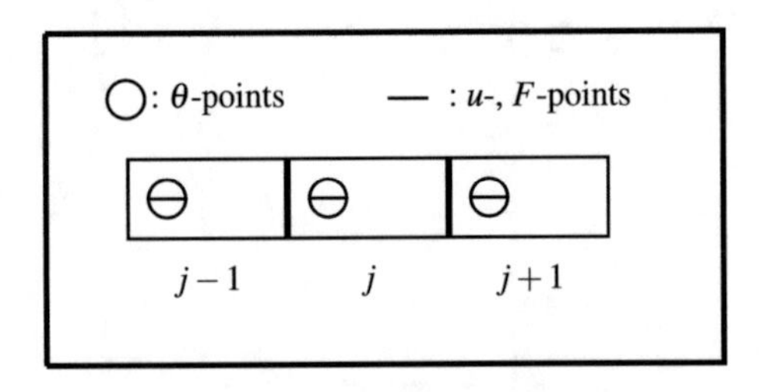

Fig. 5.10 Cell structure of lattice A in Mesinger and Arakawa (1976) (non-staggered) in one space dimension. *Circles* are associated with θ-points and *horizontal bars* with u- and F-points within the same cells. Cells are counted using the counter j. The distance between two adjacent θ-points (and hence also between adjacent u- and F-points) is Δx

C. Assume that $u = u_0$ is a constant, and show that the scheme (5.140) displays numerical diffusion with diffusion coefficient κ^* given by

$$\kappa^* = \frac{1}{2}|u_0|(\Delta x - |u_0|\Delta t) . \tag{5.143}$$

Part 2

The advection equation (5.139) allows two integration constants, one in time and one in space. We thus require one boundary condition in time (the initial condition) and one boundary condition in space to solve it. The initial tracer distribution is assumed to be a Gaussian bell

$$\theta|_{t=0} = \theta_0 e^{-(x/\sigma)^2} , \tag{5.144}$$

where θ_0 is the maximum tracer fraction at $x = 0$ and σ measures the width of the Gaussian bell. As in the computer problem of Sect. 5.19.1, make use of the cyclic condition (5.135) to mimic an infinite domain.

D. Solve (5.139) numerically for $x \in \langle -L, L]$ and within the time range $t \in \langle 0, T_{25} \rangle$, where T_{25} corresponds to the time it takes to perform the integration for 25 cycles. The initial condition at $t = 0$ is as given in (5.144). Use the scheme (5.140). Let $L = 50\,\text{km}$, $\sigma = 2L/10$, $\Delta x = \sigma/10$, $\theta_0 = 10\,°\text{C}$, $u_0 = 1\,\text{m/s}$, and $t_n = n\Delta t$, where n is the time counter and Δt is the time step. Show first that the number of time steps needed to reach the time T_{25} is $N_{25} = 50L/u_0\Delta t$. Plot the results after 0 (initial distribution), 5, 10, 15, 20, and 25 cycles.

Describe and discuss what you observe by comparing the evolution of the tracer distribution with the initial tracer distribution. Explain what happens.

E. As described in Sect. 5.16, the inherent numerical diffusion in the scheme (5.140) can be counteracted by using the flux corrective method MPDATA (Smolarkiewicz and Margolin 1998). Solve (5.139) using this method. Use at least two iteration steps, then the simple method of scaling the anti-diffusive velocity. Let the parameters and initial condition be as presented in (D). Use the scaling factor $S_c = 1.3$. When computing the anti-diffusive velocity, use a centered-in-space finite difference approximation and, as suggested by Smolarkiewicz (1983), be sure to add a small number $\varepsilon = 10^{-15}$ to the denominator.

F. Why do we have to add the small number ε to the denominator?

G. Carry out experiments for different values of the scaling factor S_c. Try out other finite difference approximations to the anti-diffusive velocity. Discuss the results.

References

Asselin RA (1972) Frequency filter for time integrations. Mon Weather Rev 100:487–490

Griffies SM (2004) Fundamentals of ocean climate models. Princeton University Press, Princeton. ISBN 0-691-11892-2

Grotjhan R, O'Brien JJ (1976) Some inaccuracies in finite differencing hyperbolic equations. Mon Weather Rev 104:180–194

Haidvogel DB, Arango H, Budgell PW, Cornuelle BD, Curchitser E, Lorenzo ED, Fennel K, Geyer WR, Hermann AJ, Lanerolle L, Levin J, McWilliams JC, Miller AJ, Moore AM, Powell TM, Shchepetkin AF, Sherwood CR, Signell RP, Warner JC, Wilkin J (2008) Ocean forecasting in terrain-following coordinates: formulation and skill assessment of the regional ocean modeling system. J Comput Phys 227(7):3595–3624. https://doi.org/10.1016/j.jcp.2007.06.016

Haltiner GJ, Williams RT (1980) Numerical prediction and dynamic meteorology, 2nd edn. Wiley, New York, 477 pp

Harten A (1997) High resolution schemes for hyperbolic conservation laws. J Comput Phys 135:260–278. https://doi.org/10.1006/jcph.1997.5713

Lax P, Wendroff B (1960) Systems of conservation laws. Commun Pure Appl Math 13:217–237

Lister M (1966) The numerical solution of hyperbolic partial differential equations by the method of characteristics. In: Ralston A, Wilf HS (eds) Mathematical methods for digital computers. Wiley, New York

Mesinger F, Arakawa A (1976) Numerical methods used in atmospheric models. GARP publication series, vol 17. World Meteorological Organization, Geneva, 64 pp

O'Brien JJ (ed) (1986) Advanced physical oceanographic numerical modelling. NATO ASI series C: mathematical and physical sciences, vol 186. D. Reidel Publishing Company, Dordrecht

Richtmyer RD, Morton KW (1967) Difference methods for initial value problems. Interscience, New York, 406 pp

Robert A (1981) A stable numerical integration scheme for the primitive meteorological equations. Atmos Ocean 19:35–46

Robert AJ (1966) The integration of a low order spectral form of the primitive meteorological equations. J Meteorol Soc Jpn 44:237–245

Røed LP, O'Brien JJ (1983) A coupled ice-ocean model of upwelling in the marginal ice zone. J Geophys Res 29(C5):2863–2872

Smolarkiewicz PK (1983) A simple positive definite advection scheme with small implicit diffusion. Mon Weather Rev 111:479. https://doi.org/10.1175/1520-0493(1983)1112.0.CO;2

Smolarkiewicz PK (1984) A fully multidimensional positive definite advection transport algorithm with small implicit diffusion. J Comput Phys 54(2):325–362. https://doi.org/10.1016/0021-9991(84)90121-9

Smolarkiewicz PK, Margolin LG (1998) MPDATA: a finite difference solver for geophysical flows. J Comput Phys 140:459–480. https://doi.org/10.1006/jcph.1998.5901

Stoker JJ (1957) Water waves: the mathematical theory with applications. Pure and applied mathematics: a series of texts and monographs, vol IV. Interscience Publishers, Inc, New York

Zalesak ST (1979) Fully multidimensional flux-corrected transport algorithms for fluids. J Comput Phys 31:335–362. https://doi.org/10.1016/0021-9991(79)90051-2

6 Shallow Water Problem

The purpose of this chapter is to learn how to solve a simple subset of the momentum equations (1.1) numerically. The focus is on the *shallow water equations*, and in particular their depth integrated versions (1.33) and (1.34). Despite their simplicity, the shallow water equations include the essence of the momentum equations. For instance, we retain the possibility of a geostrophic balance and the impact of nonlinear terms on the dynamics.

The chapter has two main parts, one investigating the linear versions of the shallow water equations, and the second focusing on the nonlinear effects. Consequently, Sect. 6.2 discusses the approximations that have to be made to linearize (1.33) and (1.34). Furthermore, to be able to check the numerical solutions, it is important to know what kind of physics is expected when solving the shallow water equations numerically. In Sect. 6.3, we therefore investigate various wave solutions, such as *inertia–gravity waves*, *Kelvin waves*, and *Rossby waves*, all possible solutions to the linear shallow water equations.

Since the shallow water equations include three dependent variables, they form a suitable set to illustrate the numerical solution of partial differential equations with more than one unknown. As a consequence, we introduce some new convergent schemes, e.g., the *forward–backward scheme* and the *quasi-Lagrangian scheme*, in addition to the conventional CTCS or leap-frog scheme. Moreover, to avoid over-specifying or under-specifying the number of boundary conditions, *staggered grids* are introduced in Sect. 6.5.

Both the linear and the nonlinear versions of the shallow water equations contain three important *first integrals*. A first integral is a quantity, say ψ, whose time rate of change within a material volume is zero in the absence of sources and sinks, i.e., $\mathrm{D}\psi/\mathrm{d}t = 0$, where $\mathrm{D}/\mathrm{d}t$ is the individual derivative. Typical first integrals are *mass*

L. P. Røed, *Atmospheres and Oceans on Computers*,
Springer Textbooks in Earth Sciences, Geography and Environment,
https://doi.org/10.1007/978-3-319-93864-6_6

(or volume when the fluid is a Boussinesq fluid), *mechanical energy*,[1] and *potential vorticity*. Conservation of mass is ensured through the continuity equation (1.31), and conservation of mechanical energy was discussed in Sect. 3.4. The conservation of potential vorticity will be presented in Sect. 6.3. It is important to note that these conservation properties should be respected when we construct a numerical scheme to solve the shallow water equations. For instance, if the fluid is contained within a fixed volume with no transport or fluxes through its boundaries, and if there are no internal sources or sinks, then the total amount of a first integral, say potential vorticity, should be conserved within that volume. Numerically, this means that, when adding up the individual contributions within a fixed volume, the total sum should not change from one time step to the next, when we neglect fluxes through the boundaries and internal sources and sinks.

Interestingly, it will be shown that the stability condition is once again $C \leq 1$, where C is the Courant number (Sect. 6.4). However, in contrast to the advection problem, the velocity that enters the Courant number is not the advection velocity, but the phase speed of gravity waves. Since such waves normally have phase speeds at least one order of magnitude greater than the wind speed or ocean current speed, the latter has a dramatic impact on the maximum allowed time step.

So far, the stability of a numerical scheme has been analyzed using the von Neumann method. However, this method is not valid for nonlinear systems, and hence not applicable to the nonlinear shallow water equations. The reason is that the nonlinear terms enable exchange of energy between wavelengths. Thus, the growth of the amplitude of each wavenumber cannot be investigated separately. It is therefore no straightforward matter to investigate the stability of a particular scheme chosen to solve the nonlinear shallow water equations, and we shall put this off until Sect. 10.3. In summary, nonlinearity creates what is known as *nonlinear instability*. To avoid this, the stability condition is as a rule more rigorous than its linear counterpart.

6.1 Shallow Water Equations

As in the previous chapter, we investigate the simplest form of the momentum equations. We thus focus is on the shallow water equations as given in (1.33) and (1.34). For a one-layer model of uniform density ρ_0, they take the form

$$\partial_t \mathbf{u} + \mathbf{u} \cdot \nabla_{\mathrm{H}} \mathbf{u} + f \mathbf{k} \times \mathbf{u} = -\nabla_{\mathrm{H}} \phi + \frac{\boldsymbol{\tau}_{\mathrm{s}} - \boldsymbol{\tau}_{\mathrm{b}}}{\rho_0 h} + \frac{\mathbf{X}}{h}, \tag{6.1}$$

$$\partial_t \phi + \mathbf{u} \cdot \nabla_{\mathrm{H}} \phi = -\phi \nabla_{\mathrm{H}} \cdot \mathbf{u}, \tag{6.2}$$

where $\mathbf{u}$ is the lateral velocity component and ϕ is the geopotential, as illustrated in Fig. 6.1. Recall that $\phi = gh = g(H + \eta)$, where g is the gravitational acceleration

[1] Mechanical energy is the sum of the kinetic energy and the available potential energy, as outlined in Sect. 3.4.

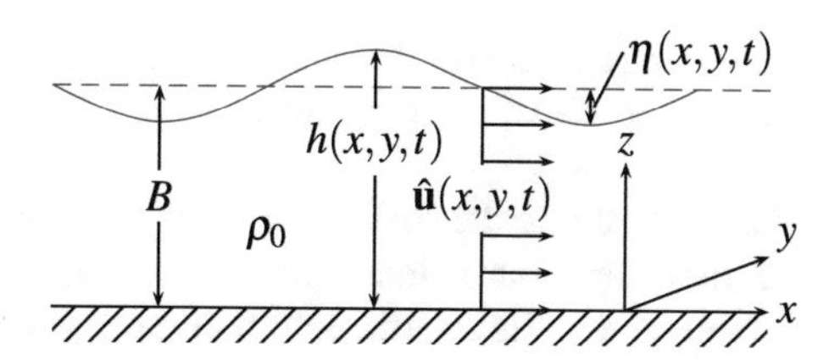

Fig. 6.1 One-layer model of the atmosphere or ocean. Here, $h = \eta + B$ is the total depth of a fluid column, where $B = B(x, y)$ is the equilibrium depth, while $\eta = \eta(x, y, t)$ is the deviation of the top reference surface away from its (level) equilibrium position. The mean velocity of the fluid column is $\hat{\mathbf{u}}$, and the uniform density of the layer is ρ_0

and h is the geopotential height. In a one-layer model, like the present one, the geopotential height $h = H + \eta$ is simply the depth (or height) of a fluid column, where $H = H(x, y)$ is the equilibrium depth and $\eta = \eta(x, y, t)$ is the deviation of the top surface away from the equilibrium depth. Furthermore, $\boldsymbol{\tau}_\mathrm{s}$ is the stress at the top surface, while $\boldsymbol{\tau}_\mathrm{b}$ is the stress at the bottom. In most cases, the stress at the top acts as an energy source. In the ocean, this is caused by the traction the atmosphere exerts on the ocean surface and is referred to as the wind stress. In contrast, the bottom stress almost always acts as an energy sink and is referred to as friction. Finally, $\mathbf{X}$ contains lateral (horizontal) viscosity arising from lateral stirring, mixing, and turbulence. Since these processes act mainly to diffuse or filter out small scale motion, this term is more often than not parameterized as a diffusion term in numerical models. Numerically, and in a prognostic equation like (6.1), the FDA of this term should follow the guidelines for constructing a stable and consistent scheme given in Chap. 4 on the diffusion equation. We shall return to this point in Sect. 10.2.

Inspection of the shallow water equations (6.1) and (6.2) reveals that they do indeed include (i) the possibility of having a steady state solution that differs from the trivial solution of a fluid at rest, and (ii) nonlinear terms enabling instabilities which in turn generate high- and low-pressure systems.[2] The former is feasible due to the effect of Earth's rotation giving rise to the so-called Coriolis force,[3] while the latter makes it possible for two waves of different wavelengths and amplitudes to exchange energy and thereby generate a new wave of wavelength and amplitude different from both of the original waves. The steady-state solution is attained by balancing the Coriolis force and the pressure force, giving rise to the so-called *geostrophic balance*[4]

[2]In the ocean, the high- and low-pressure systems are commonly referred to as mesoscale eddies.

[3]Gaspard-Gustave de Coriolis or Gustave Coriolis (1792–1843) was a French mathematician, mechanical engineer, and scientist. He is best known for his work on the extra forces that are detected in a rotating frame of reference, and one of those forces now bears his name (Source: Wikipedia). Note that this force is virtual in the sense that its presence is caused by the choice of coordinate system, viz., a frame fixed to the rotating Earth.

[4]This balance is often referred to as the thermal wind equation. The rationale is that, using the hydrostatic approximation, the pressure gradient may be written in terms of the density gradients, which contain gradients in temperature and humidity through the equation of state.

(Sect. 1.6), while the exchange of energy between waves of different wavelengths allows instability processes to transfer energy from large-scale motion to smaller scale motion.

A further motivation for studying the shallow water equations is the fact that the fully three-dimensional barotropic/baroclinic equations of motion may be described in terms of an infinite sum of so-called vertical normal modes, where each mode is governed by a set of shallow water equations similar to (6.1) and (6.2). This was first shown in the appendix to Lighthill (1969). It has since been used by scores of authors, for example, to investigate coastal upwelling and currents (Gill and Clarke 1974) and storm surges in stratified seas (Røed 1979). In addition, a stratified atmosphere and/or ocean may be discretized into a number of moving vertical layers of uniform density in which the layer interfaces are Lagrangian surfaces moving up and down in the vertical in response to the dynamics[5] (Charney 1955; Shi and Røed 1999). In either case, the end result is that we have to solve a set of shallow water equations. In particular, each mode and layer is associated with an "equivalent depth" (or equivalent geopotential height). This height corresponds roughly to the height of the coordinate surface above the ground/bottom. Hence, by learning how to solve the shallow water equations using numerical methods, we are in a sense learning how to solve the fully three-dimensional barotropic/baroclinic equations governing the motion of any geophysical fluid.

When we consider the shallow water equations (6.1) and (6.2), the focus is on the *dynamics*. Nevertheless, the *thermodynamics* includes important processes that impact the dynamics through the active tracers that control the density via the equation of state, e.g., potential temperature and humidity in the atmosphere and temperature and salinity in the ocean. The time evolution of these tracers is in turn determined by the tracer equations, that is, by the advection–diffusion equations treated in Chaps. 3–5.

Since the external forcing terms, e.g., wind stress, mixing, and friction, are energy sources and/or sinks, these terms may be neglected without loss of generality (Sect. 3.4). Under these circumstances, (6.1) and (6.2) take the form

$$\partial_t \mathbf{u} + \mathbf{u} \cdot \nabla_{\mathrm{H}} \mathbf{u} + f \mathbf{k} \times \mathbf{u} = -\nabla_{\mathrm{H}} \phi \,, \tag{6.3}$$

$$\partial_t \phi + \mathbf{u} \cdot \nabla_{\mathrm{H}} \phi = -\phi \nabla_{\mathrm{H}} \cdot \mathbf{u} \,. \tag{6.4}$$

These equations contain the three independent variables x, y, and t, and three dependent variables, namely, the two horizontal, depth integrated, velocity components u, v, and the geopotential $\phi = gh$. Moreover, (6.3) and (6.4) are complete in the sense that they constitute three equations for the three unknowns u, v, and ϕ.

To repeat, the shallow water equations do not only describe important and fundamental aspects of atmosphere–ocean dynamics. They are also a convenient set of equations whereby the numerical treatment of equations containing more than one dependent variable, or several independent variables, may be introduced in a geophysical fluid context.

[5] We return to this in Chap. 8, where various vertical coordinate systems are considered.

6.2 Linear Shallow Water Equations

In the following, both the linear and the nonlinear versions of (6.3) and (6.4) are solved numerically. We thus require an analysis of the approximations needed to linearize the shallow water equations. We begin by separating the variables into a basic state plus a perturbation, that is,

$$\mathbf{u} = \overline{\mathbf{u}} + \mathbf{u}' \quad \text{and} \quad \phi = \overline{\phi} + \phi' , \tag{6.5}$$

where the basic (unperturbed) state is denoted by an overline and the perturbations by a prime. The first approximation we make is to assume that the basic state varies slowly in time and space, compared to the motion characterized by the perturbations. Mathematically, this is expressed by

$$\begin{aligned} &|\partial_t \overline{\mathbf{u}}| \ll |\partial_t \mathbf{u}'| , \quad |\nabla_{\mathrm{H}} \overline{\mathbf{u}}| \ll |\nabla_{\mathrm{H}} \mathbf{u}'| , \quad |\nabla_{\mathrm{H}} \cdot \overline{\mathbf{u}}| \ll |\nabla_{\mathrm{H}} \cdot \mathbf{u}'| , \\ &|\partial_t \phi| \ll |\partial_t \phi'| , \quad |\nabla_{\mathrm{H}} \overline{\phi}| \ll |\nabla_{\mathrm{H}} \phi'| . \end{aligned} \tag{6.6}$$

This implies that, for short time and space scales compared to the time and space scales over which the basic state changes, the basic state may be treated as constant. Thus, the overline involves an average over several wavelengths or periods,[6] whence $\overline{\mathbf{u}'} = 0$ and $\overline{\phi'} = 0$.

The basic state is also a solution to the fully nonlinear shallow water equations (6.3) and (6.4), even if there are no perturbations. One such state is the geostrophic balance

$$\overline{\mathbf{u}} = \frac{1}{f} \mathbf{k} \times \nabla_{\mathrm{H}} \overline{\phi} \quad \Longrightarrow \quad \overline{u} = -\frac{1}{f} \partial_y \overline{\phi} \quad \text{and} \quad \overline{v} = \frac{1}{f} \partial_x \overline{\phi} . \tag{6.7}$$

Another trivial example is a basic a state at rest in which $\overline{\mathbf{u}} = 0$ and $\overline{\phi}$ is constant, say $\overline{\phi} = \overline{\phi}_0 = g H_0 = c_0^2$, where H_0 is a constant geopotential height, g is the gravitational acceleration, and c_0 is the phase speed (propagation speed) of gravity waves.

Substituting (6.5) into (6.3)–(6.4) and recalling that the basic state is governed by (6.3)–(6.4), we obtain

$$\partial_t \mathbf{u}' + \overline{\mathbf{u}} \cdot \nabla_{\mathrm{H}} \mathbf{u}' + \mathbf{u}' \cdot (\nabla_{\mathrm{H}} \overline{\mathbf{u}} + \nabla_{\mathrm{H}} \mathbf{u}') + f \mathbf{k} \times \mathbf{u}' = -\nabla_{\mathrm{H}} \phi' , \tag{6.8}$$

$$\partial_t \phi' + \overline{\mathbf{u}} \cdot \nabla_{\mathrm{H}} \phi' + \mathbf{u}' \cdot (\nabla_{\mathrm{H}} \overline{\phi} + \nabla_{\mathrm{H}} \phi') + \overline{\phi} \nabla_{\mathrm{H}} \cdot \mathbf{u}' = -\phi' (\nabla_{\mathrm{H}} \cdot \overline{\mathbf{u}} + \nabla_{\mathrm{H}} \cdot \mathbf{u}') . \tag{6.9}$$

Using the inequalities in (6.6), the relations (6.8) and (6.9) reduce to

$$\partial_t \mathbf{u}' + \overline{\mathbf{u}} \cdot \nabla_{\mathrm{H}} \mathbf{u}' + \mathbf{u}' \cdot \nabla_{\mathrm{H}} \mathbf{u}' + f \mathbf{k} \times \mathbf{u}' = -\nabla_{\mathrm{H}} \phi' , \tag{6.10}$$

$$\partial_t \phi' + \overline{\mathbf{u}} \cdot \nabla_{\mathrm{H}} \phi' + \mathbf{u}' \cdot \nabla_{\mathrm{H}} \phi' + \overline{\phi} \nabla_{\mathrm{H}} \cdot \mathbf{u}' = -\phi' \nabla_{\mathrm{H}} \cdot \mathbf{u}' . \tag{6.11}$$

The next approximation is to assume that the perturbations are "small" in a dimensionless sense. This implies that products of the perturbations are small compared to the perturbations themselves. Mathematically, this entails

$$\left| \frac{\mathbf{u}'^2}{\overline{\mathbf{u}}^2} \right| \ll \left| \frac{\mathbf{u}'}{\overline{\mathbf{u}}} \right| , \quad \left| \frac{\phi'^2}{\overline{\phi}^2} \right| \ll \left| \frac{\phi'}{\overline{\phi}} \right| . \tag{6.12}$$

[6] In this respect, the basic state is sometimes referred to as the ensemble mean.

Thus, (6.10) and (6.11) reduce further to

$$\partial_t \mathbf{u} + \overline{\mathbf{u}} \cdot \nabla_{\mathrm{H}} \mathbf{u} + f\mathbf{k} \times \mathbf{u} = -\nabla_{\mathrm{H}} \phi \,, \tag{6.13}$$

$$\partial_t \phi + \overline{\mathbf{u}} \cdot \nabla_{\mathrm{H}} \phi + \overline{\phi} \nabla_{\mathrm{H}} \cdot \mathbf{u} = 0 \,, \tag{6.14}$$

where primes on the perturbations are dropped for clarity. Equations (6.13) and (6.14) describe the time evolution of the perturbations and are henceforth referred to as the linear shallow water equations. If, for instance, the basic state is one at rest, they take the form

$$\partial_t \mathbf{u} + f\mathbf{k} \times \mathbf{u} = -\nabla_{\mathrm{H}} \phi \,, \tag{6.15}$$

$$\partial_t \phi + c_0^2 \nabla_{\mathrm{H}} \cdot \mathbf{u} = 0 \,. \tag{6.16}$$

For later reference, the two sets (6.13)–(6.16) are decomposed into component form. This leads to the following two sets of linear shallow water equations:

$$\begin{aligned} \partial_t u + \overline{u}\partial_x u + \overline{v}\partial_y u - fv &= -\partial_x \phi \,, \\ \partial_t v + \overline{u}\partial_x v + \overline{v}\partial_y v + fu &= -\partial_y \phi \,, \\ \partial_t \phi + \overline{u}\partial_x \phi + \overline{v}\partial_y v &= -\overline{\phi} \left(\partial_x u + \partial_y v \right) \,, \end{aligned} \tag{6.17}$$

and

$$\begin{aligned} \partial_t u - fv &= -\partial_x \phi \,, \\ \partial_t v + fu &= -\partial_y \phi \,, \\ \partial_t \phi + c_0^2 \partial_x u &= -c_0^2 \partial_y v \,. \end{aligned} \tag{6.18}$$

One-Dimensional Shallow Water Equations

To keep things as simple as possible, but no simpler, it is assumed that all the dependent variables are functions of x, t alone. This is achieved by setting $\partial_y = 0$ in the nonlinear shallow water equations (6.3) and (6.4), and in the linear versions (6.13)–(6.14) and (6.15)–(6.16). Note that, to ensure that the variables are indeed independent of y, we require $\partial_y f = 0$ as well.[7]

The nonlinear one-dimensional and nonlinear shallow water equations (6.3) and (6.4) then take the form

$$\partial_t \mathbf{u} + u\partial_x \mathbf{u} + f\mathbf{k} \times \mathbf{u} = -\partial_x \phi \,, \tag{6.19}$$

$$\partial_t \phi + u\partial_x \phi = -\phi \partial_x u \,, \tag{6.20}$$

or in component form

$$\begin{aligned} \partial_t u + u\partial_x u - fv + \partial_x \phi &= 0 \,, \\ \partial_t v + u\partial_x v + fu &= 0 \,, \\ \partial_t \phi + u\partial_x \phi + \phi \partial_x u &= 0 \,. \end{aligned} \tag{6.21}$$

[7] If the coordinate system is oriented in such a way that x points in the zonal direction, the latter relation implies that f does not change with latitude, an assumption leading to the so-called f-plane approximation. Note that this approximation excludes Rossby waves [see (6.48)].

Similarly, the two linear sets (6.13) and (6.14) and (6.15) and (6.16) take the form

$$\begin{aligned} \partial_t u + \overline{u}\partial_x u - fv + \partial_x \phi &= 0 \,, \\ \partial_t v + \overline{u}\partial_x v + fu &= 0 \,, \\ \partial_t \phi + \overline{u}\partial_x \phi + \overline{\phi}\partial_x u &= 0 \,, \end{aligned} \tag{6.22}$$

and

$$\begin{aligned} \partial_t u - fv + \partial_x \phi &= 0 \,, \\ \partial_t v + fu &= 0 \,, \\ \partial_t \phi + c_0^2 \partial_x u &= 0 \,, \end{aligned} \tag{6.23}$$

respectively. Finally, it should be stressed that, in order for the variables to be independent of y, the basic state must also be independent of y, that is, $\overline{u} = \overline{u}(x,t)$ and $\overline{\phi} = \overline{\phi}(x,t)$.

6.3 Analytic Considerations

It is instructive at this stage to consider some of the characteristics of the true solutions to the shallow water equations, neglecting sources and sinks. Since it is almost impossible to derive analytic solutions to the nonlinear shallow water equations, only free solutions (homogeneous solutions) to the two linear sets (6.17) and (6.18) are considered. As shown in Sect. 2.3, the relations (6.18) may be set in the form

$$\left(\partial_t^2 + f^2\right)\partial_t \phi - c_0^2 \nabla_{\mathrm{H}}^2 \partial_t \phi = 0 \,. \tag{6.24}$$

Thus, the shallow water equations form a hyperbolic system in the time versus x, y space.[8] Moreover, (6.24) tells us that only seven boundary conditions are allowed. Of these, three are initial conditions, and four are in space (two in x and two in y). If one-dimensional systems are considered, in which case (6.24) takes the form

$$\left(\partial_t^2 + f^2\right)\partial_t \phi - c_0^2 \partial_x^2 \partial_t \phi = 0 \,, \tag{6.25}$$

only five conditions are allowed, i.e., three in time and two in space.

Inertia–Gravity Waves

Under the assumption that all solutions to the sets (6.17) and (6.18) are well behaved functions, they may be represented by an infinite number of waves of different wavelengths and amplitudes (Sect. 2.10). We therefore investigate solutions in the form of a unidirectional wave in the x direction, such as

$$\mathbf{X} = \mathbf{X}_0 \mathrm{e}^{\mathrm{i}\alpha(x-ct)} \,, \tag{6.26}$$

[8]If there is no time dependence, (6.24) is elliptic in x, y.

where α is the wavenumber in the x direction and c is the phase speed.[9] Here, $\mathbf{X}$ denotes a vector consisting of the three dependent variables, that is,

$$\mathbf{X} = \begin{bmatrix} u \\ v \\ \phi \end{bmatrix}, \quad \text{with amplitude} \quad \mathbf{X}_0 = \begin{bmatrix} u_0 \\ v_0 \\ \phi_0 \end{bmatrix}. \tag{6.27}$$

Substituting (6.27) into (6.22), we obtain

$$\begin{aligned} \mathrm{i}\alpha\,(\overline{u} - c)\,u_0 - f v_0 + \mathrm{i}\alpha\phi_0 &= 0\,, \\ f u_0 + \mathrm{i}\alpha\,(\overline{u} - c)\,v_0 &= 0\,, \\ \overline{\phi} u_0 + (\overline{u} - c)\,\phi_0 &= 0\,. \end{aligned} \tag{6.28}$$

Written in matrix form, the set (6.28) takes the form

$$\begin{bmatrix} \mathrm{i}\alpha(\overline{u} - c) & -f & \mathrm{i}\alpha \\ f & \mathrm{i}\alpha & 0 \\ \overline{\phi} & 0 & \overline{u} - c \end{bmatrix} \begin{bmatrix} u_0 \\ v_0 \\ \phi_0 \end{bmatrix} = 0\,. \tag{6.29}$$

A nontrivial solution different from zero is only possible if the determinant of the matrix is zero, resulting in the dispersion relation

$$(\overline{u} - c)\left[-\alpha^2\,(\overline{u} - c)^2 + f^2 + \alpha^2\overline{\phi}\right] = 0\,. \tag{6.30}$$

Thus, there are three possible solutions for the phase speed c, viz.,

$$c_1 = \overline{u}\,, \tag{6.31}$$

and

$$c_{2,3} = \overline{u} \pm c_0\sqrt{1 + \left(\frac{1}{\alpha L_\mathrm{R}}\right)^2}\,, \tag{6.32}$$

where $c_0 = \sqrt{\overline{\phi}} = \sqrt{g H_0}$ and

$$L_\mathrm{R} = c_0/f \tag{6.33}$$

is Rossby's deformation radius. The first solution is an infinitely long wave in which the motion is in geostrophic balance, commonly referred to as the Rossby mode. This is easily derived by substituting $c_1 = \overline{u}$ from (6.31) into the first equation in the set (6.28). The latter then takes the form

$$-f v + \mathrm{i}\alpha\phi = 0\,, \quad \text{or} \quad v = \frac{1}{f}\mathrm{i}\alpha\phi\,. \tag{6.34}$$

Using the Fourier solution backwards, we recover the geostrophic balance (1.40), or

$$v = \frac{1}{f}\partial_x\phi\,. \tag{6.35}$$

[9]In general, a wave in an arbitrary horizontal direction is written $\mathbf{X} = \mathbf{X}_0 e^{\mathrm{i}(\alpha x + \beta y - \omega t)}$, where ω is the frequency. Nevertheless, assuming a unidirectional wave, it is always possible to rotate the coordinate axes so that the x-axis is in the direction of motion of the wave.

The two remaining solutions are gravity waves riding on the basic currents $\overline{u}$ and modified by the Earth's rotation. These waves are commonly referred to as *inertia–gravity waves*. The gravity wave part is associated with the first term in (6.32) (wave speeds $c = \pm c_0$), while the inertia part is associated with the second term (oscillating frequencies $\omega = f$, or wave speeds $c = f/\alpha$).

Since the inertia–gravity modes have a much higher wave speed than the Rossby mode ($|c_0| \gg |\overline{u}|$), they dominate over the Rossby mode c_1. So if the basic state is a rest state, only the inertia–gravity waves remain. The dispersion relation for the set (6.23) is therefore obtained by substituting $\overline{u} = 0$ into (6.32). Moreover, if the motion is dominated by waves with wavelengths much shorter than Rossby's deformation radius ($\alpha L_{\mathrm{R}} \gg 1$), then (6.32) shows that it is unaffected by rotation. Under these circumstances, the solution behaves like ordinary gravity waves ($f = 0$). Thus, only motion on scales of order the Rossby deformation radius or greater is affected by rotation.

Coastally Trapped Kelvin Waves

Another interesting wave solution is the so-called *coastally trapped Kelvin wave*. Two main requirements must be satisfied for a coastal Kelvin wave to exist. One is the existence of a boundary at which the normal velocity is zero, for instance, in the form of an impermeable wall or coast. The second requirement is that the alongshore velocity component should be in geostrophic balance. Assuming that the coast is at $y = 0$, the former implies that $v = 0$ at the coast, while the latter implies that the Coriolis force fu balances the pressure force $\partial_y \phi$ in the second equation of the set (6.18). Under these circumstances, the set (6.18) reduces to

$$
\begin{aligned}
\partial_t u - fv + \partial_x \phi &= 0\,, \\
fu + \partial_y \phi &= 0\,, \\
\partial_t \phi + c_0^2 \partial_x u + c_0^2 \partial_y v &= 0\,.
\end{aligned}
\tag{6.36}
$$

For waves propagating along the coast with a phase speed c, the solution must be of the form

$$
\mathbf{Y} = \mathbf{Z}(y) \mathrm{e}^{\mathrm{i}\alpha(x-ct)}\,, \tag{6.37}
$$

where α is the wavenumber in the x direction and $\mathbf{Y}$ denotes a vector consisting of the three dependent variables, that is,

$$
\mathbf{Y} = \begin{bmatrix} u \\ v \\ \phi \end{bmatrix}\,, \quad \text{with amplitude} \quad \mathbf{Z}(y) = \begin{bmatrix} U(y) \\ V(y) \\ \Phi(y) \end{bmatrix}\,. \tag{6.38}
$$

Substituting (6.38) into (6.36) leads to

$$
\begin{aligned}
-\mathrm{i}\alpha c U - fV + \mathrm{i}\alpha \Phi &= 0\,, \\
fU + \partial_y \Phi &= 0\,, \\
-\mathrm{i}\alpha c \Phi + c_0^2 \mathrm{i}\alpha U + c_0^2 \partial_y V &= 0\,.
\end{aligned}
\tag{6.39}
$$

After some manipulation, we obtain

$$
\partial_y^2 \Phi - \frac{1}{L_{\mathrm{R}}^2} \Phi = 0\,. \tag{6.40}
$$

For Φ to remain bounded as $y \to \infty$, the solution is

$$\Phi = \Phi_0 \mathrm{e}^{-y/L_\mathrm{R}} . \tag{6.41}$$

The alongshore velocity $U(y)$ is obtained by means of the second equation in the set (6.39), which leads to

$$U = c_0^{-1} \Phi . \tag{6.42}$$

Finally, substituting these results into the first equation in the set (6.39), we obtain

$$V = \frac{\mathrm{i}\alpha}{f}\left(1 - \frac{c}{c_0}\right) . \tag{6.43}$$

Since V is independent of the distance from the coast y, the boundary condition of no flow across the coastline implies $V = 0$, and hence $c = c_0$. This in turn implies that v is zero everywhere, while the two remaining variables ϕ and u take the form

$$\phi = \Phi_0 \mathrm{e}^{-y/L_\mathrm{R}} \mathrm{e}^{\mathrm{i}\alpha(x-c_0 t)} , \quad u = \frac{\Phi_0}{c_0} \mathrm{e}^{-y/L_\mathrm{R}} \mathrm{e}^{\mathrm{i}\alpha(x-c_0 t)} . \tag{6.44}$$

In summary, the coastal Kelvin wave balances the Coriolis force against a coastline, in such a way that its amplitude and the alongshore velocity component are both maximum at the coast and decrease exponentially offshore, with a length scale determined by the Rossby deformation radius. The wave always propagates with the shoreline to the right (in the northern hemisphere). A coastal Kelvin wave can also be thought of as a wave which moves eastward along a coast and which is deflected to the right in the northern hemisphere. However, the presence of the coast prevents it from turning right and instead water piles up on the coast. Thus, pressure is created with a gradient directed offshore resulting in a geostrophic current that is directed alongshore.

Equation (6.44) implies that the Kelvin wave amplitude is negligible at distances offshore that are large compared to L_R, the Rossby deformation radius. For midlatitude barotropic Kelvin waves, the deformation radius is about 200 km. Due to this relatively rapid decay, coastal Kelvin waves appear to be trapped close to the coast. Finally, it is worth mentioning that Kelvin waves belong to a class of waves called *edge waves*. Waves belonging to this class all depend on the existence of a coast and/or a stepwise continental shelf. Similar trapped waves are also found along the equator, but they are then referred to as *equatorially trapped Kelvin waves*.

Rossby Waves

Let us suppose for the moment that the time rate of change of the geopotential in (6.16) is so small that the motion is effectively divergence-free. Under these circumstances, (6.15) and (6.16) reduce to

$$\partial_t \mathbf{u} + f\mathbf{k} \times \mathbf{u} = -\nabla_\mathrm{H} \phi , \tag{6.45}$$

$$\nabla_\mathrm{H} \cdot \mathbf{u} = 0 , \tag{6.46}$$

where the second equation allows us to introduce a stream function ψ by defining

$$\mathbf{u} = \mathbf{k} \times \nabla_\mathrm{H} \psi , \quad \text{or} \quad u = -\partial_y \psi \quad \text{and} \quad v = \partial_x \psi . \tag{6.47}$$

Operating on (6.45) with the operator $\mathbf{k} \cdot \nabla_{\mathrm{H}} \times$ and remembering that the Coriolis parameter is a function of the latitude y, we have

$$\partial_t \zeta + \beta \partial_x \psi = 0 \,, \tag{6.48}$$

where $\zeta = \mathbf{k} \cdot \nabla_{\mathrm{H}} \times \mathbf{u} = \nabla_{\mathrm{H}}^2 \psi$ is the vorticity and $\beta = \partial_y f$ is the rate of change of the Coriolis parameter with latitude (sometimes referred to as the Rossby parameter). To arrive at the barotropic vorticity equation (6.48), we have used (6.47) and the approximation $\nabla_{\mathrm{H}} \cdot \mathbf{u} = 0$.

To study possible wave solutions to (6.48), we investigate solutions of the form

$$\psi = \psi_0 \mathrm{e}^{\mathrm{i}(\alpha x + l y - \omega t)} \,, \tag{6.49}$$

where α is the zonal wave number and l is the meridional wave number. Substituting (6.49) into (6.48), we obtain the well-known dispersion relation for Rossby waves, i.e.,

$$c = \frac{\omega}{\alpha} = -\frac{\beta}{(\alpha^2 + l^2)} \,, \tag{6.50}$$

where c is the zonal phase speed of the wave. The zonal group velocity, as defined by (5.32), then takes the form

$$c_{\mathrm{g}} = \alpha \partial_\alpha c + c = \frac{\beta(\alpha^2 - l^2)}{(\alpha^2 + l^2)^2} \,. \tag{6.51}$$

This implies that, for zonal wave numbers greater than the meridional wave numbers, the group velocity is positive. Under these circumstances the energy propagates eastward ($c_{\mathrm{g}} > 0$), while the Rossby wave itself always propagates westward ($c < 0$).

Potential Vorticity Conservation

In addition to the mechanical energy (see Sect. 3.4), a powerful first integral of the nonlinear barotropic shallow water equations is the *potential vorticity* defined by

$$P = \frac{\zeta + f}{h} \,, \tag{6.52}$$

where $\zeta = \mathbf{k} \cdot \nabla_{\mathrm{H}} \times \mathbf{u}$ is the relative vorticity, f is the Coriolis parameter (or planetary vorticity), and h is the geopotential height, so that $\phi = gh$. Operating on (6.3) using the operator $\mathbf{k} \cdot \nabla_{\mathrm{H}} \times$ and then substituting (6.4) into the result, we find

$$\frac{\mathrm{D_H} P}{\mathrm{d}t} = 0 \,, \quad \text{along} \quad \frac{\mathrm{D_H} \mathbf{r}}{\mathrm{d}t} = \mathbf{u} \,, \tag{6.53}$$

where the operator

$$\frac{\mathrm{D_H}}{\mathrm{d}t} = \partial_t + \mathbf{u} \cdot \nabla_{\mathrm{H}} \tag{6.54}$$

is the two-dimensional material or individual time rate of change. The interpretation of (6.53) is that P is conserved in the lateral directions $\mathrm{D_H}\mathbf{r} = \mathbf{u}\mathrm{d}t$, where $\mathbf{r} = x\mathbf{i} + y\mathbf{j}$ is the lateral position vector. Thus each fluid column conserves its initial potential vorticity as it moves around in the horizontal space.

For instance, consider a fluid initially at rest, but with an initial geopotential height $h(x, 0) = H(x)$. Then (6.53) leads to

$$\frac{\zeta + f}{h} = \frac{f}{H}, \quad \text{or} \quad \zeta = \partial_y u - \partial_x v = -f\left(1 - \frac{h}{H}\right). \tag{6.55}$$

As the fluid column moves around, the easiest way to conserve the potential vorticity is to stay at the same height contour, in which case the relative vorticity is also conserved. In contrast, if by moving around, the height of the fluid column becomes shallower or deeper than its initial height, the relative vorticity has to change in accord with (6.55). For instance, if squeezed, the relative vorticity becomes negative and gives rise to a cyclonic motion, and if stretched the relative vorticity becomes positive and an anticyclonic motion is generated. As an example, consider a fluid column in the atmosphere. By crossing a mountain chain, it tends to move to the left as it climbs the hill (in the northern hemisphere) and to the right as it descends on the other side. The conservation of potential vorticity was used, for instance, by Carl-Gustaf Rossby[10] to explain the large-scale motions of the atmosphere in terms of fluid mechanics. He was also the first to consider, and solve, the geostrophic adjustment problem (Rossby 1937, 1938) (see the computer problem in Sect. 6.11.1).

6.4 Finite Difference Forms: Linear Equations

To discretize the shallow water equations, we use the methods discussed in the previous chapters to treat the diffusion and advection problems. All terms in the shallow water equations containing derivatives of the various variables are therefore replaced by finite difference approximations using Taylor series. The latter ensures consistency of the scheme. To be convergent, it is also of vital importance that the scheme should be stable. In this regard, recall that the shallow water equations form a hyperbolic system. Schemes that worked well for the hyperbolic advection problem should therefore be tried first (Chap. 5). Recall that application of a convergent scheme is a prerequisite to ensure that the numerical solution approaches the true solution when the time step and space increments tend to zero independently of one another. It is therefore essential that the constructed schemes should be consistent and stable. This should at least be enforced for those terms in the equation that form the dominant balance.

[10]Carl-Gustaf Arvid Rossby (1898–1957) was a Swedish–U.S. meteorologist. Rossby was introduced to meteorology and oceanography while studying under Vilhelm Bjerknes in Bergen in 1919, where Bjerknes' group was developing the concept of a polar front (the Bergen School of Meteorology). His name is associated with various quantities and phenomena in meteorology and oceanography, e.g., the Rossby number, the Rossby deformation radius, and Rossby waves.

Since there is a fundamental difference between linear and nonlinear systems, it is convenient for pedagogical reasons to start with the two common finite difference forms of the linear shallow water equations.

CTCS Scheme

One scheme that worked well for the advection equation was the second-order CTCS (leap-frog) scheme (Sect. 5.3). Consequently, replacing all derivatives in the linear one-dimensional shallow water equations (6.23) with second-order centered finite difference approximations leads to the set

$$\begin{aligned} u_j^{n+1} - u_j^{n-1} &= 2f\Delta t v_j^n - \frac{\Delta t}{\Delta x}\left(\phi_{j+1}^n - \phi_{j-1}^n\right) , \\ v_j^{n+1} - v_j^{n-1} &= -2f\Delta t u_j^n , \\ \phi_j^{n+1} - \phi_j^{n-1} &= -c_0^2\frac{\Delta t}{\Delta x}\left(u_{j+1}^n - u_{j-1}^n\right) . \end{aligned} \tag{6.56}$$

This discretization provides a consistent scheme since all the FDAs are based on Taylor series. Even though these equations consist of more than one variable and include rotation, they are expected to be numerically stable, since they form a hyperbolic system.

We use the von Neumann method to investigate stability. Hence, the variables in the set (6.56) are replaced by their discrete Fourier components, i.e.,

$$u_j^n = U_n e^{i\alpha j\Delta x} , \quad v_j^n = V_n e^{i\alpha j\Delta x} , \quad \phi_j^n = \Phi_n e^{i\alpha j\Delta x} . \tag{6.57}$$

Substituting (6.57) into the set (6.56) leads to

$$\begin{aligned} U_{n+1} - U_{n-1} &= 2f\Delta t V_n - 2i\gamma\Phi_n , \\ V_{n+1} - V_{n-1} &= -2f\Delta t U_n , \\ \Phi_{n+1} - \Phi_{n-1} &= -2i\gamma c_0^2 U_n , \end{aligned} \tag{6.58}$$

where once again

$$\gamma = \frac{\Delta t}{\Delta x}\sin\alpha\Delta x . \tag{6.59}$$

Since n is a dummy index, the first equation of the set (6.58) takes the form

$$U_{n+2} - U_n = 2f\Delta t V_{n+1} - 2i\gamma\Phi_{n+1} . \tag{6.60}$$

Similarly, we have

$$U_n - U_{n-2} = 2f\Delta t V_{n-1} - 2i\gamma\Phi_{n-1} . \tag{6.61}$$

Subtracting the second from the first, and using the second equation of the set (6.58), we have

$$U_{n+2} - 2\lambda U_n + U_{n-2} = 0 , \tag{6.62}$$

where

$$\lambda = 1 - 2\gamma^2 c_0^2 - 2f^2\Delta t^2 . \quad (6.63)$$

Defining the growth factor as $G \equiv U_{n+2}/U_n$ then leads to

$$G^2 - 2\lambda G + 1 = 0 , \quad (6.64)$$

which has two complex conjugate solutions

$$G_{1,2} = \lambda \pm \mathrm{i}\sqrt{1-\lambda^2} , \quad (6.65)$$

provided that the radical is a positive-definite quantity.[11] As expected, $|G_{1,2}| = 1$. Hence, the CTCS scheme is neutrally stable under the condition $\lambda^2 \leq 1$. The latter is satisfied if and only if

$$\gamma^2 c_0^2 + f^2\Delta t^2 \leq 1 , \quad \text{or} \quad c_0^2 \left(\frac{\Delta t}{\Delta x}\right)^2 \sin^2 \alpha\Delta x + f^2 \Delta t^2 \leq 1 . \quad (6.66)$$

Thus, the CFL criterion for stability becomes

$$\Delta t \leq \frac{\Delta x}{c_0}\left[1 + \left(\frac{\Delta x}{L_\mathrm{R}}\right)^2\right]^{-1/2} , \quad (6.67)$$

where L_R is the Rossby deformation radius, as defined in (6.33). The impact of rotation is hidden in the second term in the radical. If the Coriolis parameter tends to zero (nonrotating flow), then $L_\mathrm{R} \to \infty$, which implies $\Delta x/L_\mathrm{R} \to 0$. Under these circumstances, (6.67) reduces to

$$\Delta t \leq \frac{\Delta x}{c_0} , \quad \text{or} \quad C = c_0 \frac{\Delta t}{\Delta x} \leq 1 , \quad (6.68)$$

where C is the Courant number.

Note that if Δx is chosen such that the Rossby deformation radius is well resolved, say $\Delta x = L_\mathrm{R}/10$, in which case $\Delta x/L_\mathrm{R} \ll 1$, then the second term in the radical of (6.67) is negligible. For all practical purposes, the stability condition is therefore as shown by (6.68). Moreover, even though today's computers are powerful enough to satisfy the condition $\Delta x/L_\mathrm{R} \ll 1$ for atmospheric models, this is not the case for ocean models. This is especially true for the ocean models used in global climate modeling, in which more often than not $\Delta x > L_\mathrm{R}$. Under these circumstances, the second term in the radical must also be taken into account, and this implies that the CFL condition becomes more stringent.

The Courant number defined by (6.68) contains the phase speed of the gravity waves. This is in contrast to the Courant number defined by (5.29), in relation to the advection equation. This has a decisive impact on the maximum time step allowed by the CFL criterion. As an example, the phase speed of internal baroclinic inertia–gravity waves in the atmosphere is $c_0 \sim 100$ m/s, which is one order of magnitude greater than the typical wind speed of $\sim$10 m/s. Using a grid size of say $\Delta x \sim$

[11] If the radical is negative, then the growth factor becomes real, and its absolute value becomes greater than one.

10 km, the time step according to (6.68) is restricted by the condition $\Delta t \leq 100$, or about 2 min. However, in the atmosphere, inertia–gravity waves carry little energy compared to Rossby waves, for instance. Inertia–gravity waves thus have no impact on the evolution of the weather (they are not part of the signal) and are mainly regarded as noise. They are therefore commonly neglected by treating the inertia–gravity terms semi-implicitly (Sect. 6.8). The restrictive condition on the time step stipulated by the CFL condition (6.68) is thereby avoided when considering the atmosphere.

In the ocean, in contrast to the atmosphere, inertia–gravity waves contain energetic barotropic tidal motions and the storm surge signal. Thus, oceanic models must treat these waves explicitly to avoid excessive numerical dissipation. They are therefore restricted by the CFL condition (6.68). Since the typical barotropic phase speed is $c_0 \sim 100$ m/s, and the corresponding midlatitude barotropic deformation radius is $L_{\mathrm{R}} \sim 1000$ km, the situation for the barotropic part is about the same as in the previous paragraph. In summary, regarding the ocean, the CFL condition limits the time steps to a few minutes and sometimes to seconds.

Note that the growth factor is normally defined by $G' \equiv U_{n+1}/U_n$, whereas here it was defined by $G \equiv U_{n+2}/U_n$. In that case, $G' = \pm\sqrt{G}$ and the equation (6.64) for the growth factor leads to the four solutions

$$G'_{m,n} = (-1)^m \sqrt{G_n} = (-1)^n \sqrt{\lambda - (-1)^n \mathrm{i}\sqrt{1-\lambda^2}} \,, \quad m, n = 1, 2 \,. \tag{6.69}$$

Recalling the formula for the square root of a complex number,[12] then yields

$$|G'_{m,n}| = \sqrt{G_n} = \sqrt{\frac{1+\lambda}{2} + \frac{1-\lambda}{2}} = 1 \,, \tag{6.70}$$

as expected. Thus, whenever we have to solve a fourth-order equation for the growth factor, it may be reduced to a second-order equation by defining the growth factor by U_{n+2}/U_n, instead of U_{n+1}/U_n. The result in terms of the stability condition is always the same.

Finally, recall from Sects. 5.5 and 5.7 that the leap-frog scheme applied to the advection equation contained numerical dispersion and computational modes. The question therefore arises as to whether these somewhat disadvantageous properties carry over when applying the leap-frog scheme to the linear rotating shallow water equations. To investigate this, let

$$u_j^n = U_0 \mathrm{e}^{\mathrm{i}\alpha(j\Delta x - cn\Delta t)} \quad \text{and} \quad \phi_j^n = \Phi_0 \mathrm{e}^{\mathrm{i}\alpha(j\Delta x - cn\Delta t)} \,. \tag{6.71}$$

[12] $\sqrt{a + \mathrm{i}b} = \pm(a' + \mathrm{i}b')$, where

$$a' = \sqrt{\frac{a + \sqrt{a^2+b^2}}{2}} \,, \quad b' = \mathrm{sgn}(b)\sqrt{\frac{-a + \sqrt{a^2+b^2}}{2}} \,.$$

Substituting (6.71) into the set (6.56), we obtain

$$\begin{bmatrix} \sin(\alpha c \Delta t) & -if\Delta t & -\gamma \\ if\Delta t & \sin(\alpha c \Delta t) & 0 \\ -c_0^2\gamma & 0 & \sin(\alpha c \Delta t) \end{bmatrix} \begin{bmatrix} U_0 \\ V_0 \\ \Phi_0 \end{bmatrix} = 0 \, . \tag{6.72}$$

For there to be a solution different from the trivial one, the determinant must be zero. This leads to

$$\sin^2(\alpha c \Delta t) = c_0^2\gamma^2 \left[1 + \left(\frac{f\Delta t}{c_0\gamma} \right)^2 \right] = C^2 \sin^2 \alpha \Delta x \left[1 + \left(\frac{\Delta x}{L_R \sin \alpha \Delta x} \right)^2 \right] , \tag{6.73}$$

where $C = c_0\Delta t/\Delta x$ is the Courant number. Hence, the numerical phase speed has two solutions

$$c_{1,2} = \pm \frac{1}{\alpha \Delta t} \arcsin \left[C \sin \alpha \Delta x \sqrt{1 + \left(\frac{\Delta x}{L_R \sin \alpha \Delta x} \right)^2} \right] . \tag{6.74}$$

It does therefore contain numerical dispersion similar to what is deduced for the advection equation (Sect. 5.5), but modified by rotation.[13] The two main differences are that the advection velocity is replaced by the phase speed, and the numerical solution (6.74) displays two waves propagating in opposite directions. The latter is in line with the above analytical result (Sect. 6.3).

Regarding the computational mode, this may be investigated in much the same way as in Sect. 5.7, with the same result. Thus, the leap-frog scheme applied to the linear rotating shallow water equations contains unphysical or artificial modes. These modes may be avoided, e.g., by using the Asselin filter as discussed in Sect. 5.8.

Forward–Backward Scheme

As long as the system is linear, we may use the so-called *forward–backward scheme*, credited to Sielecki (1968) (see also Martinsen et al. 1979). The scheme is essentially a forward-in-time centered-in-space (FTCS) scheme, but with one difference. This is that, as soon as one variable is updated, it is used implicitly in the remaining equations, hence the name forward–backward scheme. Thus, replacing all time tendencies in the set (6.23) by a forward-in-time FDA, all spatial derivatives by a centered-in-space FDA, and treating all updated variables implicitly, leads to the scheme

$$\begin{aligned} u_j^{n+1} &= u_j^n + f\Delta t v_j^n - \frac{\Delta t}{2\Delta x}(\phi_{j+1}^n - \phi_{j-1}^n) \, , \\ v_j^{n+1} &= v_j^n - f\Delta t u_j^{n+1} \, , \\ \phi_j^{n+1} &= \phi_j^n - c_0^2 \frac{\Delta t}{2\Delta x}(u_{j+1}^{n+1} - u_{j-1}^{n+1}) \, . \end{aligned} \tag{6.75}$$

[13]If $f = 0$ ($L_R \to \infty$), then the right-hand side of (6.74) reduces to (5.36).

Notice that the sequence in which the equations in the set (6.23) are solved numerically is random. Thus, an equally valid forward–backward scheme may be constructed by first using the first equation in the set (6.23), i.e.,

$$\begin{aligned}
\phi_j^{n+1} &= \phi_j^n - c_0^2 \frac{\Delta t}{2\Delta x}(u_{j+1}^n - u_{j-1}^n) \,, \\
u_j^{n+1} &= u_j^n + f \Delta t v_j^n - \frac{\Delta t}{2\Delta x}(\phi_{j+1}^{n+1} - \phi_{j-1}^{n+1}) \,, \\
v_j^{n+1} &= v_j^n - f \Delta t u_j^{n+1} \,.
\end{aligned} \tag{6.76}$$

Once again, note that the updated values of ϕ in the second equation and u in the third equation are treated implicitly. Similarly, we may use

$$\begin{aligned}
v_j^{n+1} &= v_j^n - f \Delta t u_j^n \,, \\
\phi_j^{n+1} &= \phi_j^n - c_0^2 \frac{\Delta t}{2\Delta x}(u_{j+1}^n - u_{j-1}^n) \,, \\
u_j^{n+1} &= u_j^n + f \Delta t v_j^{n+1} - \frac{\Delta t}{2\Delta x}(\phi_{j+1}^{n+1} - \phi_{j-1}^{n+1}) \,.
\end{aligned} \tag{6.77}$$

Note that here only ϕ is treated implicitly in the third equation. The point to make is nonetheless that, as soon as one variable is updated, it is used implicitly in the remaining equations.

In contrast to the CTCS scheme, which is a three-level scheme, the forward–backward scheme is a two-level scheme. Moreover, the scheme is consistent since the FDAs are based on the truncation of a Taylor series. The question that remains is whether the scheme is stable. If so, the scheme is convergent, implying that, in the limit $\Delta x \to 0$ and $\Delta t \to 0$, the numerical solution approaches the analytic solution. Employing von Neumann's method for this purpose, the variables in the set (6.75) are first replaced by their discrete Fourier components as defined in (6.57). Next V_n and Φ_n are eliminated by performing a similar manipulation to the one in Sect. 6.4. This leads to

$$U_{n+2} - 2\left(1 - \gamma^2 c_0^2 - f^2 \Delta t^2\right) U_{n+1} + U_n = 0 \,. \tag{6.78}$$

Finally, defining the growth factor by $G = U_{n+2}/U_n$, it takes the form

$$G^2 - 2\lambda' G + 1 = 0 \,, \tag{6.79}$$

where

$$\lambda' = 1 - \frac{1}{2}\left(f^2 \Delta t^2 + \gamma^2 c_0^2\right) \,. \tag{6.80}$$

Once again, $|G_{1,2}| = 1$ as long as $\lambda' \leq 1$. The scheme is therefore neutrally stable as long as

$$\Delta t \leq \frac{2\Delta x}{c_0}\left[1 + \left(\frac{\Delta x}{L_R}\right)^2\right]^{-1/2} \tag{6.81}$$

is satisfied. Again, for most realistic cases $\Delta x \ll L_R$. For all practical purposes the criterion for stability is therefore

$$\Delta t < \frac{2\Delta x}{c_0} , \quad \text{or} \quad C < 2 . \tag{6.82}$$

Note that this stability criterion is less stringent by a factor of 2 than the criterion for the CTCS scheme in (6.68).

Quasi-Lagrangian Scheme

In Sect. 5.12, we showed how to solve the advection equation numerically using the semi-Lagrangian method. The essence of the method was first to derive the compatibility and characteristic equations. This led to a scheme that was consistent and unconditionally stable. The same technique may be used for numerical solution, not only of the linear shallow water equations but also their nonlinear version (Sect. 6.6).

Recall that the solution to the advection equation is composed of single gravity waves traveling at the advection speed u_0. When solving the advection equation numerically using the semi-Lagrangian method, the gravity wave solution could be derived by following a characteristic in the x, t plane, with a slope given by $1/u_0$, along which the variable was conserved. The conservation property was derived from the compatibility equation.

In contrast, the solutions to the linear shallow water equations are composed of two inertia–gravity waves (Sect. 6.3) traveling at phase speeds approximately equal to c_0, as in (6.32), but in opposite directions. Thus, applying the semi-Lagrangian technique, we expect to derive at least two characteristics: one with a positive slope $1/c_0$, associated with the inertia–gravity wave propagating in the positive direction, and one with a negative slope $-1/c_0$, associated with the inertia–gravity wave propagating in the opposite direction. Similarly, we expect to derive two compatibility equations.

To obtain the compatibility and characteristic equations, recall that the linear and rotating shallow water equations are

$$\partial_t u - fv + \partial_x \phi = 0 , \tag{6.83}$$

$$\partial_t \phi + c_0^2 \partial_x u = 0 , \tag{6.84}$$

$$\partial_t v + fu = 0 . \tag{6.85}$$

Multiplying (6.84) by an as yet unknown function λ, adding the result to (6.83), and rearranging terms, leads to

$$\left(\partial_t + \lambda c_0^2 \partial_x\right) u + \lambda \left(\partial_t + \frac{1}{\lambda}\partial_x\right) h = fv . \tag{6.86}$$

Defining λ by requiring the two operators $\partial_t + \lambda c_0^2 \partial_x$ and $\partial_t + (1/\lambda)\partial_x$ to be equal leads to

$$\partial_t + \lambda c_0^2 \partial_x = \partial_t + \frac{1}{\lambda}\partial_x . \tag{6.87}$$

Thus, there are two possibilities:

$$\lambda_{1,2} = \pm \frac{1}{c_0} . \tag{6.88}$$

Defining the operator

$$\frac{\mathrm{D}^*}{\mathrm{d}t} = \partial_t + \frac{\mathrm{D}^* x}{\mathrm{d}t} \partial_x , \quad \text{where} \quad \frac{\mathrm{D}^* x}{\mathrm{d}t} = \lambda c_0^2 , \tag{6.89}$$

we find that there are two operators which take the form

$$\frac{\mathrm{D}^*_{1,2}}{\mathrm{d}t} = \partial_t + \frac{\mathrm{D}^*_{1,2} x}{\mathrm{d}t} \partial_x = \partial_t \pm c_0 \partial_x , \tag{6.90}$$

respectively. As expected, (6.90) also shows that there are two characteristics given by the two characteristic equations

$$\frac{\mathrm{D}^*_{1,2} x}{\mathrm{d}t} = \pm c_0 . \tag{6.91}$$

Since c_0 is constant, the two characteristics form straight lines in the x, t plane, with slopes $1/c_0$ and $-1/c_0$, respectively. Substituting (6.90) into (6.86), we thus find the two compatibility equations

$$\frac{\mathrm{D}^*_1 \mathscr{R}^+}{\mathrm{d}t} = fv , \quad \text{along} \quad \frac{\mathrm{D}^*_1 x}{\mathrm{d}t} = +c_0 , \tag{6.92}$$

$$\frac{\mathrm{D}^*_2 \mathscr{R}^-}{\mathrm{d}t} = fv , \quad \text{along} \quad \frac{\mathrm{D}^*_2 x}{\mathrm{d}t} = -c_0 , \tag{6.93}$$

respectively, where

$$\mathscr{R}^{\pm} = u \pm \frac{1}{c_0} \phi \tag{6.94}$$

are called the Riemann invariants. The term "invariant" is used because, in a nonrotating system ($f = 0$), the right-hand sides of (6.92) and (6.93) are zero, and under these circumstances, (6.92) and (6.93) state that the Riemann invariants are conserved along their respective characteristics.[14] Finally, (6.85) may also be written in the compatibility form

$$\frac{\mathrm{D}^*_3 v}{\mathrm{d}t} = fu , \quad \text{along} \quad \frac{\mathrm{D}^*_3 x}{\mathrm{d}t} = 0 , \tag{6.95}$$

where the operator $\mathrm{D}^*_3/\mathrm{d}t$ is defined by

$$\frac{\mathrm{D}^*_3}{\mathrm{d}t} = \partial_t + \frac{\mathrm{D}^*_3 x}{\mathrm{d}t} \partial_x = \partial_t , \tag{6.96}$$

[14] These invariants are named after Georg Friedrich Bernhard Riemann (1826–1866), an influential German mathematician who made lasting contributions to analysis, number theory, and differential geometry, some of them enabling the later development of general relativity. He first obtained the invariants in his work on plane waves in gas dynamics.

with an associated characteristic equation

$$\frac{\mathrm{D}_3^* x}{\mathrm{d}t} = 0 \,. \tag{6.97}$$

The solution in the x, t plane is therefore a straight line vertically upwards with an infinitely steep slope. Note that the solutions to (6.92), (6.93), and (6.95) are also solutions to (6.83)–(6.85). This is the reason why (6.92), (6.93), and (6.95) are referred to as the compatibility equations, Hence, the shallow water equations may be solved using either.

As in Sect. 5.12, the compatibility equations are exploited here by solving them numerically. We refer to this hereafter as the *quasi-Lagrangian method*. The method is similar to the semi-Lagrangian method discussed in Sect. 5.12, except that there are three characteristics, two with slopes equal to $\pm 1/c_0$, and one with an infinite slope in the x, t plane. As in Sect. 5.12, the finite difference approximations are forward in time. Using (6.91) and (6.97), we obtain three characteristics through the grid point (x_j, t^{n+1}) that cross the previous time step at the points P, Q, and R, as shown in Fig. 6.2. Their locations are found numerically by replacing $\mathrm{D}^*/\mathrm{d}t$ in (6.91) and (6.97) by a forward-in-time FDA, leading to

$$x_P = x_j - c_0 \Delta t \,, \quad x_Q = x_j + c_0 \Delta t \,, \quad x_R = x_j \,. \tag{6.98}$$

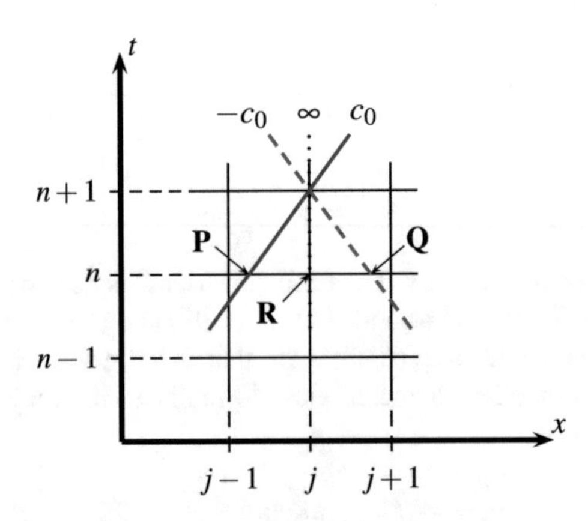

Fig. 6.2 Quasi-Lagrangian technique for the linear rotating shallow water equations, assuming $c_0 > 0$. The distances between the grid points are Δt in the vertical and Δx in the horizontal. The *sloping solid blue line* (marked c_0) is the positive characteristic through the grid point x_j, t^{n+1}, while the *dashed red line* (marked $-c_0$) is the negative characteristic through the same grid point. They are derived from (6.91) and (6.97), respectively, and the slopes are given by $\pm 1/c_0$ and infinity, respectively. The points labeled P, Q, and R mark where the characteristics cross the previous time level n. The point P is therefore a distance $c_0 \Delta t$ to the left of x_j, while the point Q is a distance $c_0 \Delta t$ to the right of x_j. The point R sits at the grid point (x_j, t^n). As long as $c_0 \Delta t \leq \Delta x$, then P is located between x_{j-1} and x_j and Q between x_{j+1} and x_j. However, if $c_0 \Delta t > \Delta x$, then the points Q and P are located to the left and right of x_{j-1} and x_{j+1}, respectively

Similarly, a forward-in-time finite difference approximation to the compatibility equations (6.92), (6.93), and (6.95) along the characteristics leads to

$$u_j^{n+1} + \frac{\phi_j^{n+1}}{c_0} = u_P^n + \frac{\phi_P^n}{c_0} + f\Delta t v_P^n \,, \tag{6.99}$$

$$u_j^{n+1} - \frac{\phi_j^{n+1}}{c_0} = u_Q^n - \frac{\phi_Q^n}{c_0} + f\Delta t v_Q^n \,, \tag{6.100}$$

$$v_j^{n+1} = v_R^n - f\Delta t u_R^n \,. \tag{6.101}$$

Solving them with respect to the three unknowns u_j^{n+1}, v_j^{n+1}, and ϕ_j^{n+1}, they take the form

$$u_j^{n+1} = \frac{1}{2}\left(u_P^n + u_Q^n\right) + \frac{1}{2c_0}\left(\phi_P^n - \phi_Q^n\right) + \frac{1}{2} f\Delta t\left(v_P^n + v_Q^n\right) \,, \tag{6.102}$$

$$v_j^{n+1} = v_R^n - f\Delta t u_R^n \,, \tag{6.103}$$

$$\phi_j^{n+1} = \frac{1}{2}\left(\phi_P^n + \phi_Q^n\right) + \frac{1}{2} c_0\left(u_P^n - u_Q^n\right) + \frac{1}{2} c_0 f\Delta t\left(v_P^n - v_Q^n\right) . \tag{6.104}$$

Since the integration is along the characteristics, they contain u_P^n, u_Q^n, u_R^n, v_R^n, v_Q^n, ϕ_P^n, and ϕ_Q^n on their right-hand sides. These are the values of u, v, and ϕ at the points P, Q, and R where the characteristics cross the previous time level (Fig. 6.2). The locations of these points are given by (6.98). The third characteristic crosses the previous time level at x_j, so $u_R^n = u_j^n$ and $v_R^n = v_j^n$. As in Sect. 5.12, the values of the remaining variables at P and Q are found by interpolation. For instance, a linear two-point Newton interpolation leads to

$$u_P^n = (1-C)u_j^n + Cu_{j-1}^n \,, \quad u_Q^n = (1-C)u_j^n + Cu_{j+1}^n \,, \tag{6.105}$$

$$v_P^n = (1-C)v_j^n + Cv_{j-1}^n, \quad v_Q^n = (1-C)v_j^n + Cv_{j+1}^n, \tag{6.106}$$

$$\phi_P^n = (1-C)\phi_j^n + C\phi_{j-1}^n \,, \quad \phi_Q^n = (1-C)\phi_j^n + C\phi_{j+1}^n \,, \tag{6.107}$$

as long as $C \leq 1$, where $C = c_0\Delta t/\Delta x$ is the Courant number. Nevertheless, as long as we keep track of the locations of the points P, Q, and R with respect to the grid points by making use of (6.98), and perform the interpolation using only the adjacent grid point values, the method also works for $C > 1$. The quasi-Lagrangian scheme is therefore unconditionally stable. Since the forward-in-time FDA is based on Taylor series, it is also consistent. It is thus a convergent scheme.

6.5 Staggered Grids

As mentioned at the beginning of Sect. 6.3, the system (6.23) contains five integration constants, namely, three in time and two in space, no more no fewer. So focusing on space, only two boundary conditions are allowed. For instance, solving the set

(6.23) in the domain $x \in \langle 0, L\rangle$, there are three options for the boundary conditions. The first is to specify one condition at $x = 0$ and another at $x = L$. The second is to specify two conditions at $x = 0$ and the third is to specify two conditions at $x = L$. Assuming that there are impermeable walls at $x = 0$ and $x = L$, the natural condition is no flow through them, i.e.,

$$u|_{x=0} = 0 \,, \quad u|_{x=L} = 0 \,, \quad \forall t \,. \tag{6.108}$$

Letting $x_j = (j-1)\Delta x$, with $x_1 = 0$ and $x_J = L$, this translates numerically to

$$u_1^n = 0 \,, \quad u_J^n = 0 \,, \quad \forall n \,. \tag{6.109}$$

No boundary condition on ϕ is allowed, since we choose to specify two for u. Thus, ϕ_1^n and ϕ_J^n are unknown. This causes a problem. It is perhaps best illustrated using the CTCS scheme (6.56), neglecting rotation. Setting $f = 0$ in (6.56) leads to

$$u_j^{n+1} = u_j^{n-1} - \frac{\Delta t}{\Delta x}\left(\phi_{j+1}^n - \phi_{j-1}^n\right) , \tag{6.110}$$

$$\phi_j^{n+1} = \phi_j^{n-1} - c_0^2 \frac{\Delta t}{\Delta x}\left(u_{j+1}^n - u_{j-1}^n\right) . \tag{6.111}$$

These equations are to be solved[15] for all "wet" points $j = 2(1)J-1$ and $n = 1(1)\infty$. At an arbitrary time level $n+1$, the values at the first "wet" point $j = 2$, updated using (6.110) and (6.111), are

$$u_2^{n+1} = u_2^{n-1} - \frac{\Delta t}{\Delta x}\left(\phi_3^n - \phi_1^n\right) , \tag{6.112}$$

$$\phi_2^{n+1} = \phi_2^{n-1} - c_0^2 \frac{\Delta t}{\Delta x}\left(u_3^n - u_1^n\right) . \tag{6.113}$$

Here, the term ϕ_1^n in (6.112) poses a problem simply because it is unknown, and there is no boundary condition specifying it. A similar problem also arises at the other boundary $x = L$. To get the values of u and ϕ at the last "wet" point, j is replaced by $J-1$ in (6.110) and (6.111), giving

$$u_{J-1}^{n+1} = u_{J-1}^{n-1} - \frac{\Delta t}{\Delta x}\left(\phi_J^n - \phi_{J-2}^n\right) , \tag{6.114}$$

$$\phi_{J-1}^{n+1} = \phi_{J-1}^{n-1} - c_0^2 \frac{\Delta t}{\Delta x}\left(u_J^n - u_{J-2}^n\right) . \tag{6.115}$$

Once again, the term ϕ_J^n poses a problem since it is unknown.

Trying to remedy the problem by using option (ii) above, that is, specifying u and ϕ only at $x = 0$, the problem gets worse at $x = L$. It is therefore tempting to specify more than the two allowed conditions in space. However, specifying ϕ at the boundaries $x = 0$ and L in addition to u leads to a system that is over-specified, which is not recommended. The computer will still produce numbers, but they will be false, even though they may look reasonable. The present author strongly advises against exploring such an avenue.

[15] As mentioned in Sect. 5.6, an Euler scheme has to be used to arrive at the solution for the first time level.

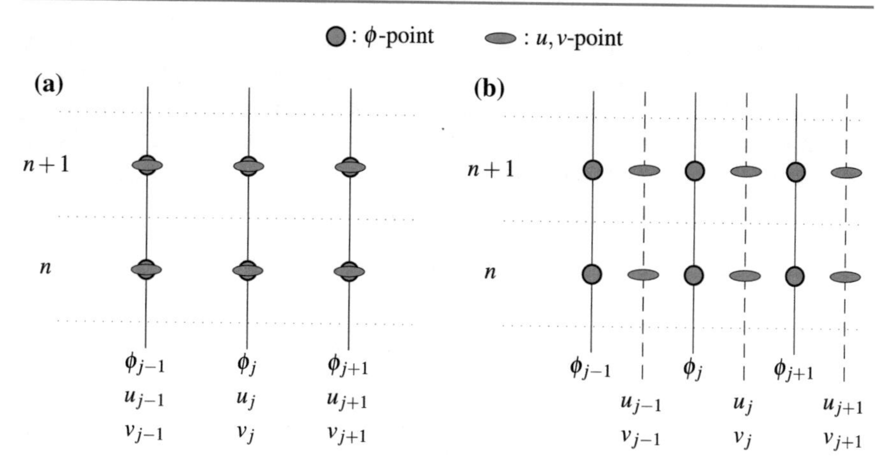

Fig. 6.3 Comparison of the arrangement of **a** a non-staggered and **b** a staggered grid in one spatial dimension for the linear shallow water equations. *Circles* are associated with ϕ-points, while *ellipses* are associated with u, v-points. The staggering illustrated in **b** is such that the ϕ-points and the u, v-points are one half-grid distance apart, but at the same time level

To avoid the problem, Mesinger and Arakawa (1976) suggested using *staggered grids*. Instead of calculating the two variables u and ϕ at the same points in space, they suggested staggering one of them with respect to the other, say one half-grid length along the x-axis (Fig. 6.3b). With this arrangement, or grid structure, u is calculated at $x_{j+1/2}$ points, while ϕ is calculated at x_j points. As an example, consider the CTCS scheme (6.110) and (6.111). Using the notation in Fig. 6.3 to avoid using the cumbersome $j + 1/2$ notation, the finite difference approximation form of (6.110) and (6.111) is

$$u_j^{n+1} = u_j^{n-1} - \frac{2\Delta t}{\Delta x}\left(\phi_{j+1}^n - \phi_j^n\right) , \tag{6.116}$$

$$\phi_j^{n+1} = \phi_j^{n-1} - c_0^2\frac{2\Delta t}{\Delta x}\left(u_j^n - u_{j-1}^n\right) . \tag{6.117}$$

Note the appearance of the factor of 2 in the second term on the right-hand sides of (6.116) and (6.117). This appears because, in the staggered formulation, the distance between two adjacent points in the finite difference approximation of ∂_x is Δx rather than $2\Delta x$, while the centered-in-time scheme still carries $2\Delta t$. It should be stressed that (6.116) and (6.117) are correct if and only if the two grids are staggered by exactly one half-grid length. In principle, any finite staggering length different from zero will avoid the problem of over-specifying the number of boundary conditions. In particular, solving the linear shallow water equations using the forward–backward scheme does not help. Staggering is necessary to avoid over-specification of boundary conditions.

It remains to see whether the staggering has any impact on the numerical stability. If we investigate this using the von Neumann method, care has to be taken when constructing the discrete Fourier components. Once again, we use the nonrotating case as an example. Since u is staggered by one half-grid length with respect to ϕ, the discrete Fourier components are

$$\phi_j^n = \Phi_n e^{i\alpha j \Delta x} , \quad u_j^n = U_n e^{i\alpha (j+1/2)\Delta x} . \tag{6.118}$$

Substituting these expressions into (6.116) and (6.117) leads to

$$U_{n+1} - U_{n-1} = -4i\gamma' \Phi_n , \tag{6.119}$$

$$\Phi_{n+1} - \Phi_{n-1} = -4i\gamma' c_0^2 U_n , \tag{6.120}$$

where

$$\gamma' = \frac{\Delta t}{\Delta x} \sin \frac{\alpha \Delta x}{2} . \tag{6.121}$$

Eliminating U_n leads to

$$\Phi_{n+2} - 2\Phi_n + \Phi_{n-2} = -16\gamma'^2 c_0^2 \Phi_n . \tag{6.122}$$

Moreover, defining a growth factor by $G \equiv \Phi_{n+2}/\Phi_n$, we obtain

$$G^2 - 2\lambda' G + 1 = 0 , \tag{6.123}$$

where

$$\lambda' = 1 - 8c_0^2 \left(\frac{\Delta t}{\Delta x}\right)^2 \sin^2 \frac{\alpha \Delta x}{2} . \tag{6.124}$$

The growth factor therefore has two complex conjugate solutions given by

$$G_{1,2} = \lambda' \pm i\sqrt{1 - \lambda'^2} . \tag{6.125}$$

Thus, as long as $\lambda'^2 \le 1$, the magnitudes of the growth factors are $|G_{1,2}| = 1$. The staggered scheme is therefore neutrally stable, as expected. Nonetheless, this is only true as long as $\lambda'^2 \le 1$. Hence,

$$\Delta t \le \frac{\Delta x}{2c_0} , \quad \text{or} \quad C \le \frac{1}{2} . \tag{6.126}$$

Note that (6.126) is a more stringent condition than the CFL condition (6.68) associated with the non-staggered and nonrotating grid. This should be no surprise. When the grid is staggered, the distance between two adjacent points decreases to one-half the original grid length. Thus, the distance between two adjacent points in the staggered grid, say Δx_{stagg}, is simply $\Delta x_{\text{stagg}} = \Delta x/2$. Using Δx_{stagg} instead of Δx, the CFL condition becomes

$$\Delta t \le \frac{\Delta x_{\text{stagg}}}{c_0} , \tag{6.127}$$

as expected. It is worth mentioning that the more stringent constraint on the time step caused by the staggering is also valid for other schemes, e.g., the forward–backward scheme.

6.6 Finite Difference Forms: Nonlinear Equations

In most applications and models of atmospheres and oceans, the nonlinear terms play an important role. For instance, they are responsible for the physical instabilities that lead to meanders, jets, and eddies in the ocean and low- and high-pressure systems in the atmosphere. The question therefore arises as to how to include them appropriately in a numerical model. To investigate this, the shallow water equations are expanded to the fully nonlinear and rotating equations (6.3) and (6.4). Assuming a one-dimensional system by neglecting all terms differentiated with respect to y in these equations, we obtain

$$\partial_t u + u\partial_x u = fv - \partial_x \phi \,, \tag{6.128}$$

$$\partial_t v + u\partial_x v = -fu \,, \tag{6.129}$$

$$\partial_t \phi + u\partial_x \phi = -\phi\partial_x u \,. \tag{6.130}$$

We retain three equations with u, v, and ϕ as the three dependent variables, and the independent variables are time t and the horizontal direction x.

Defining the volume fluxes $U = hu$ and $V = hv$, and remembering that $\phi = gh$, the set (6.129) yields

$$\partial_t U = fV - \partial_x \left(\frac{U^2}{h}\right) - \frac{1}{2} g \partial_x h^2 \,, \tag{6.131}$$

$$\partial_t V = -fU - \partial_x \left(\frac{UV}{h}\right) , \tag{6.132}$$

$$\partial_t h = -\partial_x U \,. \tag{6.133}$$

The advantage of using the latter set compared to the former is that the continuity equation (6.133) becomes linear. Moreover, the nonlinear terms in (6.131) and (6.132) are written in flux form. This is beneficial, when the nonlinear shallow water equations are later turned into finite difference form. Under these circumstances the flux form has better conservation properties, e.g., energy conservation, than turning the set (6.128)–(6.130) into finite difference form.

Nonlinear CTCS Scheme

The system is still hyperbolic, so it is natural to first try out a CTCS (leap-frog) scheme. Using centered-in-time and centered-in-space FDAs to replace the derivatives in (6.131)–(6.133), and employing first a non-staggered grid (Fig. 6.3), we obtain

$$U_j^{n+1} = U_j^{n-1} + 2f\Delta t V_j^n + A_j^n + P_j^n \,, \tag{6.134}$$

$$V_j^{n+1} = V_j^{n-1} - 2f\Delta t U_j^n + B_j^n \,, \tag{6.135}$$

$$h_j^{n+1} = h_j^{n-1} - \frac{\Delta t}{\Delta x}\left(U_{j+1}^n - U_{j-1}^n\right) , \tag{6.136}$$

where the nonlinearity is hidden in the terms

$$A_j^n = -\frac{\Delta t}{\Delta x}\left[\frac{\left(U_{j+1}^n\right)^2}{h_{j+1}^n} - \frac{\left(U_{j-1}^n\right)^2}{h_{j-1}^n}\right] , \tag{6.137}$$

$$B_j^n = -\frac{\Delta t}{\Delta x}\left[\left(\frac{U_{j+1}^n V_{j+1}^n}{h_{j+1}^n}\right) - \left(\frac{U_{j-1}^n V_{j-1}^n}{h_{j-1}^n}\right)\right], \tag{6.138}$$

$$P_j^n = -\frac{g\Delta t}{2\Delta x}\left[\left(h_{j+1}^n\right)^2 - \left(h_{j-1}^n\right)^2\right]. \tag{6.139}$$

To avoid using more boundary conditions in space than allowed, the grid must be staggered (Sect. 6.5). Using the cell arrangement shown in Fig. 6.3b, we find

$$U_j^{n+1} = U_j^{n-1} + 2f\Delta t V_j^n + A{*}_j^n + P{*}_j^n, \tag{6.140}$$

$$V_j^{n+1} = V_j^{n-1} - 2f\Delta t U_j^n + B{*}_j^n, \tag{6.141}$$

$$h_j^{n+1} = h_j^{n-1} - \frac{2\Delta t}{\Delta x}\left(U_j^n - U_{j-1}^n\right), \tag{6.142}$$

where $A{*}_j^n$, $B{*}_j^n$, and $P{*}_j^n$ are

$$A{*}_j^n = -\frac{\Delta t}{2\Delta x}\left[\frac{\left(U_{j+1}^n + U_j^n\right)^2}{h_{j+1}^n} - \frac{\left(U_j^n + U_{j-1}^n\right)^2}{h_j^n}\right], \tag{6.143}$$

$$B{*}_j^n = -\frac{\Delta t}{2\Delta x}\left[\frac{\left(U_{j+1}^n + U_j^n\right)\left(V_{j+1}^n + V_j^n\right)}{h_{j+1}^n} - \frac{\left(U_j^n + U_{j-1}^n\right)\left(V_j^n + V_{j-1}^n\right)}{h_j^n}\right], \tag{6.144}$$

$$P{*}_j^n = -g\frac{\Delta t}{\Delta x}\left[\left(h_{j+1}^n\right)^2 - \left(h_j^n\right)^2\right]. \tag{6.145}$$

As is evident by comparing the two sets of equations (6.143)–(6.145) and (6.137)–(6.139), the terms containing the nonlinearity become more complex due to the staggering. The reason is that the FDA of the terms on the right-hand side of (6.131) must be evaluated at a u-point, all terms on the right-hand side of (6.132) at a v-point, and all terms on the right-hand side of (6.133) at an h-point. Since the u, v-points are staggered one half-grid length with respect to the h-points, a linear interpolation must be performed to find U and V at the respective h-points.

To illustrate the last point consider the FDA of the first term $\partial_x\left(U^2/h\right)$ on the right-hand side of (6.131). Since it is to be evaluated at the U-point in cell j, the centered-in-space FDA is constructed by taking the difference between the value of U^2/h at the h-point in cell $j+1$ and the value of U^2/h at the h-point in cell j (Fig. 6.4). This leads to

$$\left[\partial_x\left(\frac{U^2}{h}\right)\right]_j^n = \frac{1}{\Delta x}\left[\frac{\left(U{*}_{j+1}^n\right)^2}{h_{j+1}^n} - \frac{\left(U{*}_j^n\right)^2}{h_j^n}\right], \tag{6.146}$$

where $U{*}_{j+1}^n$ and $U{*}_j^n$ are the values of U at the h-point in cells number $j+1$ and j, respectively. Since they are not evaluated at h-points, they have to be established

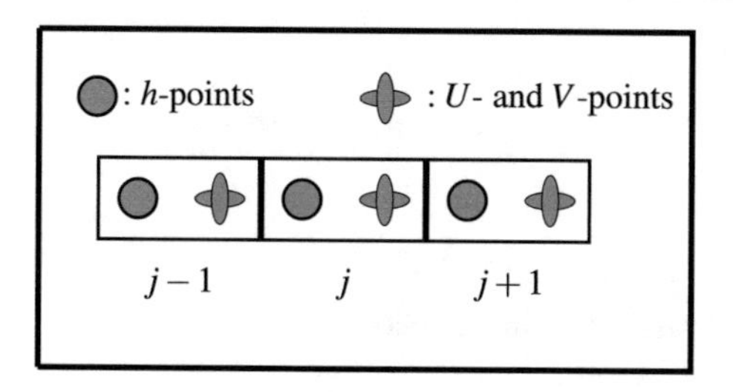

Fig. 6.4 Cell arrangement of Lattice B in Mesinger and Arakawa (1976) (staggered) in one space dimension. *Filled circles* are associated with h-points, *horizontal ellipses* with U-points, and *vertical ellipses* with V-points in the same cells. The cell counter is j. The distance between two adjacent h-points, and hence also between adjacent U- and V-points, is Δx

by performing a linear interpolation. The staggering is exactly one half-grid length, so the interpolation leads to

$$U^{*n}_{j+1} = \frac{1}{2}\left(U^n_{j+1} + U^n_j\right) , \quad U^{*n}_j = \frac{1}{2}\left(U^n_j + U^n_{j-1}\right) . \tag{6.147}$$

Substituting these expressions into the corresponding terms in (6.146) then yields (6.143). Similarly, the FDA of the term $\partial_x (UV/h)$ on the right-hand side of (6.132) is

$$\left[\partial_x \left(\frac{UV}{h}\right)\right]^n_j = \frac{1}{\Delta x}\left[\frac{\left(U^{*n}_{j+1}\right)\left(V^{*n}_{j+1}\right)}{h^n_{j+1}} - \frac{\left(U^{*n}_j\right)\left(V^{*n}_j\right)}{h^n_j}\right] , \tag{6.148}$$

where

$$V^{*n}_{j+1} = \frac{1}{2}\left(V^n_{j+1} + V^n_j\right) , \quad V^{*n}_j = \frac{1}{2}\left(V^n_j + V^n_{j-1}\right) . \tag{6.149}$$

Quasi-Lagrangian Scheme

Although the forward–backward scheme cannot be employed in the nonlinear case, the quasi-Lagrangian scheme can. For this purpose, it is convenient to replace the geopotential by the new variable $c = \sqrt{\phi}$. The one-dimensional nonlinear rotating shallow water equations then take the form

$$\partial_t u + u\partial_x u = fv - 2c\partial_x c , \tag{6.150}$$

$$\partial_t c + u\partial_x c = -\frac{1}{2}c\partial_x u , \tag{6.151}$$

$$\partial_t v + u\partial_x v = -fu , \tag{6.152}$$

where a division by $2c$ has been performed in the continuity equation to arrive at (6.151).

As in the linear case, the first task is to establish the compatibility and characteristic equations. The starting point is to multiply (6.151) by an as yet unknown function λ. Adding the result to (6.150) and rearranging terms, we find

$$\left[\partial_t + \left(u + \frac{1}{2}\lambda c\right)\partial_x\right]u + \lambda\left[\partial_t + \left(u + \frac{2c}{\lambda}\right)\partial_x\right]c = fv . \tag{6.153}$$

Next, requiring that

$$u + \frac{1}{2}\lambda c = u + \frac{2c}{\lambda} = \frac{\mathrm{D}^* x}{\mathrm{d}t} , \tag{6.154}$$

we obtain

$$\lambda_{1,2} = \pm 2 , \tag{6.155}$$

whence the characteristic equations become

$$\frac{\mathrm{D}_1^* x}{\mathrm{d}t} = u + c , \quad \frac{\mathrm{D}_2^* x}{\mathrm{d}t} = u - c . \tag{6.156}$$

Thus, (6.153) may be written

$$\left(\partial_t + \frac{\mathrm{D}_{1,2}^* x}{\mathrm{d}t} \partial_x \right) u \pm 2 \left(\partial_t + \frac{\mathrm{D}_{1,2}^* x}{\mathrm{d}t} \partial_x \right) c = f v . \tag{6.157}$$

Finally, defining the two operators

$$\frac{\mathrm{D}_{1,2}^*}{\mathrm{d}t} = \partial_t + \frac{\mathrm{D}_{1,2}^* x}{\mathrm{d}t} \partial_x = \partial_t + (u \pm c) \partial_x , \tag{6.158}$$

and substituting them into (6.157), we obtain the two compatibility equations

$$\frac{\mathrm{D}_{1,2}^* \mathscr{R}^+}{\mathrm{d}t} = f v , \quad \text{along} \quad \frac{\mathrm{D}_1^* x}{\mathrm{d}t} = u + c , \tag{6.159}$$

$$\frac{\mathrm{D}_{1,2}^* \mathscr{R}^-}{\mathrm{d}t} = f v , \quad \text{along} \quad \frac{\mathrm{D}_2^* x}{\mathrm{d}t} = u - c , \tag{6.160}$$

where the two Riemann invariants are

$$\mathscr{R}^{\pm} = u \pm 2c . \tag{6.161}$$

As in the linear case, the last equation (6.152) may also be transformed into a compatibility form. Defining the third characteristic equation by

$$\frac{\mathrm{D}_3^* x}{\mathrm{d}t} = u , \tag{6.162}$$

and substituting it into (6.152), the latter takes the form

$$\frac{\mathrm{D}_3^* v}{\mathrm{d}t} = -f u , \quad \text{along} \quad \frac{\mathrm{D}_3^* x}{\mathrm{d}t} = u . \tag{6.163}$$

Thus, three characteristics are obtained, with slopes $u + c$, $u - c$, and u, as depicted in Fig. 6.5.

Applying simple forward-in-time finite difference approximations to the terms in (6.159), (6.160), and (6.163) along their respective characteristics, we find

$$u_j^{n+1} + 2c_j^{n+1} = u_P^n + 2c_P^n + f \Delta t v_P^n , \tag{6.164}$$

$$u_j^{n+1} - 2c_j^{n+1} = u_Q^n - 2c_Q^n + f \Delta t v_Q^n , \tag{6.165}$$

$$v_j^{n+1} = v_R^{n+1} - f \Delta t u_R^n , \tag{6.166}$$

where the subscripts P, Q, and R refer to the evaluation of the variables in question at the points P, Q, and R, where the three characteristics cross the previous time

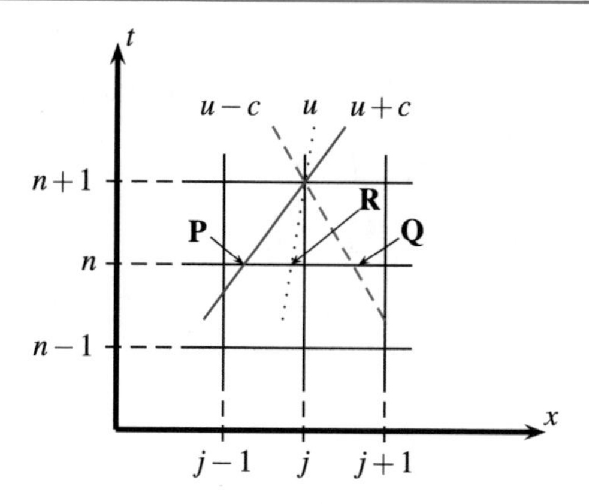

Fig. 6.5 The *blue solid line* is the positive sloping characteristic through the open boundary grid point x_{j+1}, t^{n+1} with slope equal to $(u+c)^{-1}$, while the *red dashed line* is the negative sloping characteristic through the same point with slope equal to $(u-c)^{-1}$. The distance between the grid points is Δt in the vertical and Δx in the horizontal. These are derived from (6.156). The last characteristic with slope $1/u$ is the *dotted black line* derived from (6.163). Provided that $u \geq 0$, the point labeled P is a distance $(u+c)\Delta t$ to the left of x_j, while the point Q is a distance $(u_j^n - c_j^n)\Delta t$ to the right of x_j, hence the asymmetry. Finally, the point labeled R is located a distance $u\Delta t$ to the left of x_j. As long as $(u+c)\Delta t \leq \Delta x$ and $u > 0$, then P and R are located between x_{j-1} and x_j, and Q between x_{j+1} and x_j. However, if $(u+c)\Delta t > \Delta x$, then the points Q and P are located to the left and right of x_{j-1} and x_{j+1}, respectively

level as shown in Fig. 6.5. Solving (6.164)–(6.166) with respect to the unknowns u_j^{n+1}, c_j^{n+1}, and v_j^{n+1}, we obtain

$$u_j^{n+1} = \frac{1}{2}\left[\left(u_P^n + u_Q^n\right) + 2\left(c_P^n - c_Q^n\right) + f\Delta t\left(v_P^n + v_Q^n\right)\right], \tag{6.167}$$

$$v_j^{n+1} = v_R^{n+1} - f\Delta t u_R^n, \tag{6.168}$$

$$c_j^{n+1} = \frac{1}{4}\left[\left(u_P^n - u_Q^n\right) + 2\left(c_P^n + c_Q^n\right) + f\Delta t\left(v_P^n - v_Q^n\right)\right]. \tag{6.169}$$

It remains to find u_P^n, u_Q^n, u_R^n, v_P^n, v_Q^n, v_R^n, c_P^n, and c_Q^n. To this end, we perform an interpolation using adjacent grid points. To accomplish this, we seek the places where the characteristics cross the previous time level. Let these points be denoted by x_P, x_Q, and x_R, respectively. As illustrated in Fig. 6.5, the point P is associated with the first $u+c$ characteristic, while the point Q is associated with the second $u-c$ characteristic defined by (6.158). The point R is associated with the third u characteristic defined by (6.162). Integrating these characteristic equations using a forward-in-time finite difference approximation, we obtain

$$x_P = x_j - (u_j^n + c_j^n)\Delta t, \quad x_Q = x_j - (u_j^n - c_j^n)\Delta t, \quad \text{and} \quad x_R = x_j - u_j^n\Delta t\,. \tag{6.170}$$

Let u be positive. Since $c \gg u$, it follows that $x_P, x_R \leq x_j$ and $x_Q \geq x_j$. Assuming in addition that $x_P, x_R \geq x_{j-1}$ and $x_Q \leq x_{j+1}$, a two-point linear interpolation leads to

$$u_P^n = \left(1 - C_{Pj}^n\right) u_j^n + C_{Pj}^n u_{j-1}^n \, , \tag{6.171}$$

$$u_Q^n = \left(1 - C_{Qj}^n\right) u_j^n + C_{Qj}^n u_{j+1}^n \, , \tag{6.172}$$

$$u_R^n = \left(1 - C_{Rj}^n\right) u_j^n + C_{Rj}^n \begin{cases} u_{j-1}^n \, , & \text{if } u_j^n \geq 0 \, , \\ u_{j+1}^n \, , & \text{if } u_j^n < 0 \, , \end{cases} \tag{6.173}$$

where

$$C_{Pj}^n = \frac{\Delta t}{\Delta x} |u_j^n + c_j^n| \, , \quad C_{Qj}^n = \frac{\Delta t}{\Delta x} |u_j^n - c_j^n| \, , \quad C_{Rj}^n = \frac{\Delta t}{\Delta x} |u_j^n| \, , \tag{6.174}$$

are local Courant numbers modified by nonlinearity. The remaining values v_P^n, v_Q^n, v_R^n, c_P^n, and c_Q^n are found by a similar procedure. Once again, (6.171)–(6.173) are only valid as long as the local Courant numbers are less than one. This is equivalent to requiring that $x_{j-1} \leq x_P, x_R \leq x_j$, and $x_j \leq x_Q \leq x_{j+1}$. As in the linear case, these are not conditions that have to be satisfied. One just has to keep track of where the characteristics cross with respect to the grid points, and then make an interpolation like (6.171) using the two nearest grid points. Another option is to resort to higher order interpolation. Either way one ends up with three equations to solve for the three unknowns u_j^{n+1}, v_j^{n+1}, and c_j^{n+1}.

In the nonlinear case the characteristics are no longer straight lines in the x, t plane. Thus, the locations of x_P, x_Q, and x_R obtained by use of (6.170), and the values found for u_j^{n+1}, v_j^{n+1}, and c_j^{n+1}, may be viewed as the first guess in an iterative procedure. Improved locations may then be computed, e.g., using the formulas

$$x_P^{(\nu+1)} = x_j - \frac{1}{2}\left[\left(u^{(\nu)}{}_j^{n+1} + c^{(\nu)}{}_j^{n+1}\right) + \left(u_P^{(\nu)} + c_P^{(\nu)}\right)\right] \Delta t \, , \tag{6.175}$$

$$x_Q^{(\nu+1)} = x_j - \frac{1}{2}\left[\left(u^{(\nu)}{}_j^{n+1} - c^{(\nu)}{}_j^{n+1}\right) + \left(u_Q^{(\nu)} - c_Q^{(\nu)}\right)\right] \Delta t \, , \tag{6.176}$$

$$x_R^{(\nu+1)} = x_j - \frac{1}{2}\left(u^{(\nu)}{}_j^{n+1} + u_R^{(\nu)}\right) \Delta t \, , \tag{6.177}$$

where $u^{(\nu)}{}_j^{n+1}$, $v^{(\nu)}{}_j^{n+1}$, and $c^{(\nu)}{}_j^{n+1}$ are given by

$$u^{(\nu)}{}_j^{n+1} = \frac{1}{2}\left[\left(u_P^{(\nu-1)} + u_Q^{(\nu-1)}\right) + 2\left(c_P^{(\nu-1)} - c_Q^{(\nu-1)}\right) + f\Delta t\left(v_P^{(\nu-1)} + v_Q^{(\nu-1)}\right)\right] , \tag{6.178}$$

$$v^{(\nu)}{}_j^{n+1} = v_R^{(\nu-1)} - f \Delta t u_R^{(\nu-1)} \, , \tag{6.179}$$

$$c^{(\nu)}{}_j^{n+1} = \frac{1}{4}\left[\left(u_P^{(\nu-1)} - u_Q^{(\nu-1)}\right) + 2\left(c_P^{(\nu-1)} + c_Q^{(\nu-1)}\right) + f\Delta t\left(v_P^{(\nu-1)} - v_Q^{(\nu-1)}\right)\right] . \tag{6.180}$$

The iteration may be repeated until a satisfactory accuracy is reached.

6.7 Nonlinear Instability

The quasi-Lagrangian scheme is always stable, whether applied to the linear or the nonlinear, rotational shallow water equations. The linear CTCS scheme, on the other hand, is stable if and only if the CFL condition is satisfied. The question therefore arises as to whether the nonlinear CTCS scheme is stable, and if so under what conditions. Since the equations are nonlinear, the analysis is not as straightforward as for a linear system. In fact, throwing in the nonlinear terms adds to the complexity of the possibilities for an unstable solution. The reason is, as mentioned in Sects. 3.2 and 4.9, that the nonlinear dynamics allows energy to be exchanged between waves (see, e.g., Phillips 1977). Thus, the nonlinear terms redistribute energy among the different wavelengths present, something that is impossible in a linear system. Moreover, as time goes by, the nonlinear terms act to cascade the energy progressively toward shorter and shorter wavelengths, which are eventually lost through the effects of viscosity.

Wavelengths shorter than $2\Delta x$ remain unresolved when we solve nonlinear problems numerically using finite difference approximations to replace the differential operators in the PDEs. In the real world, and in accordance with the theories explained in detail by Phillips (1977), the energy cascade continues beyond this wavelength, and progressively toward shorter and shorter waves. The numerical solution inhibits this cascading across the $2\Delta x$ resolution limit, and energy therefore tends to accumulate in the $2\Delta x < \lambda < 4\Delta x$ wavelength band. Eventually, when enough energy has accumulated, a scheme that works well for a linear system will blow up. This is called *nonlinear instability* and is treated in more detail in Sect. 10.3.

The processes responsible for the energy cascade across and beyond the grid resolution wavelength act on scales that are shorter than the $2\Delta x$ grid resolution. Such processes are commonly referred to as *sub-grid scale (SGS) processes* (e.g., Griffies 2004, Chap. 7), and in numerical models, the SGS processes have to be parameterized. Since diffusion acts to smooth out noise, as discussed in Sect. 4.3, a common parameterization is to add a diffusion term in the form of a harmonic operator, say $\kappa \nabla_{\mathrm{H}}^2 u$, or biharmonic operator, say $-\kappa \nabla_{\mathrm{H}}^4 u$, where κ is a diffusion coefficient. As mentioned in Chap. 4, these operators are scale selective, and the biharmonic even more so, and they therefore help to effectively damp out the shortest waves without diminishing the energy of the longer waves. In the relatively simple one-dimensional problems presented here and in the previous chapters, the addition of a harmonic diffusion term leads to the amplitude of an individual Fourier component being proportional to $\mathrm{e}^{-\kappa \alpha^2 t}$, which implies that the highest wavenumbers are damped more than the lowest. The results are nonetheless sensitive to the choice of diffusion coefficient. If it is too high, the energy is damped too fast. This results in more energy being absorbed than is necessary and the solution will be too smooth. If it is too low, the energy contained in the low wavenumber band still increases in time, in which case the nonlinear instability will kick in sooner or later.

6.8 Semi-implicit and Time-Splitting Methods

The above analysis reveals that the introduction of a pressure force (in addition to advection) results in a much more stringent CFL criterion (shorter time step). It is therefore tempting to treat terms responsible for this behavior implicitly, while treating the remaining terms explicitly. Such a method is commonly referred to as a *semi-implicit method.*

As an example, consider the nonlinear rotating one-dimensional shallow water equations (6.128)–(6.130), which for the present purpose take the form

$$\partial_t u = A_u - \partial_x \phi \ , \tag{6.181}$$

$$\partial_t v = A_v \ , \tag{6.182}$$

$$\partial_t \phi = A_\phi - \Phi \partial_x u \ . \tag{6.183}$$

Here, ϕ is the deviation away from the (constant) equilibrium geopotential Φ, and

$$A_u = fv - u\partial_x u \ , \tag{6.184}$$

$$A_v = -fu - u\partial_x v \ , \tag{6.185}$$

$$A_\phi = -u\partial_x \phi - \phi\partial_x u \ , \tag{6.186}$$

include all the nonlinear terms as well as the Coriolis terms. The terms responsible for the stringent CFL criterion are $\partial_x \phi$ and $\Phi \partial_x u$, and are kept as is. As mentioned in Sect. 4.8, treating the terms on the right-hand sides of (6.181)–(6.183) implicitly avoids any restriction on the time step. It is therefore tempting to treat the terms $\partial_x \phi$ and $\Phi \partial_x u$ implicitly, while integrating the remaining terms explicitly. Applying a centered-in-time finite difference approximation to the time tendency terms and treating the pressure terms implicitly, we obtain

$$\frac{u^{n+1} - u^{n-1}}{2\Delta t} = A_u^n - \partial_x \left(\phi^{n+1}\right) \ , \tag{6.187}$$

$$\frac{v^{n+1} - v^{n-1}}{2\Delta t} = A_v^n \ , \tag{6.188}$$

$$\frac{\phi^{n+1} - \phi^{n-1}}{2\Delta t} = A_\phi^n - \Phi \partial_x \left(u^{n+1}\right) \ . \tag{6.189}$$

Notice that only the time tendency terms are given a finite difference form, so that each individual term is still a function of x. Solving (6.187) with respect to u^{n+1} and differentiating the result with respect to x, we find

$$\partial_x \left(u^{n+1}\right) = \partial_x \left(u^{n-1}\right) + 2\Delta t \partial_x \left(A_u^n\right) - 2\Delta t \partial_x^2 \left(\phi^{n+1}\right) \ . \tag{6.190}$$

Inserting this expression in (6.189) and rearranging terms, we obtain the Helmholtz equation

$$\partial_x^2 \left(\phi^{n+1}\right) - \frac{1}{4\Phi \Delta t^2} \phi^{n+1} = B^n \ , \tag{6.191}$$

where

$$B^n = \partial_x\left(A_u^n\right) + \frac{1}{2\Delta t}\partial_x\left(u^{n-1}\right) - \frac{\phi^{n-1}}{4\Phi\Delta t^2} - \frac{A_\phi^n}{2\Phi\Delta t} \tag{6.192}$$

contains known quantities at time levels $n, n-1, \ldots$.

With proper boundary conditions (ϕ or its normal derivative at lateral boundaries), these equations may easily be solved by standard numerical methods called elliptic solvers. One such elliptic solver is the direct elliptic solver treated in Sect. 4.11 (Gauss elimination), but there are also a host of iterative (non-direct) solvers to choose from. A widely used non-direct elliptic solver is the *successive over-relaxation* or SOR method. Having obtained ϕ^{n+1}, the values of u^{n+1} and v^{n+1} are found using (6.187) and (6.188), respectively. This method is commonly used in atmospheric models, since it avoids the restrictive CFL condition when estimating an upper bound for the time step. Nevertheless, the time step should not be excessively large. If it is too large, the numerical solution may be heavily damped, and energy is lost.

Recall that this method cannot be employed in the ocean. While the barotropic mode (inertia–gravity waves) carries very little energy in the atmosphere, things are quite different in the ocean. Thus, treating the fast barotropic waves implicitly would ruin the inertia–gravity waves, which carry information about such important signals as tides and storm surges. To speed up the computations of ocean models, it is common to resort to so-called time-splitting (see, e.g., Griffies 2004, p. 246). The idea is to treat the barotropic and baroclinic parts of the motion separately. The procedure is to first integrate the slow baroclinic modes forward from time t using a time step Δt_{bc}. Next, the barotropic mode, which is influenced by the baroclinic modes, is integrated forward in time over the time span $t + \Delta t_{\text{bc}}$ using a much shorter time step, say Δt_{bt}. Commonly, $\Delta t_{\text{bc}} = N\Delta t_{\text{bt}}$, where N is of order 50.

6.9 Summary and Remarks

In this chapter, the discussion focused on various schemes for solving the linear and nonlinear shallow water equations by numerical methods. We stressed the importance of choosing a scheme that is both stable and consistent, since this ensures that the scheme will be convergent. Since both the shallow water equations and the advection equation belong to the class of hyperbolic problems, some of the characteristics of the latter are expected to carry over to the shallow water equations. A key point was therefore also to determine whether this was true.

After introducing the full nonlinear rotating shallow water equations in Sects. 6.1, 6.2 presented a detailed exposition of what it takes to linearize them, followed by an overview of the one-dimensional linear and nonlinear shallow water equations.

Section 6.3 discussed what kind of wave solutions to expect, deriving inertia–gravity waves, coastally trapped Kelvin waves, and Rossby waves, and discussing first integrals like mass, energy, and potential vorticity. We stressed the importance of choosing schemes that conserve these first integrals. For example, a scheme that is losing mass (or volume in the case of a Boussinesq fluid) is of little value.

To show some of the common finite difference forms used to solve the shallow water equations, its linear and nonlinear forms were treated separately. For instance, we presented the application of the leap-frog or CTCS scheme to the linear shallow water equations in Sect. 6.4. We also presented the forward–backward scheme, originally formulated by Sielecki (1968), in Sect. 6.4. This is a one-level scheme, very fast and efficient on the computer, and was used in many ocean models to forecast tides and storm surges in the 1970 and 1980s.

The von Neumann stability analysis was used to investigate the stability of the two linear schemes. It was found that the leap-frog and forward–backward schemes were both neutrally and conditionally stable under the condition that the Courant number is less than or equal to one. However, it was emphasized that the velocity entering the Courant number is the phase speed of the waves, as discussed in Sect. 6.3. Since the phase speed is at least one order of magnitude faster than the advection velocity, this puts a much stricter constraint on the time step. Section 6.4 ends by showing how the semi-Lagrangian scheme presented for the advection equation could be expanded to obtain a new scheme called the quasi-Lagrangian scheme for the shallow water equations. The prefix "quasi" is used to remind us that the integration is along characteristics whose slopes are determined by the phase speed, rather than the advection velocity. The advantage of the quasi-Lagrangian scheme is that it is unconditionally stable. Hence, there is no constraint on the time step other than those dictated by variations in the forcing.

As emphasized several times, the number of boundary conditions must exactly match the number of integration constants in the PDE. Section 6.5 highlights this mantra. The one-dimensional shallow water equations admit only two spatial boundary conditions. Thus, solving the shallow water equations within a closed basin, the two natural boundary conditions are, for instance, no flow through the walls. Hence, there are no conditions for specifying the thickness or geopotential at the boundaries. To avoid over-specifying the number of conditions, Mesinger and Arakawa (1976) suggested staggering the grid, so that the grid in which the velocity components are evaluated is staggered with respect to the grid where the geopotential is evaluated. There are several ways in which this may be done, and this led (Mesinger and Arakawa 1976) to suggest staggered grids, later named Arakawa B, C, and D-grids (or lattice B, C, and D). Later, grids named E and F were also suggested. An unstaggered grid is referred to as an A-grid.

Moving on to the nonlinear part (Sect. 6.6), the forward–backward scheme had to be abandoned. The leap-frog and quasi-Lagrangian schemes could still be applied, but the nonlinearity of the equations has some implications. Since energy may now be interchanged between wavenumbers, the von Neumann stability condition no longer applies, and a formal stability analysis is compromised. Moreover, as outlined in Sect. 6.7, this interchange means that energy tends to be cascaded toward higher and higher wavenumbers (shorter and shorter wavelengths). Since the solutions to the finite difference equations are band-limited in wavenumber space, this tends to produce an accumulation of energy in the low wavelength band. If it is allowed to accumulate, the result is that the model will sooner or later blow up due to *nonlinear instability*. Nonetheless, if the *linear* version is unstable, the nonlinear version will definitely be unstable as well. Thus, one has to ensure that the linear version of the nonlinear system is stable before discussing the effect of the nonlinear terms on the stability.

As mentioned in Sect. 6.8, the barotropic inertia–gravity waves carry very little energy in the atmosphere, in contrast to the ocean, where they are highly significant, e.g., in the analysis of tides and storm surges. In the atmosphere, they are therefore regarded more or less as noise. In addition, they are fast and hence put an unnecessarily strict constraint on the time step. To avoid treating them explicitly, it is common to deal with those terms responsible for the inertia–gravity waves implicitly, while treating the remaining terms explicitly. This is the idea behind the so-called *semi-implicit methods*. When implementing them on the computer, the implication is that for every time step one has to solve an elliptic equation to determine the geopotential, and this adds to the computational burden. This is somewhat lessened, however, by the gain allowed by the increased time step.

6.10 Exercises

1. Consider the nonlinear rotating shallow water equations

$$\partial_t \mathbf{U} + \nabla_{\mathrm{H}} \cdot \left(h^{-1}\mathbf{U}\mathbf{U}\right) + f\mathbf{k} \times \mathbf{U} = -gh\nabla_{\mathrm{H}}\zeta + \rho_0^{-1}(\boldsymbol{\tau}_{\mathrm{s}} - \boldsymbol{\tau}_{\mathrm{b}}) \,, \tag{6.193}$$

$$\partial_t h + \nabla_{\mathrm{H}} \cdot \mathbf{U} = 0 \,, \tag{6.194}$$

where

$$\mathbf{U}(x, y, t) = \int_{-H(x,y)}^{\zeta(x,y,t)} \mathbf{u}(x, y, z, t)\, \mathrm{d}z \tag{6.195}$$

with components (U, V) along the x, y-axes, is the volume transport component in a water column of depth $h = H + \zeta$, with ζ the deviation of the sea level from the equilibrium depth H, $\boldsymbol{\tau}_{\mathrm{s}}$ and $\boldsymbol{\tau}_{\mathrm{b}}$ are the wind and bottom stresses, respectively, with components $(\tau_{\mathrm{s}}^x, \tau_{\mathrm{s}}^y)$ and $(\tau_{\mathrm{b}}^x, \tau_{\mathrm{b}}^y)$, g is the gravitational acceleration, and ρ_0 is the density (uniform in time and space). The Coriolis parameter is $f = 2\Omega \sin\phi$, where Ω is the Earth's rotation rate and ϕ is the latitude (Fig. 6.6). Solutions to the storm surge problem are sought in the presence of a straight coast

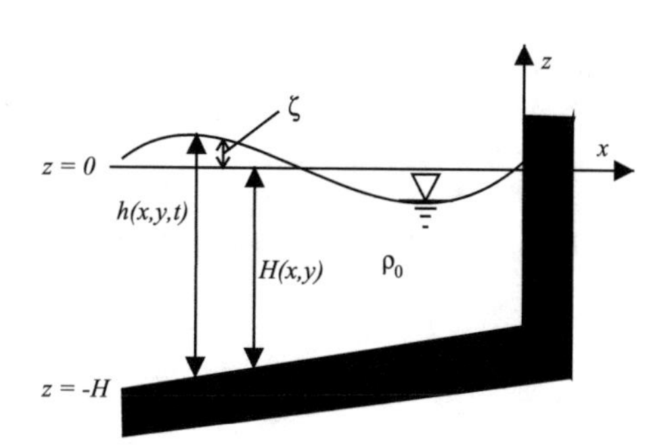

Fig. 6.6 Storm surge model along a straight coast, showing some of the notation

along the y-axis, as sketched in Fig. 6.6. Hence, it may be assumed that the solution is one-dimensional ($\partial_y = 0$), as long as the wind stress is also independent of y. In addition, we assume that changes in the equilibrium depth H are insignificant and that $|\mathbf{U}|^2 \ll |\mathbf{U}|$. It is also safe to neglect the variation with latitude in the effect of the Earth's rotation, using the well known f-plane approximation. Hence, the Coriolis parameter f is assumed to be constant in time and space. In addition, changes in the equilibrium depth are assumed to be small enough to ensure that H may to a good approximation be considered constant as well ($H = H_0$). Furthermore, the initial condition is an ocean at rest and in equilibrium. Finally, the coast is impermeable, so that the natural boundary condition is no flow through the coast. Under these circumstances, show that

$$\partial_t U - fV = -gH\partial_x h + \rho_0^{-1}(\tau_s^x - \tau_b^x) , \tag{6.196}$$

$$\partial_t V + fU = \rho_0^{-1}(\tau_s^y - \tau_b^y) , \tag{6.197}$$

$$\partial_t h + \partial_x U = 0 , \tag{6.198}$$

follow by linearizing (6.193) and (6.194).

2. Even the system (6.196)–(6.198) is hard to solve analytically. The problem is therefore simplified even further by neglecting the term $\partial_t U$ in (6.196). Show that the inertial oscillations (of frequency f) are avoided if this term is neglected.
3. Assume that there is no bottom friction, that is, $\tau_b^x = \tau_b^y = 0$, and that there is no wind stress in the x-direction ($\tau_s^x = 0$). Show that, under these assumptions and the assumption that $\partial_t U$ is small compared to the Coriolis acceleration, the motion along the coast (in the y-direction) is in geostrophic balance. Furthermore, under these assumptions, show that the analytic solutions to (6.196)–(6.198) are

$$U = U_\mathrm{E}\left(1 - \mathrm{e}^{x/L_\mathrm{R}}\right) , \tag{6.199}$$

$$V = ftU_\mathrm{E}\mathrm{e}^{x/L_\mathrm{R}} , \tag{6.200}$$

$$h = H\left(1 + \frac{tU_\mathrm{E}}{L_\mathrm{R}H}\mathrm{e}^{x/L_\mathrm{R}}\right) , \tag{6.201}$$

where $L_\mathrm{R} = \sqrt{gH}/f$ is the Rossby radius of deformation and

$$U_\mathrm{E} = \frac{\tau_s^y}{\rho_0 f} , \tag{6.202}$$

is the Ekman transport.

4. To investigate the effect of bottom friction, solve (6.196)–(6.198) analytically under the assumption that $\tau_b^x = 0$, $\tau_b^y = \rho_0 RV/H$, and that the term $\partial_t U$ in (6.196) can be neglected.
Hint: Make use of Laplace transforms.

6.11 Computer Problems

6.11.1 Geostrophic Adjustment

A fundamental question is how the atmosphere and ocean actually adjust from an unbalanced state to one in geostrophic balance under gravity. This problem, known as the problem of geostrophic adjustment, was first raised and solved back in the 1930s by Rossby (1937, 1938). It is nicely summarized in Blumen (1972) and Gill (1982), Sect. 7.2, p. 191. This will be the background for this computer problem.

One of the strongest and most important balances in the atmosphere and ocean, confirmed again and again by observation, is the geostrophic balance. When the fluid motion is in geostrophic balance, the Coriolis acceleration and the pressure forcing balance one another, i.e.,

$$f\mathbf{k} \times \mathbf{u}_{\mathrm{g}} = -\frac{1}{\rho_0}\nabla_{\mathrm{H}} p \,, \quad \text{or} \quad v_{\mathrm{g}} = \frac{1}{\rho_0 f}\partial_x p \,, \quad u_{\mathrm{g}} = -\frac{1}{\rho_0 f}\partial_y p \,, \tag{6.203}$$

where $f = 2\Omega \sin\phi$ is the Coriolis parameter, $\mathbf{k}$ is the unit vector along the vertical z-axis, $\mathbf{u}_{\mathrm{g}}$ is the (horizontal) geostrophic velocity with components u_{g}, v_{g} along the x-axis and y-axis, respectively, ρ_0 is the density, $\nabla_{\mathrm{H}} = \mathbf{i}\partial_x + \mathbf{j}\partial_y$ is the horizontal component of the three-dimensional del operator, and p is the pressure. Note that (6.203) contains three unknowns, namely, p, u_{g}, and v_{g}, but only two equations. The solution is therefore underdetermined, but by specifying one of them, say the pressure p, the other two variables may be found.

In the following, we consider the one-dimensional (1D) shallow water equations, which conveniently serve the purpose of illustrating the wave modes of the solution, the role of initial conditions, and the use of an open boundary condition (FRS).

Recall that the shallow water equations assume a hydrostatic balance, $p = \rho_0 g h$, where h is the geopotential height. Thus, the inherently nonlinear governing equations are

$$\partial_t h = -\nabla_{\mathrm{H}} \cdot (h\mathbf{u}) \,, \tag{6.204}$$

$$\partial_t \mathbf{u} = -f\mathbf{k} \times \mathbf{u} - \mathbf{u} \cdot \nabla_{\mathrm{H}} \mathbf{u} - g\nabla_{\mathrm{H}} h \,, \tag{6.205}$$

where the Coriolis parameter is $f = 1.26 \times 10^{-4}\mathrm{s}^{-1}$ (corresponding to its value at 60° N). The variable h is the geopotential height of a pressure surface in the atmosphere, and the depth of a water column in the ocean. The equilibrium height of h in the atmosphere is associated with a pressure surface of ~900 hPa, while the equilibrium depth in the ocean is ~1 km.

Part 1

A. Show that by introducing $\mathbf{U} = h\mathbf{u}$ and $h = h$ as new variables, (6.204) and (6.205) become

$$\partial_t h = -\nabla_{\mathrm{H}} \cdot \mathbf{U} \,, \tag{6.206}$$

$$\partial_t \mathbf{U} = -f\mathbf{k} \times \mathbf{U} - \nabla_{\mathrm{H}} \cdot \left(\frac{\mathbf{U}\mathbf{U}}{h}\right) - \frac{1}{2} g\nabla_{\mathrm{H}} h^2 \,. \tag{6.207}$$

B. Show that (6.204) and (6.205) may be combined to yield the vorticity equation

$$(\partial_t + \mathbf{u} \cdot \nabla_H) P_v = 0 \,, \tag{6.208}$$

where P_v is the potential vorticity defined by

$$P_v = \frac{\zeta + f}{h} \,, \tag{6.209}$$

with $\zeta = \mathbf{k} \cdot \nabla_H \times \mathbf{u}$ the relative vorticity.
C. Assume that the motion is independent of y ($\partial_y = 0$). Show that under these circumstances (6.204) and (6.205) reduce to

$$\partial_t h = -u\partial_x h - h\partial_x u \,, \tag{6.210}$$

$$\partial_t u = fv - u\partial_x u - g\partial_x h \,, \tag{6.211}$$

$$\partial_t v = -fu - u\partial_x v \,. \tag{6.212}$$

Moreover, show that the steady-state solution to (6.210)–(6.212) is in geostrophic balance, i.e.,

$$u = 0 \quad \text{and} \quad v = \frac{g}{f}\partial_x h \,. \tag{6.213}$$

D. Use (6.208) and (6.213) to show that the steady-state solution to (6.210)–(6.212) is a solution to the ordinary differential equation

$$\partial_x^2 h - \frac{f P_{v0}}{g} h = -\frac{f^2}{g} \,, \tag{6.214}$$

where $P_{v0} = P_{v0}(x)$ is the initial distribution of the potential vorticity.
E. Assume that the initial condition is one at rest, and that the geopotential height is given by a Heaviside function, i.e.,

$$u = v = 0 \,, \quad h = H_0 - \mathrm{sgn}(x - x_m)\Delta H \,, \quad \text{at } t = 0, \ \forall x \,, \tag{6.215}$$

where $\mathrm{sgn}(\psi) = +1$ if $\psi \geq 0$ and $\mathrm{sgn}(\psi) = -1$ if $\psi < 0$ (Fig. 6.7). Show that, under these circumstances, the solution to (6.214) is

$$h = H_0 + \Delta H \begin{cases} 1 - \dfrac{2\lambda_-}{\lambda_- + \lambda_+} \exp\left(\dfrac{x - x_m}{\lambda_-}\right) \,, & \text{if } x < x_m \,, \\ -1 + \dfrac{2\lambda_+}{\lambda_- + \lambda_+} \exp\left(-\dfrac{x - x_m}{\lambda_+}\right) \,, & \text{if } x \geq x_m \,, \end{cases} \tag{6.216}$$

where

$$\lambda_\pm = \frac{1}{f}\sqrt{g(H_0 \mp \Delta H)}$$

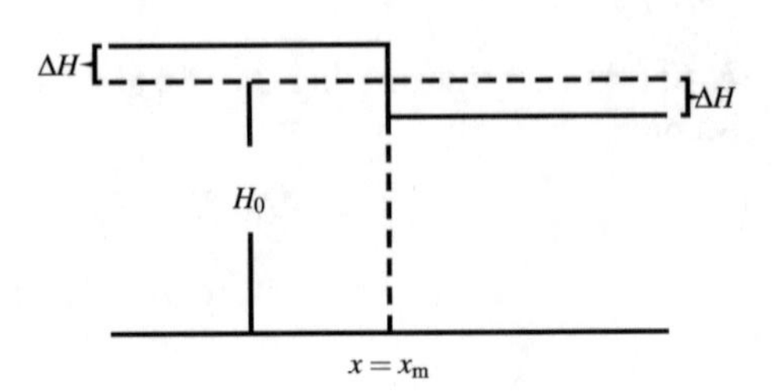

Fig. 6.7 Initial geopotential height according to (6.215)

is the Rossby deformation radius, $H_0 = 1000$ m, $\Delta H = 15$ m, $u_g = 0$ ms^{-1}, and $x_m = D/2$ is the midpoint of the domain of length D. Is the height anomaly $h - H_0$ negative or positive at $x = x_m$? Discuss the solution.

F. If you were to solve the system (6.210)–(6.212), how many boundary and initial conditions would there be? Explain how you derive the number of conditions.

Part 2

Solve the system (6.210)–(6.212) using numerical methods for the limited domain $x \in \langle 0, D \rangle$, assuming that the motion starts from rest, and where the geopotential height (or ocean surface) is given by (6.215). To this end, impose boundary conditions at $x = 0, D$, as entered under items (J) and (K) below.

G. Construct a CTCS (leap-frog) scheme to solve (6.210)–(6.212). Describe in some detail how you derive the various terms. Explain your choices. Add a filter to remove the two grid-length noises as explained in Sect. 5.8.

H. Is the CTCS scheme consistent? Derive condition(s) under which the scheme you have constructed is stable.[16] How long a time step Δt may be used?

I. Construct a quasi-Lagrangian scheme to solve (6.210)–(6.212). Describe in some detail how you derive the finite difference analogue and the choices you make.

J. Let the grid length be $\Delta x = 100$ km and $D = 62\Delta x$. Solve the above equations numerically using first the leap-frog (CTCS) scheme and then the semi-Lagrangian scheme constructed for the domain $x \in \langle 0, D \rangle$. Assume that the variables u, v, and h retain their initial values at the boundaries $x = 0$ and $x = D$. Plot h hourly, including the initial time, for the first 10 hours. Plot also the solution after 300 hours. Discuss the solutions. Try to make a movie spanning $t \in [0, 300]$ hours. What kind of waves do you observe?

K. Repeat the above computation using the FRS method explained in Sect. 7.5 to relax the inner solution toward the externally specified values

$$(\hat{u}, \hat{v}, \hat{h}) = (0, 0, H_0 + \Delta H) , \quad \text{at } x = 0 ,$$
$$(\hat{u}, \hat{v}, \hat{h}) = (0, 0, H_0 - \Delta H) , \quad \text{at } x = D .$$

Let the buffer zone be seven points wide, where the relaxation parameter λ_j is as given in Table 6.1. Compare the two solutions, for instance, by plotting the difference between them, and discuss any differences. Compare the two solutions to the steady-state solution derived in (E). Explain and discuss any differences you observe.

L. Compute the geostrophic component of the velocity, viz.,

$$v_g = \frac{g}{f}\partial_x h , \tag{6.217}$$

using the solution for h at $t = 6$ hours. Compare v_g and v at $t = 10$ hours, and describe and discuss what you observe. What do you think has happened?

M. Finally, replace the initial condition and boundary conditions for v in (6.215) by one in geostrophic balance, that is,

$$v = \frac{g}{f}\partial_x h_s , \tag{6.218}$$

[16] Hint: We always neglect the nonlinear terms when performing the stability analysis.

Table 6.1 Values of the relaxation parameter used in Part 2 (J). The left-hand column refers to the j numbers in the left-hand FRS zone, where $x = 0$ corresponds to $j = 1$. The right-hand column refers to the right-hand FRS zone, where $x = D$ corresponds to $j = j_{max}$

j	λ_j	j
1	1.0	j_{max}
2	0.69	$j_{max} - 1$
3	0.44	$j_{max} - 2$
4	0.25	$j_{max} - 3$
5	0.11	$j_{max} - 4$
6	0.03	$j_{max} - 5$
7	0.0	$j_{max} - 6$

where h_s is the steady-state solution of (6.216). Redo the computations and discuss the solutions you get.

6.11.2 Storm Surges

Here, we consider the storm surge problem. The aim is to gain experience in constructing numerical solutions to geophysical problems that include more than one dependent variable.

The water level of the ocean changes due to three main factors. To begin with, it responds to astronomical forcing, which generates tides, a periodic deterministic response in the water level. It also reacts to atmospheric forcing through wind stress and atmospheric forcing via sea level pressure. Changes in water level due to atmospheric forcing are referred to as storm surges. From time to time, the joint occurrence of high tides and high storm surges may lead to devastatingly high coastal water levels, and combined with ocean waves, these may threaten life and cause severe damage to coastal infrastructures. This was exemplified in Oslo Harbor, Norway, in mid-October 1987, where the water level reached 1.96 m above the normal sea level. More Norwegian examples are given in Gjevik (2009). A more recent example is the storm surge in combination with high waves during the hurricane "Sandy". This storm surge hit New York City on October 29, 2012, flooding streets, tunnels, and subway lines, and cutting power in and around the city. The cost of the damage in the United States amounted to $65 billion (2012 USD). The water level of the ocean may also change due to expansion by heating. Because of global warming, the oceans are heating up and hence also expanding. This rise in water level is a major concern, since it threatens to submerge low-lying islands and coastal regions.

Early on, many countries developed forecasting services for tides and storm surges to secure life and property. When numerical ocean models were being developed in the late 1960s and early 1970s, one of the first of these was to forecast tides and storm surges. For instance, the National Oceanography Center, Liverpool, UK (formerly the Proudman Oceanographic Laboratory) was founded in the late 1960s to forecast tides and storm surges in British waters, and has since been one of the leading institutes working in this field. Since the late 1970s and early 1980s, the Norwegian Meteorological Institute has also developed numerical models to forecast sea level changes due to storm surges and tides, using models developed through the work of Gjevik and Røed (1976), Martinsen et al. (1979), and Røed (1979).

Many of the earlier studies of storm surges have shown that they are mainly a barotropic response, more or less independent of the vertical baroclinic structure. A storm surge model may therefore be successfully constructed by assuming that the density is uniform (constant) in time and space. The equations to be considered therefore reduce to the well known shallow water equations, as given by (6.193) and (6.194) in Exercise 1.

The second term on the left-hand side of (6.193) and the first term on its right-hand side contain nonlinear terms. The main effect of these terms is to get the interaction of tides and storm surges correct down to centimeter precision. In many instances, this effect is small, and in the remainder of this computer problem, they are neglected. Furthermore, linearizing and neglecting variations with respect to y, (6.193) and (6.194) take the form given in (6.196)–(6.198). It is also safe to neglect the latitudinal variation in the effect of the Earth's rotation, and hence assume that the Coriolis parameter f is constant in time and space, the well-known f-plane approximation. In addition, changes in the equilibrium depth are assumed to be small enough so that H may also be considered constant to a good approximation ($H = H_0$).

Below we investigate numerical solutions to (6.196)–(6.198). The initial condition is an ocean at rest and in equilibrium. The coast is impermeable, and hence the natural boundary condition is no flow through the coast. Otherwise, the domain is assumed to be unlimited. Hence, far away from the coast, the solution is assumed to approach the well known Ekman solution, which follows from (6.196)–(6.198) by letting $\partial_x = 0$ and $\partial_t = 0$ and assuming that $\tau_b^x = \tau_b^y = 0$.

Note that the analytic solutions to (6.196)–(6.198) listed in Exercise 3 may be used to check the numerical solutions. This checking is part of the so-called quality assurance procedures (Chap. 11).

A. Let the coast be located at $x = 0$ with the ocean extending infinitely in the negative x direction. Explain why the Ekman solution is the natural open boundary condition to use far away from the coast. Show that mathematically the condition is equivalent to

$$h = H \quad \text{or} \quad \partial_x h = 0\,, \quad \text{at} \quad x \to -\infty\,, \quad \forall t\,. \tag{6.219}$$

Show also that the condition of no flow through the coast at $x = 0$ is

$$U = 0\,, \quad \text{at} \quad x = 0\,, \quad \forall t\,, \tag{6.220}$$

and that the initial condition may be formulated as

$$U = V = 0\,, \quad h = H_0\,, \quad \text{at} \quad t = 0\,, \quad \forall x\,. \tag{6.221}$$

B. Plot the analytical solutions of h, U, and V as derived in Exercises 3 and 4 in an x, t diagram (Hovmöller diagram).

C. Let the parameters appearing in (6.196)–(6.198) be

$$g = 10 \text{ ms}^{-1}, \quad \tau_s^x = 0, \quad \tau_s^y = 0.1 \text{ Pa}, \quad \rho_0 = 10^3 \text{ kg m}^{-3}, \quad f = 10^{-4} \text{ s}^{-1},$$
$$H_0 = 300 \text{ m}, \quad \tau_b^x = \rho_0 R \frac{U}{H_0}, \quad \tau_b^y = \rho_0 R \frac{V}{H_0}, \quad R = 2.4 \times 10^{-3} \text{ m s}^{-1},$$

unless explicitly stated. Since one of the spatial boundaries is at $x \to -\infty$, the domain is infinite. On the computer, however, the domain has to be finite. Thus, the computations have to be stopped at a finite distance away from the coast, say $x = -L$. The wording "far away from the coast" is equivalent to saying that L is large compared to some typical dynamical length scale of the problem. Note that the boundary at $x = -L$ is an open boundary. Thus, Eqs. (6.196)–(6.198) are still valid there, and we have to construct a condition that does not violate them.

Show that the dominant length scale in this problem is the Rossby deformation radius $L_R = c_0/f$, and by assuming $|x| \gg L_R$, show that the dynamics may safely be assumed to be independent of x. Discuss why under these circumstances the natural open boundary condition at $x = -L$ is the Ekman solution at $x = -L$.

D. Construct a forward–backward scheme (Sect. 6.4) on a staggered grid (Sect. 6.5). The staggered grid is the one-dimensional version of Lattice B in Mesinger and Arakawa (1976), p. 47, as sketched in Fig. 6.4. Thus, the U- and V-points are staggered one half-grid length with respect to the h-points.

E. Show using the von Neumann method that the scheme constructed in (D) is neutrally stable under the condition

$$\Delta t \le \frac{\Delta x}{c_0 \sqrt{1 + (\Delta x / 2L_R)^2}}, \tag{6.222}$$

where $c_0 = \sqrt{gH}$. Discuss situations where the simpler condition $C \le 1$ is valid, where $C = c_0 \Delta t / \Delta x$.

Hint: When analyzing the instability, neglect all forcing (stress) terms.

F. Show that the scheme constructed in (D) is consistent.

G. Solve the storm surge problem (6.196)–(6.198) numerically, including bottom stresses, using the scheme constructed in (D) and the parameters listed in (C). Choose Δx such that $\Delta x \ll L_R$, say $\Delta x = 0.1 L_R$. Explain why the latter choice is important.

H. Plot the numerical solution of the dependent variables h, U, and V in a Hovmöller diagram. Compare the numerical solution with those derived analytically in Exercises 3 and 4. Discuss differences and similarities.

6.11.3 Quasi-Lagrangian Method Applied to a Nonlinear System

Here, we consider solutions to the nonlinear rotating shallow water equations using the quasi-Lagrangian method explained in Sect. 6.6. This method was outlined in the appendix to Røed and O'Brien (1983), where it was referred to as the method of characteristics. It is also associated with the semi-Lagrangian method. We will assume that the variables only depend on one independent variable in space, that

is, we let $\partial_y = 0$. Furthermore, we note that $\mathbf{U} = h\mathbf{u}$, and that the pressure term in (6.193) may be written $gh\partial_x h$, under the assumption of a constant equilibrium depth.
A. Show that, if we define $c = \sqrt{gh}$, then (6.193) may be rewritten to yield the compatibility equations

$$\frac{D_1^*}{dt}(u + 2c) = fv + \frac{\tau_s^x - \tau_b^x}{\rho h}\,, \quad \text{along} \quad \frac{D_1^* x}{dt}\,, \tag{6.223}$$

$$\frac{D_2^*}{dt}(u - 2c) = fv + \frac{\tau_s^x - \tau_b^x}{\rho h}\,, \quad \text{along} \quad \frac{D_2^* x}{dt}\,, \tag{6.224}$$

$$\frac{D_3^* v}{dt} = -fu + \frac{\tau_s^y - \tau_b^y}{\rho h}\,, \quad \text{along} \quad \frac{D_3^* x}{dt}\,, \tag{6.225}$$

where the operators $D_{1,2,3}^*/dt$ are defined by

$$\frac{D_{1,2,3}^*}{dt} = \partial_t + \frac{D_{1,2,3}^* x}{dt}\partial_x\,, \tag{6.226}$$

and where the characteristic equations are[17]

$$\frac{D_1^* x}{dt} = u + c\,, \qquad \frac{D_2^* x}{dt} = u - c\,, \qquad \frac{D_3^* x}{dt} = u\,. \tag{6.227}$$

B. Solve (6.223)–(6.225) numerically using the method of characteristics, with fixed space increments Δx and time steps Δt. Disregard the wind and bottom stress, and look for solutions for $t > 0$, assuming that the initial condition at $t = 0$ is given by

$$h|_{t=0} = H + \Delta H \tanh(\kappa x)\,, \quad u|_{t=0} = v|_{t=0} = 0\,. \tag{6.228}$$

The solution domain is $x \in \langle L/2, L/2 \rangle$, where the boundaries at $x = \pm L/2$ are open. Furthermore, we let $\kappa = 10/L$, $\Delta H = H/2$, $H = 100$ m, and $L = 2000$ km. As open boundary condition, we recommend the gradient condition or the "flow relaxation scheme" (see Chap. 7).
C. Display the solution in graphical form by plotting the time evolution of h, u, and v as isolines in the x, t space (Hovmøller diagram).

References

Blumen W (1972) Geostrophic adjustment. Rev Geophys Space Phys 10:485–528
Charney JG (1955) The generation of ocean currents by wind. J Mar Res 14:477–498
Gill AE (1982) Atmosphere–ocean dynamics. International geophysical series, vol 30. Academic, London

[17] Hint: The problem is nonlinear and hence the characteristics are not straight lines. Since we use fixed Δx and Δt, they may from time to time cross the previous time level outside the domain bounded by x_{j-1} on the left-hand side and x_{j+1} on the right-hand side. Remember to check whether this is the case before interpolating.

Gill AE, Clarke AJ (1974) Wind-induced upwelling, coastal currents and sea-level changes. Deep Sea Res 21(5):325–345. https://doi.org/10.1016/0011-7471(74)90038-2
Gjevik B (2009) High and low tides along the coast of Norway and Svalbard. Farleia Forlag, Norway (in Norwegian only)
Gjevik B, Røed LP (1976) Storm surges along the western coast of Norway. Tellus 28:166–182
Griffies SM (2004) Fundamentals of ocean climate models. Princeton University Press, Princeton. ISBN 0-691-11892-2
Lighthill MJ (1969) Unsteady wind-driven ocean currents. Q J R Meteorol Soc 95:675–688
Martinsen EA, Gjevik B, Røed LP (1979) A numerical model for long barotropic waves and storm surges along the western coast of Norway. J Phys Oceanogr 9:1126–1138
Mesinger F, Arakawa A (1976) Numerical methods used in atmospheric models. GARP publication series, vol 17. World Meteorological Organization, Geneva, 64 pp
Phillips OM (1977) The dynamics of the upper ocean, 2nd edn. Cambridge University Press, Cambridge
Røed LP (1979) Storm surges in stratified seas. Tellus 31:330–339
Røed LP, O'Brien JJ (1983) A coupled ice-ocean model of upwelling in the marginal ice zone. J Geophys Res 29(C5):2863–2872
Rossby CG (1937) On the mutual adjustment of pressure and velocity distributions in certain simple current systems. J Mar Res 1:15–28
Rossby CG (1938) On the mutual adjustment of pressure and velocity distributions in certain simple current systems. Part II. J Mar Res 1:239–263
Shi XB, Røed LP (1999) Frontal instability in a two-layer, primitive equation ocean model. J Phys Oceanogr 29:948–968
Sielecki A (1968) An energy-conserving numerical scheme for the solution of the storm surge equations. Mon Weather Rev 96:150–156

7 Open Boundary Conditions and Nesting Techniques

The aim of this chapter is to discuss open boundaries and some of the techniques used to deal with them. An open boundary is defined as a computational boundary at which disturbances originating in the interior of the computational domain are allowed to leave without disturbing or deteriorating the interior solution (Røed and Cooper 1986). Even though the governing equations are still valid at these boundaries, they nonetheless constitute a boundary in a numerical sense. Hence, we focus on how to construct conditions, or *open boundary conditions* (OBCs), in such a way that disturbances originating in the interior of the computational domain are indeed allowed to leave without disturbing or deteriorating the interior solution.

Problems arising from not treating open boundaries properly were already encountered in the very first attempts at making a numerical weather forecast in the late 1940s (Charney et al. 1950) (see the historical notes in the preface of the present book). Recall that, just after the Second World War, digital computers were still in their infancy. The equation used by Charney et al. (1950) was the barotropic quasi-geostrophic vorticity equation, and its finite difference version was solved on a rectangular domain barely covering the North American continent. The model featured four "open" boundaries to the north, south, east, and west, across which the atmosphere was free to flow. They were thus forced to apply OBCs. They opted for a set in which they specified the potential vorticity at the inflowing boundaries and applied a radiation condition (Sect. 7.3) at the outflowing boundaries. Later, Platzman (1954) showed that their solution was unstable due to the choice of OBCs.

Since this first application, where the OBCs were shown to play a crucial part, there has been much research on OBCs, including techniques for nesting a high-resolution model in a low-resolution model (e.g., Davies 1976, 1985; Orlanski 1976; Sundström and Elvius 1979; Hedstrøm 1979; Røed and Smedstad 1984; Chapman 1985; Røed and Cooper 1986, 1987; Palma and Matano 2000; Blayo and Debreu 2005, 2006; Debreu and Blayo 2008; Mason et al. 2010; Debreu et al. 2012). As a consequence, most national meteorological institutes today run limited-area weather

L. P. Røed, *Atmospheres and Oceans on Computers*,
Springer Textbooks in Earth Sciences, Geography and Environment,
https://doi.org/10.1007/978-3-319-93864-6_7

forecast models nested within a global model. As an example, the limited-area model run at the Norwegian Meteorological Institute is the AROME–MetCoOp model, a 2.5 km mesh non-hydrostatic (convective scale) weather prediction model nested in the ECMWF model (Müller et al. 2017).[1] The global model is then the parent model and the limited-area model the child model, the latter covering the area of interest to that particular nation with a much higher resolution (smaller grid size). The nesting technique used is mostly one-way, simply because most nations use forecasts from a global model which is not run in-house, such as the ECMWF. This is also the case regarding ocean forecasting. For instance, the European Community recently established the Copernicus Marine Core Service, in which several European institutions collaborate to provide regional ocean weather forecasts for European Seas.[2] These include a global ocean model within which the regional models are nested.

This chapter details some of the common OBCs developed over the years. Details are presented of two of the most promising OBCs that turn out to be useful as one-way nesting techniques (Sects. 7.5 and 7.6). These are the flow relaxation scheme (e.g., Davies 1976, 1985) and the weakly reflective approach (e.g., Navon et al. 2004; Blayo and Debreu 2006; Mason et al. 2010). While the former was developed in the meteorological community, the latter comes from the study of electromagnetic fields (Berenger 1994). In addition, we shall look briefly at the two-way nesting technique (e.g., Debreu et al. 2012) in Sect. 10.5.

7.1 Open Boundaries

It is well known that computers, however large, can only hold a finite amount of data in their *random access memory* (RAM). Even the biggest computers have a limited capacity, and this constitutes a constraint on how "large" a numerical weather prediction (NWP) and/or a numerical ocean weather prediction (NOWP) model can be. For a given geographical area (or computational domain), the size of a numerical model is determined by the model's horizontal grid size and the number of vertical levels. Decreasing the horizontal grid size for a given computer implies that the computational domain has to be shrunk. Likewise, increasing the computational domain implies that the horizontal grid size has to be enlarged. Thus, the grid size limits the geographical area a numerical model is able to cover, while the finite computer capacity constrains the scale a model is able to resolve for a given geographical area. In this respect, the steady growth in computer capacity since the birth of computers back in the 1940s has resulted in ever-increasing resolution. Indeed, the capacity of modern-day supercomputers allows today's global atmospheric models to resolve the

[1]The forecast provided by this model is what you get when looking at yr (http://www.yr.no/) for the Norwegian forecast area. Outside this area, the forecast is built on the ECMWF model.
[2]Daily updated forecasts are available at http://marine.copernicus.eu/.

larger scale weather patterns more than adequately. For instance, the high-resolution global weather forecast issued twice a day with a lead time of 10 days by the European Center for Medium-Range Weather Forecasts (ECMWF) features a grid size of about 9 km in the horizontal and 137 vertical levels, a resolution one could only dream of in the 1970s.

The question therefore arises: is there a lower limit to the grid size needed? The dynamical scale of the weather systems in the atmospheres and oceans, also referred to as the synoptic scale in the atmosphere and mesoscale in the ocean, is determined by what is known as the Rossby deformation radius (Sect. 6.4). The first answer that comes to mind is therefore affirmative. Recall though that Rossby's deformation radius in the ocean differs vastly from the atmospheric one. At subpolar latitudes, the (baroclinic) deformation radius in the atmosphere is about 500–1000 km and the typical timescale is a few days. In contrast, the baroclinic deformation radius in the ocean at the same latitude is about 10–50 km, and the timescale is a few weeks to months. While the capacity of yesterday's computers was already enough to resolve the Rossby deformation radius in the atmosphere in the 1980s and 1990s for limited-area models, and today even for global models with a good margin, this is not the case for the ocean. This difference is perhaps the main reason why NOWP models are less mature than NWP models.

To illustrate the latter point consider a global model with a grid size of about 2° (Fig. 7.1 upper panel). A mesh size of 2°, or about 200 km, entails a grid size of about one-fifth of the atmospheric Rossby radius. This is a tolerable grid size for a numerical atmosphere model. Scaling this to Rossby's deformation radius in the ocean, the grid for the atmosphere model would look like the one displayed in the lower panel of Fig. 7.1. As is evident, the grid size of the latter is about three to four times the Rossby deformation radius. No meteorologist in his right mind would consider it to be an adequate grid for an NWP model. To obtain a similar tolerable resolution in ocean, grids of mesh sizes ∼2–4 km, or ∼1/200th of a degree, have to be employed. Hence, for a given geographical region, the amount of RAM required by an ocean model is much higher than in an atmosphere model. In addition, to satisfy the CFL criterion, the time step is much shorter for an ocean model, since the grid size is so much smaller. In practice, it requires a much greater computational effort to provide say a 24 h "weather" forecast for the ocean for a given area on a given computer than to provide a similar weather prediction. To enable computers to provide numerical ocean weather forecasts for the same area as fast as today's NWP models, faster computers are needed. To make things even worse recall that the timescale in the ocean is much longer than in the atmosphere. A weather prediction of say ten days corresponds to an ocean forecast of at least one month. Note that the above remarks regarding the ocean are associated with the baroclinic modes or internal part of the ocean response. Regarding the external or barotropic mode, responsible for tides and storm surges, the Rossby deformation radius is about 500–1000 km, and therefore more or less the same as the baroclinic deformation radius in the atmosphere.

In the infancy of NWP in the 1950s and 1960s, computer capacity was too limited to allow global atmospheric forecasts that resolved large-scale weather systems (synoptic scales). At the time most national meteorological institutes therefore made

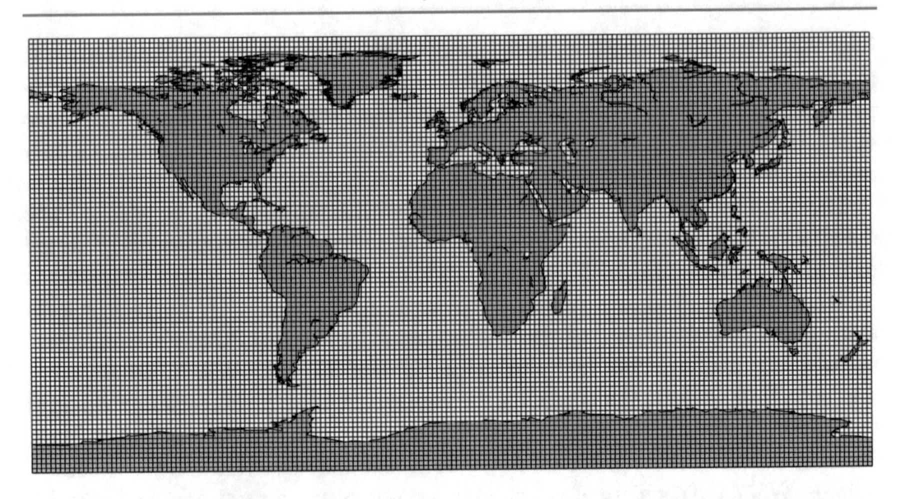

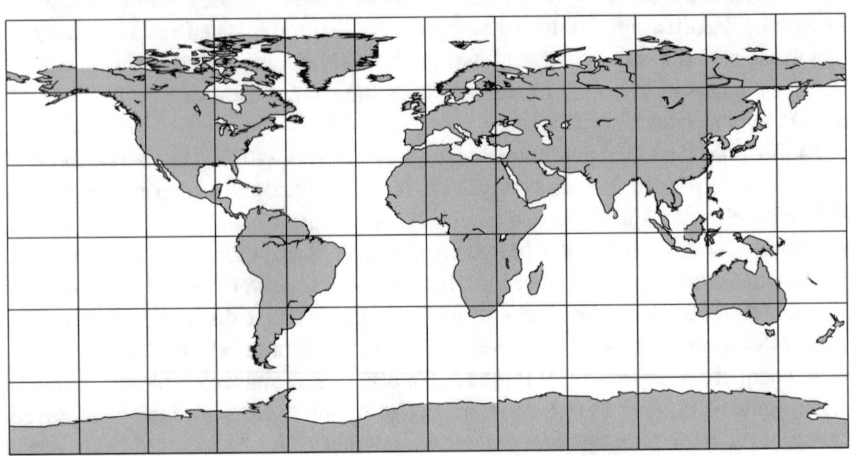

Fig. 7.1 *Upper*: Earth's surface covered by a 2° mesh. *Lower*: Similar mesh of 30° mesh size. The figure illustrates what a 2° mesh in the ocean would look like in the atmosphere scaled by the Rossby deformation radius

use of limited-area models or LAMs to produce a forecast. The boundaries of an LAM are therefore "open" in the sense that there are no natural boundaries, such as an impermeable wall, to help in specifying a boundary condition. Fluids are therefore free to flow through them, and as a consequence, the governing equations are still valid there. Hence, they are referred to as *open boundaries*. Nevertheless, in a mathematical/numerical sense, an open boundary still constitutes a boundary at which a boundary condition has to be specified, and such boundary conditions are referred to as *open boundary conditions*, or OBCs for short.

As mentioned, today's computers can run global NWP models with a resolution that is more than adequate to resolve the atmospheric synoptic scale. Nonetheless, there are still processes on much shorter scales than the synoptic scale that have a decisive impact on the local weather, notably processes associated with cloud formation (showers) and irregular topography. Processes associated with these scales, which are not yet resolved properly by the global models, impact not only local weather phenomena, but equally importantly also impact large-scale weather. Their effect is therefore parameterized in the global models.

Regarding ocean models, the situation is somewhat similar. In fact, global NOWP models resolving the oceanic deformation radius, also referred to as the mesoscale,[3] are not yet feasible at high latitudes, even on today's supercomputers. Thus, even today oceanographers have to resort to limited-area models to resolve the scales associated with oceanic weather systems. It thus appears that there is no limit in the quest for higher and higher resolution, so the second answer to the above question is negative.

As a consequence, early on, both ocean and atmosphere models had to handle open boundaries. In particular, they had to develop suitable OBCs, that is, OBCs that ensure, in a mathematical sense, that a solution to the governing equations of the child model exists and is unique. Recall that the governing equations are still valid. But since the computational domain of the child model ends at the open boundary, the governing equations must be replaced by some kind of OBC there. From a physical point of view, the solution should be as close to the "correct" solution as possible, or the solution obtained if the model were global with natural boundary conditions applied along its boundaries. However, since the solution to the governing equations is determined, not only by the equations themselves but also by the boundary conditions (Sect. 2.4), this is a dilemma. Thus, when an OBC has to be applied, there is no guarantee that the solution obtained will be the correct one. In general, it is impossible, except in special cases, even to prove that a solution exists and is unique, and then only for very simplified linear systems.

7.2 Nesting Techniques

Although to begin with global models did not resolve weather systems, they did provide information on scales larger than the weather systems. It was therefore recognized early on that the information inherent in these coarser mesh models should be exploited by the higher resolution, finer mesh limited-area models. The answer is *dynamic downscaling*, in which the results from a coarser mesh model are used to provide OBCs for the finer mesh models. These techniques are referred to as *nesting techniques*. It is common today to refer to the coarse mesh model as the "parent" model and to refer to the fine mesh model as the "child". If the parent is a "stand-alone" model, that is, there is no feedback from the child to the parent, the

[3]The reason for the difference in terminology regarding dynamical scales in the atmosphere and ocean is that the dynamical scale in the ocean measured in kilometers is called the mesoscale in the atmosphere.

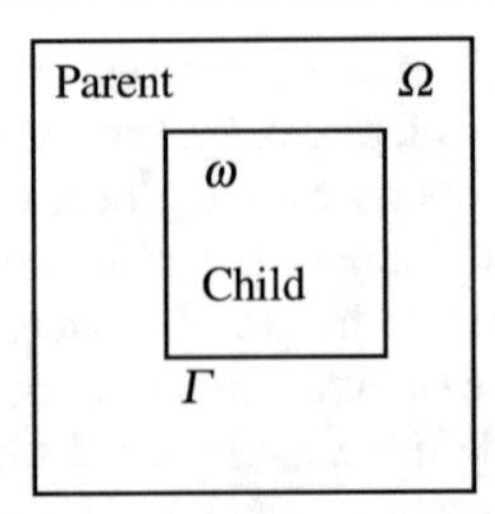

Fig. 7.2 Configuration of a child covering a domain ω, embedded in a parent covering the domain Ω. Commonly, the child has a higher resolution than the parent, with a refinement factor of 3–5. The interface, denoted by Γ, is an open boundary where an open boundary condition must be imposed

nesting is referred to as *one-way nesting*. This is in contrast to *two-way nesting*, in which the child is allowed to impact the parent. A treatment of the latter is postponed to Sect. 10.5, while the focus here is on one-way nesting.

The parent–child situation is visualized in Fig. 7.2, in which the parent covers the domain denoted by Ω and the child covers the domain denoted by ω. The open boundary or interface between them is denoted by Γ. The task is then to provide an OBC on Γ so that results from the parent are somehow transferred to the child. This then becomes the condition on the boundary Γ that must be obeyed by the child. In this way, the child provides a solution that is consistent with the larger scales of the parent, while at the same time taking into account the finer scales present in the child, e.g., through a refinement of the topography. Early on, with limited computer capacity, there were no parents around. Nonetheless, the boundary Γ was still there, but then a truly open boundary at which an OBC had somehow to be specified.

The choice of conditions at open boundaries is not arbitrary. There are certain requirements that must be satisfied. Foremost is the condition that disturbances originating in the interior of the domain and propagating toward the open boundary should be allowed to pass through to the exterior without distorting or disturbing the interior solution. Second, the chosen OBC must lead to a numerically stable solution. From a mathematical point of view, and this is perhaps the most important thing, the OBC together with the governing equations must form a well-posed problem, or at least a problem that is well-posed enough. Finally, disturbances originating in the exterior (parent domain) should be free to enter the child domain without distortion or damping. The latter is sometimes hard to achieve, since knowledge of the exterior solution may be nonexistent, or come from observation alone.

To illustrate the first requirement, consider a wave created in the child domain and propagating toward the open boundary. Whether an external parent domain exists or not, the condition imposed at the boundary should then be constructed in such a way as to let the wave pass, as if there were no boundary there. Thus, none of the energy contained in the wave should be allowed to radiate back into the child domain. Likewise, if a parent exists, a wave created in the parent domain should be free to enter the child without being distorted or damped.

7.3 Radiation Conditions

Many of the processes in atmospheres and oceans involve wave propagation in one way or another. Early attempts to develop OBCs were therefore based on simple wave equations. In its simplest form, the wave equation reads

$$\partial_t \phi + c_\phi \partial_\mathrm{n} \phi = 0 , \tag{7.1}$$

where ϕ is the dependent variable, c_ϕ is the component of the phase velocity normal the boundary associated with the variable ϕ, and ∂_n denotes the derivative normal to the open boundary. When (7.1) is used as an OBC, it is known as the *Sommerfeld radiation condition* (Schot 1992). This basically assumes that disturbances passing through the open boundary consist of waves. Note that the disturbances passing through the boundary may consist of several waves of different wavelengths, and hence that (7.1), strictly speaking, is valid only for one Fourier component. It is therefore suitable for linear problems, but less so for nonlinear problems.

One of the first obstacles in employing (7.1) as an OBC is that the phase velocity c_ϕ is unknown. Recall that c_ϕ is the slope of the characteristics (Sect. 5.13). Thus, if the choice of c_ϕ perfectly matches the slope of the characteristics, then (7.1) is a perfect open boundary condition. However, it is only for very simple physical problems, e.g., monochromatic wave problems, that the characteristics are known *a priori*, so c_ϕ is generally unknown.

Equation (7.1) contains two special cases, $c_\phi = 0$ and its opposite $c_\phi \to \infty$. Recall that in the former case the characteristics are vertical straight lines in the x, t plane. Under these circumstances, integrating (7.1) over time leads to

$$\phi = \text{constant at the open boundary} , \tag{7.2}$$

which is a Dirichlet condition (Sect. 2.4). Equation (7.2) is commonly referred to as a *clamped condition*, since the dependent variable ϕ does not change in time at the OB. In the second case, $c_\phi \to \infty$, the characteristics are horizontal straight lines in the x, t plane. If $\partial_t \phi$ should remain finite, (7.1) says that the gradient $\partial_\mathrm{n} \phi$ must be zero. Hence,

$$\partial_\mathrm{n} \phi = 0 \quad \text{at the open boundary} , \tag{7.3}$$

which is a Neumann condition (Sect. 2.4), and commonly referred to as a *gradient condition*. Notice that the first implies that the wave speed is zero, while the second implies that it is infinite. Both are obviously wrong in general. For instance, if the wave has a nonzero finite wave speed, the wave will be reflected and the solution in the interior domain will be affected. However, it may still work for simple problems, in particular elliptic problems (infinite wave speed) and stationary problems.

This nevertheless leaves the problem of how to determine or specify the phase velocity in order to avoid reflections. If the solution is in the form of waves with a known phase speed, the task is easy. For instance, if the problem is linear, the solution in most cases is composed of waves. For the shallow water equations, these may be in the form of inertia–gravity waves, coastally trapped Kelvin waves, or Rossby waves (Sect. 6.3). The phase speed is then given by the corresponding dispersion relation.

Consider a fluid contained in a channel of equilibrium depth H. Furthermore, consider frictionless motion, and let h denote the total depth or layer thickness of a fluid column and u the speed of the fluid column.[4] Moreover, consider that the motion is on a non-rotational Earth, and that the fluid has constant and uniform density. Then, the governing equations may be written

$$\partial_t u = -g\partial_x h \,, \tag{7.4}$$

$$\partial_t h = -H\partial_x u \,. \tag{7.5}$$

The classic method to solve the above set is first to differentiate (7.4) with respect to x and (7.5) with respect to t and then to add the results. The result is

$$\partial_t^2 h - c_0^2 \partial_x^2 h = 0 \,, \tag{7.6}$$

that is, a wave equation with phase speed equal to $c_0 = \sqrt{gH}$. The set (7.4) and (7.5) requires two boundary conditions in x space. Assume that the channel has two open boundaries at $x = 0$ and $x = L$. It is then natural to use the radiation condition (7.1) with a phase speed of $+c_0$ at $x = L$ and $-c_0$ at $x = 0$ as open boundary conditions.

Since the phase velocity is determined by the slope of the characteristics, the quasi-Lagrangian method (Sect. 6.4) may be used to find a useful boundary condition. The compatibility equations and the characteristic equations for this simple system follow from (6.92) and (6.93) by letting $f = 0$, which leads to

$$\partial_t \left(u + \frac{c_0}{H}h\right) + \frac{\mathrm{D}_1^* x}{\mathrm{d}t}\partial_x \left(u + \frac{c_0}{H}h\right) = 0 \,, \quad \text{along} \quad \frac{\mathrm{D}_1^* x}{\mathrm{d}t} = c_0 \,, \tag{7.7}$$

$$\partial_t \left(u - \frac{c_0}{H}h\right) + \frac{\mathrm{D}_2^* x}{\mathrm{d}t}\partial_x \left(u - \frac{c_0}{H}h\right) = 0 \,, \quad \text{along} \quad \frac{\mathrm{D}_2^* x}{\mathrm{d}t} = -c_0 \,. \tag{7.8}$$

While (7.7) describes a wave propagating in the positive x-direction with phase velocity c_0, we observe that (7.8) describes a wave propagating in the opposite direction, but with the same phase velocity. In particular, we notice that (7.7) and (7.8) express the idea that the specific combinations of the dependent variables u and h, namely, the Riemann invariants $u \pm hc_0/H$, are conserved along their respective characteristics.

As above, we consider the solution of (7.4) for $0 < x < L$, where $x = 0$ and $x = L$ are open boundaries. Additionally, assume that motion is generated in the interior of the domain, say in the form of an initial local deviation of the layer thickness h. The question then arises: what is the correct boundary condition to impose on the two open boundaries? The two compatibility equations (7.7) and (7.8) tell us that the information about the deviation will propagate toward the boundaries along the two characteristics. Toward the right-hand boundary at $x = L$, the information will propagate along the $\mathrm{D}_1^* x/\mathrm{d}t = +c_0$ characteristic, and toward the left-hand boundary $x = 0$ along the $\mathrm{D}_2^* x/\mathrm{d}t = -c_0$ characteristic. To avoid reflections, we must impose a condition that ensures that information cannot propagate back into the interior

[4] Since frictionless motion is considered, it is safe to assume that u is independent of depth.

domain from the boundary point. This is ensured if there are no characteristics at $x = 0$ or $x = L$ that slope toward the interior, and this is equivalent to requiring

$$\frac{\mathrm{D}_2^* x}{\mathrm{d}t} = 0 \,, \quad \text{at} \quad x = L \,, \tag{7.9}$$

and

$$\frac{\mathrm{D}_1^* x}{\mathrm{d}t} = 0 \,, \quad \text{at} \quad x = 0 \,. \tag{7.10}$$

Substituting this into the left-hand sides of (7.7) and (7.8), respectively, leads to

$$\partial_t \left(u - \frac{c_0}{H} h \right) = 0 \,, \quad \text{at} \quad x = L \,, \tag{7.11}$$

and

$$\partial_t \left(u + \frac{c_0}{H} h \right) = 0 \,, \quad \text{at} \quad x = 0 \,. \tag{7.12}$$

We now integrate (7.11) and (7.12) over time to obtain

$$u = \frac{c_0}{H} h + \text{const.} \,, \quad \text{at} \quad x = L \,, \tag{7.13}$$

and

$$u = -\frac{c_0}{H} h + \text{const.} \,, \quad \text{at} \quad x = 0 \,. \tag{7.14}$$

This is in fact the radiation condition. Indeed, if we substitute the expression (7.5) for $\partial_t h$ into (7.11), we get (7.1) with $\phi = u$ and $c_\phi = c_0$.

This relatively easy derivation of a nonreflective boundary condition using the quasi-Lagrangian method provides a general way to construct open boundary conditions. For instance, on the basis of earlier work by Hedstrøm (1979), it was used by Røed and Cooper (1987) to construct a weakly reflective boundary condition for a more general problem, and this is presented in Sect. 7.6.

The next question is how to implement the radiation condition in a numerical model. To this end we turn to the one-dimensional version of the radiation condition (7.1). Consider a computational domain $x \in \langle 0, L \rangle$ and $t \in \langle 0, T \rangle$. In addition, the boundary at $x = L$ is assumed to be an open boundary, while $x = 0$ is a natural boundary in the form of an impermeable wall. We also construct a grid in the x, t coordinates, where $x_j = (j - 1)\Delta x$ and $t^n = n\Delta t$ (Fig. 7.3). The radiation condition (7.1) then takes the form

$$\partial_t \phi + c_\phi \partial_x \phi = 0 \,, \quad \text{at} \quad x = L \,, \tag{7.15}$$

while the boundary condition at $x = 0$ is $u = 0$.

Since (7.15) is an advection equation, it is natural to employ one of the stable schemes developed in Sect. 5.2. Recall the importance of applying schemes that have the same accuracy. For instance, if the interior scheme is second order accurate in time and space, the scheme used to solve (7.15) should also be a second order scheme. Under these circumstances, it is natural to employ a leap-frog scheme to solve (7.15). If, on the other hand, the interior scheme is first order in time and space, the scheme used to solve the radiation condition (7.15) should be first order accurate

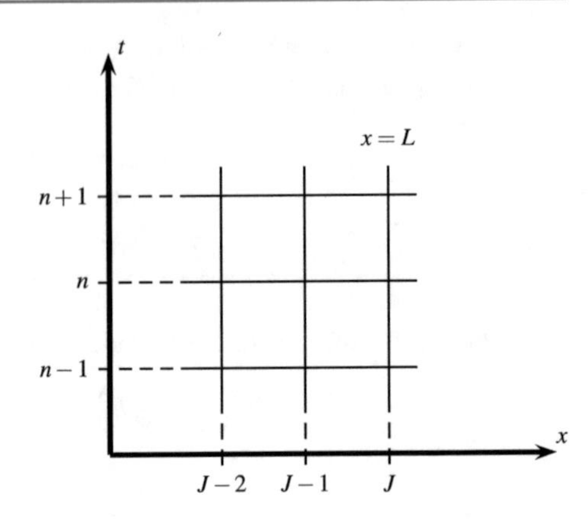

Fig. 7.3 Mesh in the x, t plane, close to the right-hand open boundary. The computational domain is to the left of $x = L$. The grid point J is associated with the boundary $x = L$, while $J - 1$ and $J - 2$ are the two first "wet" points inside the boundary. Time levels are denoted by n, $n - 1$, and $n + 1$, respectively

as well. Assuming the latter, the upstream scheme may be implemented. With the notation shown in Fig. 7.3, (7.15) then takes the finite difference form

$$\frac{\phi_J^{n+1} - \phi_J^n}{\Delta t} + c_\phi \frac{\phi_J^n - \phi_{J-1}^n}{\Delta x} = 0 \,, \tag{7.16}$$

or

$$\phi_J^{n+1} = (1 - C_\phi)\phi_J^n + C_\phi \phi_{J-1}^n \,, \tag{7.17}$$

where

$$C_\phi = c_\phi \frac{\Delta t}{\Delta x} \tag{7.18}$$

is the Courant number.

However, the problem persists. The Courant number is as unknown as the phase velocity c_ϕ. Following the suggestion by Orlanski (1976), (7.17) may be solved for the phase velocity (or C_ϕ) using grid points where the variable ϕ is known. Then, several options are available. One is to use interior points, but conserve the time levels, in which case

$$C'_\phi = -\frac{\phi_{J-1}^{n+1} - \phi_{J-1}^n}{\phi_{J-1}^n - \phi_{J-2}^n} \,. \tag{7.19}$$

A second is to use information at previous times, but conserving the spatial points:

$$C'_\phi = -\frac{\phi_J^n - \phi_J^{n-1}}{\phi_J^{n-1} - \phi_{J-1}^{n-1}} \,. \tag{7.20}$$

Both of these provide an expression for the Courant number. But which is the correct one? As recommended above and detailed in Sect. 5.12, (7.19) and (7.20) should be interpreted in terms of characteristics. From this perspective, (7.19) assumes that the slope of the characteristic through the grid point (x_J, t^{n+1}) equals the slope through the grid point (x_{J-1}, t^{n+1}). In contrast, (7.20) assumes that it equals the slope through the grid point (x_J, t^n). Following this argument, a third option is available, namely, that the characteristic through (x_J, t^{n+1}) continues backward in time and crosses the time level $n - 1$ between x_{J-1} and x_{J-2}. This is equivalent to assuming that to a first approximation the slope through (x_J, t^{n+1}) equals the slope through (x_{J-1}, t^n). Hence, the third option is

$$C'_\phi = -\frac{\phi^n_{J-1} - \phi^{n-1}_{J-1}}{\phi^{n-1}_{J-1} - \phi^{n-1}_{J-2}} \,. \tag{7.21}$$

Primes are deliberately attached to C_ϕ in (7.19)–(7.21). A negative C'_ϕ implies that c_ϕ is negative as well, implying that the wave is propagating in the negative direction. Under these circumstances there is no positive characteristic leaving the domain at $x = L$. The variable should therefore not be updated, but remain the same. This is ensured by letting $C_\phi = 0$ if $C'_\phi < 0$. Hence, the Courant number that should be substituted in (7.17) is

$$C_\phi = \begin{cases} C'_\phi \,, & C'_\phi \geq 0 \,, \\ 0 \,, & C'_\phi < 0 \,. \end{cases} \tag{7.22}$$

It is not clear which of the three approximations is the "best" to use. Nonetheless, Røed and Cooper (1987) advocate (7.21) as perhaps the better approximation of the three. Finally, note that the same procedure may be followed if the boundary at $x = 0$ is an open boundary. Under these circumstances, the inequality sign in (7.22) must be reversed. The reason is that, at the left-hand boundary, the characteristics that leave the domain are the characteristics with a negative slope.

As mentioned, it is only for problems where processes like nonlinear interactions, friction, wind forcing, and the rotation of the Earth can be neglected that the radiation condition (7.1) turns out to be useful. However, all modern-day numerical models of atmospheres and oceans are much more complex and include at least the processes just mentioned, and in most cases many more. The radiation condition is therefore far from being a perfect open boundary condition. Strictly speaking, there is no such thing as a perfect boundary condition, since the problem in a geophysical context can be shown to be ill-posed. The need to construct open boundary conditions stems from the fact that computers have a limited capacity, which limits the geographical domain that can be covered for a given resolution. Since the radiation condition is in most cases far from perfect, the meteorological and oceanographic communities have developed several optional OBCs (e.g., Chapman 1985; Røed and Cooper 1986, 1987; Palma and Matano 2000; Blayo and Debreu 2006). Some of the more popular ones are presented in the following sections.

7.4 Sponge Condition

One OBC which is still fairly popular is the so-called *sponge condition*. This will also serve as a convenient prelude to our discussion of the flow relaxation scheme (FRS) in Sect. 7.5. In essence, the method extends the computational domain outside the area of interest (the interior domain), so as to include an area where the energy leaving the interior domain is gradually decreased to avoid reflection. This is comparable to waves impinging on a sandy beach. In contrast to waves hitting a wall, there is no reflection; it is as though the beach absorbs the waves like a sponge. In practice, this is implemented by gradually increasing the relative importance of those terms associated with energy extraction, e.g., frictional processes. As the disturbance is advected or propagated into the extended exterior domain (sometimes referred to as the sponge layer), the parameter associated with energy extraction is gradually increased until all the energy is sucked up.

As an example, we study the nonrotating shallow water equations (7.4) and (7.5), and let the domain of interest be $x \in \langle -L_\mathrm{i}, L_\mathrm{i} \rangle$. However, the two boundaries $x = -L_\mathrm{i}, L_\mathrm{i}$ are not physical, but open, implying that the governing equations (7.4) and (7.5) are valid outside the domain of interest as well. As mentioned, the idea of the sponge method is to extend the *computational domain* by adding an area outside the area of interest, say to $x \in \langle -L_\mathrm{e}, L_\mathrm{e} \rangle$, where $L_\mathrm{e} > L_\mathrm{i}$, so that the domain of interest becomes a sub-domain within the computational domain. Furthermore, a friction term is added to the momentum equation, for instance, in the form of Rayleigh friction, that is, $-\gamma u$, where γ is a frictional parameter. The term should be active in the extended domain only to suck up the energy emanating from the interior domain, if any. Thus, (7.4) and (7.5) are replaced by

$$\partial_t u = -g\partial_x h - \gamma u \, , \tag{7.23}$$

$$\partial_t h = -H\partial_x u \, . \tag{7.24}$$

These equations are solved within the whole computational domain $x \in \langle -L_\mathrm{e}, L_\mathrm{e} \rangle$, including the two so-called *sponge layers* $-L_\mathrm{e} \le x \le -L_\mathrm{i}$ and $L_\mathrm{i} \le x \le L_\mathrm{e}$ to $x = L_\mathrm{e}$. By letting γ be zero within $x \in \langle -L_\mathrm{i}, L_\mathrm{i} \rangle$, (7.23) and (7.24) reduce to the original equations (7.4) and (7.5) in the interior. In the exterior domains $x \in \langle -L_\mathrm{e}, -L_\mathrm{i} \rangle$ and $x \in \langle L_\mathrm{i}, L_\mathrm{e} \rangle$, γ is gradually and monotonically increased from zero at the open boundaries $x = L_\mathrm{i}$ and $X = -L_\mathrm{i}$ to some finite value at the end of the computational domain. In particular, to avoid losing mass, Rayleigh friction is only added to the momentum equation to ensure mass (or volume) conservation. Since the solution to the equation $\partial_t u = -\gamma u$ is $u = u_0 \mathrm{e}^{-\gamma t}$ where u_0 is the initial velocity, adding Rayleigh friction has the added advantage that it is nonselective, implying that all wavelengths are equally damped.

A frictional parameter satisfying the above is

$$\gamma = \gamma_0 \begin{cases} 1 - \mathrm{e}^{-\lambda(x+L_\mathrm{i})} \, , & -L_\mathrm{e} \le x < -L_\mathrm{i} \, , \\ 0 \, , & -L_\mathrm{i} \le x \le L_\mathrm{i} \, , \\ 1 - \mathrm{e}^{\lambda(x-L_\mathrm{i})} \, , & L_\mathrm{i} < x \le L_\mathrm{e} \, . \end{cases} \tag{7.25}$$

Clearly, this increases exponentially from zero at interfaces between the interior and exterior domain to

$$\gamma_0 \big[1 - \mathrm{e}^{\lambda(L_\mathrm{e} - L_\mathrm{i})} \big]$$

at the end of the two computational domains. Any wave or disturbance, of any wavelength, created inside the interior domain will therefore start to be damped by friction as it propagates into the sponge layers, and more so as it progresses farther into the sponge. Together with the length of the sponge layers, the parameters λ and γ_0 determine how fast the waves are damped. To avoid reflection, which might deteriorate the interior solution, it is extremely important to choose these parameters so that the friction increases slowly near the interfaces.

The best way to show how the sponge method works is perhaps to derive an analytic solution to (7.23) and (7.24). Differentiating (7.23) with respect to time and then using the expression for $\partial_t h$ in (7.24), we obtain

$$\partial_t^2 u + \gamma \partial_t u = g H \partial_x^2 u \ . \tag{7.26}$$

Searching for a wave-like solution, we let

$$u = u_0 \mathrm{e}^{\omega t} \mathrm{e}^{\mathrm{i}\alpha x} \ , \tag{7.27}$$

where α is the wavenumber and ω is a complex frequency. Substituting this expression into (7.26), we obtain the dispersion relation

$$\omega^2 + \gamma\omega + gH\alpha^2 = 0 \ . \tag{7.28}$$

It has the two solutions

$$\omega_{1,2} = -\frac{1}{2}\gamma \pm \mathrm{i}\alpha c \ , \tag{7.29}$$

where

$$c = \sqrt{gH - \left(\frac{\gamma}{2\alpha}\right)^2} \tag{7.30}$$

is the phase speed of the wave entering the sponge. The solution then takes the form

$$u = u_0 \mathrm{e}^{-\gamma t/2} \mathrm{e}^{\mathrm{i}\alpha(x \pm ct)} \ , \tag{7.31}$$

where u_0 is a constant. Thus, the solution consists of two waves traveling in opposite directions, and the amplitude of the waves within the sponges, where $\gamma \neq 0$, decreases exponentially in time. Furthermore, the phase velocity (7.30) decreases with increasing γ. As the wave propagates deeper into the sponge areas, it is not only damped but also slows down.

Since application of the sponge condition requires the sponge zone to have a certain extension, it adds computer time. Including sponges therefore slows down the wall clock time. In addition, if the solution consists of forced waves (Røed and Cooper 1986), then the volume (or mass) may change within the sponge layer. This may be illustrated by adding a forcing term to the momentum equation, in which case the governing equations take the form

$$\partial_t u = -g\partial_x h - \gamma u + \tau \ , \tag{7.32}$$

$$\partial_t h = -H\partial_x u \ , \tag{7.33}$$

where τ represents forcing that is uniform in time and space. Solving the set analytically, we obtain the solution

$$u = u_0 \mathrm{e}^{-\gamma t/2} \mathrm{e}^{\mathrm{i}\alpha(x \pm ct)} + \frac{\tau}{\gamma} \ . \tag{7.34}$$

Since $\gamma = \gamma(x)$, u decreases as the forced wave progresses into the sponge. This implies that h changes as well, so the pressure forcing will also change. Sooner or later it therefore impacts the interior solution. It will be important to bear this fact in mind when discussing the flow relaxation scheme (FRS) in the next section.

7.5 Flow Relaxation Scheme

The *flow relaxation scheme* was first suggested for use as an OBC by Davies (1976) in atmospheric models. As we shall see, it has many features in common with the sponge. First, it requires us to extend the computational domain to include an exterior domain or buffer zone. Second, it is in essence a sponge in which the solution is suppressed as it progresses into the buffer zone. The method is commonly abbreviated to FRS and is also widely used in oceanographic models (Martinsen and Engedahl 1987; Cooper and Thompson 1989; Engedahl 1995; Shi et al. 1999, 2001). The last two references are particularly useful as they give a detailed description of the FRS and a nice example of its use as an OBC.

One of the advantages of the FRS, as compared, for instance, with the sponge, is that it allows us to specify an exterior solution. The FRS can therefore be used as a one-way nesting condition, in which an exterior solution is specified by, e.g., a coarser grid model (parent) covering a much larger area. In essence, this method, like the sponge method, only modifies the numerical solution in a buffer or relaxation zone outside the area of interest or interior domain, commonly referred to as the FRS zone. In the FRS zone, for each time step, the solution is relaxed toward a specified exterior solution. In general, the FRS zone is not very wide, but should contain at least 7 grid points. Since the FRS zone is an extension of the interior domain, it extends the computational domain and therefore increases the computational burden. Within the FRS zone, the solution is relaxed toward a previously specified exterior solution or outer solution. The relaxation is performed by specifying a weighting function which, for each grid point in the FRS zone, computes a weighted mean between the specified outer solution and the interior solution.

Let $\phi(x, t)$ be the dependent variable and let the interior domain be $x \in \langle -L_\text{i}, L_\text{i} \rangle$, where $x = -L_\text{i}, L_\text{i}$ are open boundaries. The FRS zones extend the interior domain by increasing the computational domain to the left and right (Fig. 7.4). The FRS zone to the left starts at $x = -L_\text{e}$ and ends at $x = -L_\text{i}$, while the FRS zone to the right starts at $x = L_\text{i}$ and ends at $x = L_\text{e}$. This is quite similar to the addition of sponge layers (Sect. 7.4). As usual, the grid points are defined by $x_j = (j - 1)\Delta x$, where the index j counts all grid points of the computational domain. The first grid point $j = 1$ is at $x = -L_\text{e}$, the leftmost boundary of the computational domain, and the last is $j = J + 1$ at $x = L_\text{e}$, the rightmost boundary. Furthermore, let $j = J_\text{l} + 1$

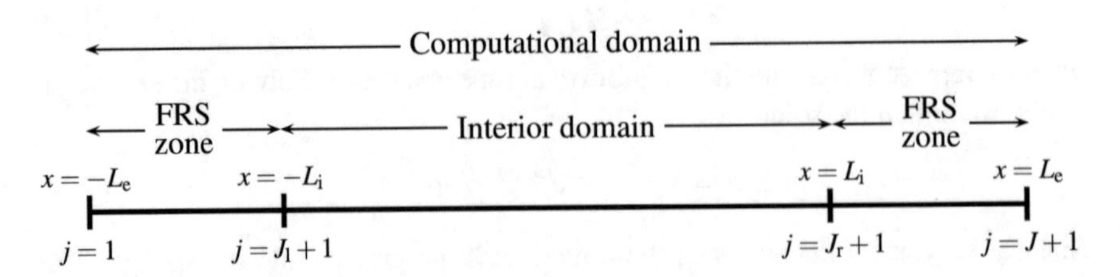

Fig. 7.4 The FRS zone, the interior domain, and the computational domain. Also shown are the appropriate indices referred to in the text

be associated with $x = -L_{\mathrm{i}}$, the left-hand open boundary, and let $j = J_{\mathrm{r}} + 1$ be associated with $x = L_{\mathrm{i}}$, the right-hand open boundary.

As mentioned, the FRS allows an outer solution to be specified, e.g., resulting from a presumably coarser grid parent model. Let the exterior solution be denoted by $\phi_{\mathrm{e}}{}_j^n$, emphasizing that the outer solution is a function of space and time. By making use of the governing equations, the dependent variable ϕ_j^n may be computed at all points within the computational domain, including the FRS zones. Assuming that this is done up to and including time level n, we may derive a predictor solution ϕ_j^* at the next time level $n+1$, except at the boundary grid points $j = 1$ and $j = J + 1$. The next step is to correct the predictor by computing the dependent variable as a weighted mean between the predicted solution ϕ_j^* and the specified outer solution $\phi_{\mathrm{e}}{}_j^n$. Thus, at all grid points including the boundaries, the final solution takes the form

$$\phi_j^{n+1} = (1 - \alpha_j)\phi_j^* + \alpha_j \phi_{\mathrm{e}}{}_j^{n+1} \ , \quad j = 1(1)J \ , \tag{7.35}$$

where $0 \leq \alpha_j \leq 1$ is a relaxation parameter defined so that

$$\alpha_j = \begin{cases} 1 \ , & j = 1 \ , \\ \alpha'_j \ , & j = 2(1)J_{\mathrm{l}} \ , \\ 0 \ , & j = J_{\mathrm{l}} + 1(1)J_{\mathrm{r}} + 1 \ , \\ \alpha'_j \ , & j = J_{\mathrm{r}} + 2(1)J \ , \\ 1 \ , & j = J + 1 \ , \end{cases} \tag{7.36}$$

and α'_j increases monotonically within the FRS zones. Hence, ϕ_j^{n+1} equals the specified outer solution at the end points of the FRS zones, while it equals the predictor in the interior. In the FRS zones, excluding the boundaries, ϕ_j^{n+1} is a weighted mean between the inner and the outer solutions.

Experiments involving the FRS, e.g., Martinsen and Engedahl (1987), Engedahl (1995), show that the solution is sensitive to the distribution of the specified weighting function α throughout the FRS zone. They find that a hyperbolic tangent function, i.e.,

$$\alpha'_j = 1 - \tanh \frac{j-1}{2} \ , \quad j = 1(1)J_{\mathrm{l}} \ , \tag{7.37}$$

works in most cases. Furthermore, as with the sponge method, they also found that the solution is sensitive to the width of the FRS zone. They concluded that, for ocean applications, the width of the FRS zone should be at least seven grid points, that is, $J_{\mathrm{l}} \geq 7$.

As with the sponge condition, one of the drawbacks of using the FRS is that the computational domain is increased, thus adding to the computational burden. Since the FRS allows an outer parent solution to be specified, this weak point is somewhat alleviated. Indeed, as will be shown in the next section, this may be used to minimize possible reflections of energy. Another limitation, and perhaps a more significant one, is that it does not conserve fundamental quantities such as volume (or mass). As with the sponge conditions, this may cause serious problems.

As an example consider the numerical solution to the continuous problem

$$\partial_t \phi = \mathscr{L}[\phi] \ , \quad x \in \langle -L_\mathrm{i}, L_\mathrm{i} \rangle \ , \tag{7.38}$$

where $\mathscr{L}$ is a spatial differential operator. For simplicity, it is assumed that the right-hand boundary $x = -L_\mathrm{i}$, associated with the counter $j = J_\mathrm{r} + 1$, is a natural boundary, but that the left-hand boundary $x = L_\mathrm{i}$, associated with the counter $j = J_\mathrm{l} + 1$, is an open boundary. As above, let ${\phi_\mathrm{e}}_j^n$ denote the specified parent solution. Applying a forward-in-time centered-in-space scheme, (7.38) takes the finite difference form

$$\phi_j^* = \phi_j^n + \Delta t \mathscr{L}_j^n \ , \quad j = 2(1)J_\mathrm{r} \ , \tag{7.39}$$

where ϕ_j^* is the predictor. To correct the predictor, the relaxation (7.35) is used, leading to

$$\phi_j^{n+1} = (1 - \alpha_j)\phi_j^* + \alpha_j(\phi_\mathrm{e})_j^{n+1} \ , \quad j = 1(1)J_\mathrm{r} \ . \tag{7.40}$$

To ensure that no corrections to the predictor are made within the interior domain, the relaxation parameter is specified as

$$\alpha_j = \begin{cases} 1 - \tanh \dfrac{j-1}{2} \ , & j = 1(1)J_\mathrm{l} \ , \\ 0 \ , & j = J_\mathrm{l} + 1(1)J_\mathrm{r} \ . \end{cases} \tag{7.41}$$

Using the expression on the right-hand side of (7.39) to substitute for ϕ_j^* in (7.40), and adding the zero $(\alpha_j \phi_j^{n+1} - \alpha_j \phi_j^{n+1})$ to the left-hand side of (7.40), we obtain

$$\frac{\phi_j^{n+1} - \phi_j^n}{\Delta t} = \mathscr{L}_j^n + \gamma_j \left({\phi_\mathrm{e}}_j^{n+1} - \phi_j^{n+1} \right) \ , \quad j = 1(1)J_\mathrm{l} \ , \tag{7.42}$$

where the coefficient γ_j is defined by

$$\gamma_j = \frac{\alpha_j}{1 - \alpha_j} \ . \tag{7.43}$$

Letting Δt and Δx tend to zero then leads to the continuous equation

$$\partial_t \phi = \mathscr{L}[\phi] + \gamma \left(\phi_\mathrm{e} - \phi \right) \ , \quad x \in \langle -L_\mathrm{e}, L_\mathrm{i} \rangle \ . \tag{7.44}$$

Except for the additional "frictional" term $\gamma\,(\phi_\mathrm{e} - \phi)$, Eq. (7.44) is the same as (7.38). Moreover, the additional term is proportional to the difference between the interior solution and the exterior solution, and the factor γ_j varies from zero at the open boundary ($x = -L_\mathrm{i}$) to infinity at the edge of the FRS zone. Thus, the relative importance of the frictional term increases as we progress into the FRS zone.

Specifying an exterior solution which is constant and zero, (7.44) becomes

$$\partial_t \phi = \mathscr{L}[\phi] - \gamma \phi \ . \tag{7.45}$$

Under these circumstances, the FRS acts like a sponge with a frictional parameter γ which gradually and monotonically increases toward infinity within the FRS zone, not unlike the exponential function specified in Sect. 7.4.

To illustrate the nonconservative properties of the FRS, we may reuse the example specified by (7.4) and (7.5), i.e.,

$$\partial_t u = -g\partial_x h \ , \tag{7.46}$$

$$\partial_t h = -H\partial_x u \ . \tag{7.47}$$

Notice that, as Δx and Δt tend to zero, (7.40) takes the form

$$\phi = (1-\alpha)\phi^* + \alpha\phi_{\mathrm{e}} \ . \tag{7.48}$$

There are two dependent variables, namely, h and u. Hence, the relaxation has to be performed on both, leading to

$$h = (1-\alpha)h^* + \alpha h_{\mathrm{e}} \ , \tag{7.49}$$

$$u = (1-\alpha)u^* + \alpha u_{\mathrm{e}} \ . \tag{7.50}$$

Substituting u from (7.50) into (7.47), we have

$$\partial_t h = -H\partial_x u = -H\partial_x u^* + H(u^* - u_{\mathrm{e}})\partial_x\alpha + \alpha H\partial_x(u^* - u_{\mathrm{e}}) \ . \tag{7.51}$$

The first term on the right-hand side of (7.51) is the term obtained without relaxation ($\alpha = 0$). However, since the relaxation parameter α is also a function of x, there appears a term containing its divergence. Unless $u^* = u_{\mathrm{e}}$, volume (or mass) conservation, as expressed by (7.47), is violated, and the varying relaxation parameter builds up an artificial pressure force in the FRS. This eventually leads to currents that may deteriorate the interior solution. To avoid this violation of mass conservation affecting the interior solution, the relaxation parameter α must be a very slowly varying function close to the open boundary at $x = -L_{\mathrm{i}}$. Moreover, the FRS zone must be wide enough for this to be realized. This is similar to the sponge, once again.

Finally, if the exterior solution is close to or equal to the true solution, the friction term in (7.45) disappears, as does the false divergence in (7.51). Under these circumstances, the FRS is close to being a perfect open boundary or nesting condition. Consequently, the usefulness of the FRS depends to a certain extent on how close the specified exterior solution is to the interior solution. This is the reason why the FRS is most successful when used as a one-way nesting condition. Then the exterior solution is a solution of the same equations that govern the interior solution, albeit for a coarser mesh parent model. Nonetheless, it is as close to the child solution as one may hope for.

7.6 A Weakly Reflective OBC

As mentioned in Sect. 7.3, the method of characteristics may also be used to construct a weakly reflective OBC for problems including nonlinearities, Coriolis effects, and forcing (see Røed and Cooper 1987).

As an example, consider the nonlinear rotating shallow water equation including forcing terms [see (6.128) in Sect. 6.4], i.e.,

$$\partial_t u + u\partial_x u - fv = -g\partial_x h + F^x \, , \tag{7.52}$$
$$\partial_t v + u\partial_x v + fu = F^y \, , \tag{7.53}$$
$$\partial_t h + \partial_x (hu) = 0 \, , \tag{7.54}$$

where F^x, F^y are the forcing terms and f is the Coriolis parameter. Solutions to (7.52)–(7.54) are sought in the domain $x \in \langle 0, L \rangle$, with an open boundary at $x = L$ and a natural boundary at $x = 0$.

Manipulating (7.52) and (7.54), as shown in Sect. 6.6, two of the three compatibility and characteristic equations take the form

$$\frac{\mathrm{D}^*_{1,2}}{\mathrm{d}t}(u \pm 2c) = fv + F^x \, , \quad \text{along} \quad \frac{\mathrm{D}^*_{1,2}x}{\mathrm{d}t} = u \pm c \, , \tag{7.55}$$

where $c = \sqrt{gh}$ replaces h as variable. Since $x = L$ is an open boundary, the compatibility and characteristic equations are valid there. Thus, there are two characteristics through any point along the vertical axis at $x = L$ (Fig. 7.5), with slopes given by (7.55). Since, $|u| \ll c$, there is always a positive sloping characteristic and a negative sloping one, but the presence of u creates an asymmetry. As long as $|u + c|\Delta t \leq \Delta x$, the point P, which is the point where the positive sloping characteristic crosses the time level n, is located between J and $J + 1$. The point Q where the negative sloping characteristic crosses the time level n lies outside the computational domain. This characteristic therefore carries information about the exterior or parent solution. If something is known about the exterior solution, this can be used as a boundary

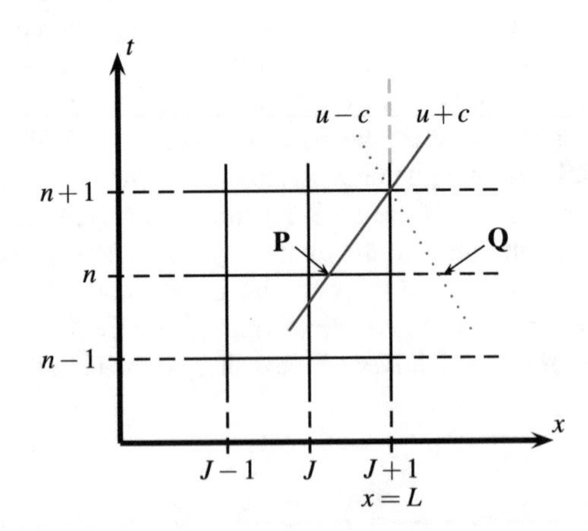

Fig. 7.5 Constructing weakly reflective open boundary conditions based on characteristics. The open boundary is at $x = L$ ($j = J + 1$). The *blue solid line* is the positive sloping characteristic through the open boundary grid point x_{J+1}, t^{n+1}, with slope equal to $(u + c)^{-1}$, while the *red dotted line* is the negative sloping characteristic through the same point, with a slope equal to $(u - c)^{-1}$. These are derived from (7.55). The distance between the grid points is Δt in the vertical and Δx in the horizontal direction. The point labeled P is a distance $|u^n_j + c^n_j|\Delta t$ to the left of the boundary, while the point Q is a distance $|u^n_j - c^n_j|\Delta t$ to the right of the open boundary, and hence outside the computational domain. The *green dashed line* corresponds to the characteristic required to formulate the weakly reflective open boundary condition

condition (e.g., Mason et al. 2010). Moreover, this characteristic passes through the boundary at $x = L$ and continues inwards. Thus, any energy reflected at the open boundary is signaled to the interior along this characteristic. To avoid reflections, it is therefore sufficient to require that the negative sloping characteristic be replaced by one with an infinite slope at the boundary, in which case it points vertically upwards along $x = L$ (see the green dashed line in Fig. 7.5). This is achieved by requiring

$$\frac{\mathrm{D}_2^* x}{\mathrm{d}t} = 0 , \quad \text{at} \quad x = L . \tag{7.56}$$

Substituting this expression into (7.55), the weakly reflective open boundary condition takes the form

$$\frac{\mathrm{D}_1^*}{\mathrm{d}t}(u + 2c) = fv + F^x , \quad \text{along} \quad \frac{\mathrm{D}_1^* x}{\mathrm{d}t} = u + c , \tag{7.57}$$

$$\partial_t (u - 2c) = fv + F^x , \quad \text{at} \quad x = L . \tag{7.58}$$

To find u and h at the boundary $x = L$ numerically, forward-in-time finite difference approximations of (7.57) and (7.58) lead to

$$u_{J+1}^{n+1} + 2c_{J+1}^{n+1} = u_P^n + 2c_P^n + \left(fv + F^x\right)_P^n \Delta t , \tag{7.59}$$

$$u_{J+1}^{n+1} - 2c_{J+1}^{n+1} = u_{J+1}^n - 2c_{J+1}^n + (fv + F^x)_{J+1}^n \Delta t , \tag{7.60}$$

where $J + 1$ is associated with the open boundary at $x = L$ and the time levels with the integer n. Solving (7.59) and (7.60) with respect to u_{J+1}^{n+1} and c_{J+1}^{n+1}, we find

$$u_{J+1}^{n+1} = \frac{1}{2}\left\{u_P^n + u_{J+1}^n + 2\left(c_P^n - c_{J+1}^n\right) + \left[\left(fv + F^x\right)_P^n + \left(fv + F^x\right)_{J+1}^n\right] \Delta t\right\} , \tag{7.61}$$

$$c_{J+1}^{n+1} = \frac{1}{4}\left\{u_P^n - u_{J+1}^n + 2\left(c_P^n + c_{J+1}^n\right) + \left[(fv + F^x)_P^n - (fv + F^x)_{J+1}^n\right] \Delta t\right\} . \tag{7.62}$$

It remains to find u_P^n, v_P^n, and c_P^n. To this end, we need the position of the point P along the x-axis, say x_P. Once again, a first approximation is found by a forward-in-time discretization of the characteristic equation (7.57), which leads to

$$x_P = x_j - (u_{J+1}^n + c_{J+1}^n)\Delta t . \tag{7.63}$$

As explained in Sect. 5.12 u_P^n, v_P^n, and c_P^n are found by a linear or higher order interpolation using the grid points adjacent to x_P. Since we make use of finite difference approximations, (7.57) and (7.58) are not perfectly solved, and some weak reflection is unavoidable. However, the accuracy may be somewhat increased by making an iteration along the lines described in Sect. 6.6.

7.7 Summary and Remarks

This chapter goes into some detail about open boundary conditions that have been developed over the years. Some are applicable as true open boundary conditions, while others may also be used as nesting techniques, whereby a child model is nested into a parent model. Two-way nesting techniques have also been put forward recently (e.g., Debreu et al. 2012), but details of this technique, including an example, will be postponed to Sect. 10.5.

After presenting some historical notes on open boundaries and nesting techniques in Sects. 7.1 and 7.2, including some reflections on whether there is a limit to how fine a grid needs to be, Sect. 7.3 introduces the *radiation condition*, one of the earliest OBCs. It is shown that this is a perfect OBC in those cases where the solution consists of waves propagating with known phase speeds. Furthermore, the OBCs referred to as the *fixed condition* and the *gradient condition* were shown to be special cases of the radiation condition, specifying the phase speed to be zero and infinite, respectively. In most cases, the phase speed is not known, and it is therefore shown in some detail how to implement the radiation condition under these circumstances. The radiation condition is a true OBC and cannot be used as a nesting technique.

The *sponge condition* is considered in Sect. 7.4. This is also a true OBC. The idea is to extend the computational area to include an area outside the domain, referred to as the sponge area, such that anything that propagates into it is progressively damped. The tricky part is to set up the sponge so that the solution inside the sponge area is not deteriorated. If a pressure force builds up in the sponge, it may easily start to build up false currents close to the interface with the sponge (Røed and Cooper 1986, 1987).

The idea of extending the domain was considered early on in the atmospheric community. As early as 1976, H.C. Davies suggested what later became known as the flow relaxation scheme (FRS) in his paper (Davies 1976). At that time, attempts were being made to develop limited-area models embedded in global scale models, and there was a need for OBCs that could make use of the information provided by the global parent model. The idea behind the FRS as a one-way nesting technique is presented in Sect. 7.5. However, if nothing is known about the exterior solution, the FRS behaves more or less like a sponge condition. As such, it shares the problem with the sponge method that a buildup of false pressure gradients may deteriorate the interior solution, even when used as a one-way nesting technique.

Finally, Sect. 7.6 proposes an OBC based on the method of characteristics. The idea is to inhibit information from propagating along the inward sloping characteristic by requiring it to have an infinite slope in the two-dimensional time versus space domain (Røed and Cooper 1987). The constructed OBC turns out to be weakly reflective. It was also mentioned that, using the method of characteristics, the weakly reflective OBC may be further developed to act as a two-way nesting technique. This idea was exploited by Mason et al. (2010), for instance.

There is a single computer problem at the end of this chapter (Sect. 7.9.1), which the reader is strongly urged to attempt. This will give an insight into the inner workings of the FRS, as well as the weakly reflective OBC. It concerns one-dimensional planetary waves in a limited domain that is open at both ends. It will also give experience in programming the quasi-Lagrangian and CTCS schemes for a nonlinear problem.

7.8 Exercises

1. Show that a one-sided finite difference scheme, in time and in space, of the radiation condition (7.1) can be written

$$\phi_B^{n+1} = \begin{cases} \phi_B^n \,, & c_\phi > 0 \,, \\ \left(1 + c_\phi \dfrac{\Delta t}{\Delta x}\right) \phi_B^n - c_\phi \dfrac{\Delta t}{\Delta x} \phi_{B+1}^n \,, & c_\phi \leq 0 \,. \end{cases} \quad (7.64)$$

The open boundary is to the left, so the subscript B denotes the values of the variables on the open boundary, and the subscript $B + 1$ the values to the right of the open boundary.

2. Using (7.64), show that the radiation condition is "simply" an interpolation of values on the inside of the computational domain.

7.9 Computer Problems

7.9.1 Planetary Waves

Recall that the shallow water equations assume a hydrostatic balance and hence that $p = \rho_0 g h$, where h is the geopotential height. Thus, the inherently nonlinear governing equations are

$$\partial_t h = -\nabla_{\mathrm{H}} \cdot (h\mathbf{u}) \,, \quad (7.65)$$

$$\partial_t \mathbf{u} = -f\mathbf{k} \times \mathbf{u} - \mathbf{u} \cdot \nabla_{\mathrm{H}} \mathbf{u} - g\nabla_{\mathrm{H}} h \,, \quad (7.66)$$

where the Coriolis parameter is $f = 1.26 \times 10^{-4}\,\mathrm{s}^{-1}$ (corresponding to its value at 60°N). It is usual to consider h as the geopotential height of a pressure surface in the atmosphere and as the depth of a water column in the ocean. The equilibrium height of h in the atmosphere is associated with a pressure surface of $\approx$900 hPa, while the equilibrium depth in the ocean is typically $\approx$1 km.

Part 1

A. By introducing $\mathbf{U} = h\mathbf{u}$ and $h = h$ as new variables, show that (7.65) and (7.66) take the form

$$\partial_t h = -\nabla_{\mathrm{H}} \cdot \mathbf{U} \,, \quad (7.67)$$

$$\partial_t \mathbf{U} = -f\mathbf{k} \times \mathbf{U} - \nabla_{\mathrm{H}} \cdot \left(\frac{\mathbf{UU}}{h}\right) - \frac{1}{2} g\nabla_{\mathrm{H}} h^2 \,. \quad (7.68)$$

B. Show that (7.67) and (7.68) may be combined to yield the vorticity equation

$$(\partial_t + \mathbf{u} \cdot \nabla_{\mathrm{H}}) P_{\mathrm{v}} = 0 \,, \quad (7.69)$$

where P_{v} is the potential vorticity defined by

$$P_{\mathrm{v}} = \frac{\zeta + f}{h} \,, \quad (7.70)$$

and $\zeta = \mathbf{k} \cdot \nabla_{\mathrm{H}} \times \mathbf{u}$ is the vorticity.

C. Assume that the motion is independent of y ($\partial_y = 0$). Show that under these circumstances (7.65) and (7.66) reduce to

$$\partial_t h = -u\partial_x h - h\partial_x u \,, \tag{7.71}$$

$$\partial_t u = fv - u\partial_x u - g\partial_x h \,, \tag{7.72}$$

$$\partial_t v = -fu - u\partial_x v \,, \tag{7.73}$$

and (7.67) and (7.68) reduce to

$$\partial_t h = -\partial_x U \,, \tag{7.74}$$

$$\partial_t U = fV - \partial_x \left(\frac{U^2}{h} + \frac{1}{2} g h^2 \right) , \tag{7.75}$$

$$\partial_t V = -fU - \partial_x \left(\frac{UV}{h} \right) . \tag{7.76}$$

Part 2

The aim here is to solve the shallow water equations using numerical methods for a limited domain $x \in \langle 0, L \rangle$. To this end boundary conditions at $x = 0, L$ and initial conditions at time $t = 0$ are needed. Assuming that the motion starts from rest and a perturbed geopotential height (or ocean surface), the initial conditions are

$$u = v = 0 \,, \quad h(x,t) = H_0 + A \exp\left[-\left(\frac{x - x_{\mathrm{m}}}{\sigma} \right)^2 \right] , \tag{7.77}$$

where $H_0 = 1000\,\mathrm{m}$, $A = 15\,\mathrm{m}$, $x_{\mathrm{m}} = L/2$ is the middle point of the domain, and σ is a measure of the width of the Gaussian bell.

D. How many boundary and initial conditions can be specified when solving the system (7.71)–(7.73)? Explain how you derive the number of conditions.
E. Construct a CTCS scheme for the two shallow water equations listed in (7.71)–(7.73) and (7.74)–(7.76). Describe in detail how you derive the finite difference approximation to the various terms. Explain your choices.
F. Under what conditions are the schemes stable? Describe in detail how you analyze the stability and consistency of the scheme. How long a time step Δt can be used? Explain your choice.
Hint: Neglect the nonlinear terms when performing the stability analysis.
G. Solve either the shallow water equations (7.71)–(7.73) or (7.74)–(7.76) using the CTCS scheme you have constructed for the domain $x \in \langle 0, L \rangle$. Let the grid length be $\Delta x = 100\,\mathrm{km}$, $L = 62\Delta x$, and $\sigma = 5\Delta x$. Use the FRS method to relax the interior solution to an exterior solution. Let the exterior solution be $(\hat{u}, \hat{v}, \hat{h}) = (0, 0, H_0)$ at the two open boundaries, and let the FRS zones be seven points wide. Let the relaxation parameter λ_j vary within the FRS zones, as specified in Table 7.1. Plot h at times $t = 0$, 1.5, 3.0, 4.5, 6.0, and 10.0 h. Discuss the solution. Try to make a movie spanning $t \in [0, 10]$ h. What kind of waves do you observe?
H. Compute the geostrophic component of the velocity

$$v_{\mathrm{g}} = \frac{g}{f} \partial_x h \,, \tag{7.78}$$

Table 7.1 Values of the relaxation parameter used in Part 2

j	λ_j	j
1	1.0	j_{max}
2	0.69	$j_{max} - 1$
3	0.44	$j_{max} - 2$
4	0.25	$j_{max} - 3$
5	0.11	$j_{max} - 4$
6	0.03	$j_{max} - 5$
7	0.0	$j_{max} - 6$

using the solution for h at $t = 6\,\text{h}$. Compare v_g and v at $t = 6.0\,\text{h}$, and describe and discuss what you observe. What do you think has happened?

I. Replace the initial condition for v in (7.77) by

$$v = \frac{g}{f}\partial_x h\ , \tag{7.79}$$

and repeat the computation in (H) using (7.79) as initial condition. Discuss the solution by comparing it with the solution obtained in (G).

J. Finally, solve the shallow water equations (7.71)–(7.73) using the quasi-Lagrangian method described in Sect. 6.6 and the weakly reflective open boundary condition described in Sect. 7.6. Compare the solution with the one obtained in (G).

References

Berenger J-P (1994) A perfectly matched layer for the absorption of electromagnetic waves. J Comput Phys 126:185–200

Blayo E, Debreu L (2005) Revisiting open boundary conditions from the point of view of characteristics. Ocean Model 9(3):234–252. https://doi.org/10.1016/j.ocemod.2004.07.001

Blayo E, Debreu L (2006) Nesting ocean models. In: Chassignet E, Verron J (eds) Ocean weather forecasting: an integrated view of oceanography. Springer, Berlin, pp 127–146. https://doi.org/10.1007/1-4020-4028-8_5

Chapman DC (1985) Numerical treatment of cross-shelf open boundaries in a barotropic coastal ocean model. J Phys Oceanogr 15:1060–1075

Charney JG, Fjørtoft R, von Neumann J (1950) Numerical integration of the barotropic vorticity equation. Tellus 2:237–254

Cooper C, Thompson JD (1989) Hurricane generated currents on the outer continental shelf. J Geophys Res 94:12, 513–12,539

Davies HC (1985) Limitation of some common lateral boundary schemes used in regional NWP models. Mon Weather Rev 111:1002–1012

Davies HC (1976) A lateral boundary formulation for multilevel prediction models. Q J R Meteorol Soc 102:405–418. https://doi.org/10.1002/qj.49710243210

Debreu L, Blayo E (2008) Two-way embedding algorithms: a review. Ocean Dyn 58:415–428. https://doi.org/10.1007/s10236-008-0150-9

Debreu L, Marchesiello P, Penven P, Cambon G (2012) Two-way nesting in split-explicit ocean models: algorithms, implementation and validation. Ocean Model 49–50:1–21

Engedahl H (1995) Use of the flow relaxation scheme in a three-dimensional baroclinic ocean model with realistic topography. Tellus 47A:365–382

Hedstrøm GW (1979) Nonreflecting boundary conditions for nonlinear hyperbolic systems. J Comput Phys 30:222–237

Martinsen EA, Engedahl H (1987) Implementation and testing of a lateral boundary scheme as an open boundary condition in a barotropic ocean model. Coast Eng 11:603–627

Mason E, Molemaker J, Shchepetkin AF, Colas F, McWilliams J (2010) Procedures for offline grid nesting in regional ocean models. Ocean Model 35:1–15. https://doi.org/10.1016/j.ocemod.2010.05.007

Müller M, Homleid M, Ivarsson K-I, Køltzow MA, Lindskog M, Andrae U, Aspelien T, Bjørge D, Dahlgren P, Kristiansen J, Randriamampianina R, Ridal M, Vignes O (2017) AROME-MetCoOp: a Nordic convective-scale operational weather prediction model. Weather Forecast 32:609–627. https://doi.org/10.1175/WAF-D-16-0099.1

Navon IM, Neta B, Hussaini MY (2004) A perfectly matched layer approach to the linearized shallow water equations models. Mon Weather Res 132(6):1369–1378

Orlanski I (1976) A simple boundary condition for unbounded hyperbolic flows. J Comput Phys 21:251–269

Palma ED, Matano RP (2000) On the implementation of open boundary conditions to a general circulation model: The 3-d case. J Geophys Res 105:8605–8627

Platzman GW (1954) The computational stability of boundary conditions in numerical integration of the vorticity equation. Arch Meteorol Geophys Bioklimatol 7:29–40

Røed LP, Cooper CK (1986) Open boundary conditions in numerical ocean models. In: O'Brien J (ed) Advanced physical oceanographic numerical modelling, series C: mathematical and physical sciences, vol 186. D. Reidel Publishing Co, Dordrecht, pp 411–436

Røed LP, Cooper CK (1987) A study of various open boundary conditions for wind-forced barotropic numerical ocean models. In: Nihoul JCJ, Jamart BM (eds) Three-dimensional models of marine and estuarine dynamics. Elsevier oceanography series, vol 45. Elsevier Science Publishers B.V., Amsterdam, pp 305–335

Røed LP, Smedstad OM (1984) Open boundary conditions for forced waves in a rotating fluid. SIAM J Sci Stat Comput 5:414–426

Schot SH (1992) Eighty years of Sommerfeld's radiation condition. Hist Math 19:385–401

Shi XB, Hackett B, Røed LP (1999) Documentation of DNMI's MICOM version, part 2: implementation of a flow relaxation scheme (FRS), Research Report 87, Norwegian Meteorological Institute [Available from Norwegian Meteorological Institute, Postboks 43 Blindern, N-0313 Oslo, Norway]

Shi XB, Røed LP, Hackett B (2001) Variability of the Denmark strait overflow: a numerical study. J Geophys Res 106:22 277–22 294

Sundström A, Elvius T (1979) Computational problems related to limited-area modeling. GARP publication series, vol 17. World Meteorological Organization, Geneva, p 11

Generalized Vertical Coordinates

8

So far, we have only used the Cartesian geopotential coordinate system consisting of three orthogonal spatial coordinates x, y, z. Relaxing the orthogonality between the vertical coordinate z and the two horizontal coordinates x, y can make it much easier to analyze phenomena in atmospheres and oceans, and devise compelling models of them. The development of such non-orthogonal coordinate systems remains at the forefront of research in numerical modeling. The purpose of this chapter is, therefore, to present the salient issues relating to these *generalized vertical coordinates*.

Most modern models employed in the meteorological and oceanographic community replace the natural geopotential vertical coordinate z with a new vertical coordinate. The reason for this is that the geopotential coordinate is quite cumbersome to work with in the presence of steep topography, such as mountains in the atmosphere and shelf breaks and sea mountains in the ocean.

In the late 1940s, at the dawn of numerical weather prediction, Sutcliffe (1947) and Eliassen (1949) were already suggesting the use of pressure surfaces to replace surfaces of geopotential height as the vertical coordinate, a method successfully tested by Charney and Phillips (1953) in a quasi-geostrophic model (Sect. 1.6). The pressure coordinate has several advantages over ordinary geopotential height models. For instance, it reduces the mass conservation equation to a diagnostic equation, and this eases the analysis of large scale (hydrostatic) motions.

The pressure coordinate, however, has certain computational disadvantages, in particular near mountains, since the ground is not a pressure surface. To remedy this, Phillips (1957) suggested using terrain-following surfaces as the vertical coordinate. This coordinate system, which has become quite popular in ocean models, is now commonly referred to as the σ-coordinate system (Blumberg and Mellor 1987; Haidvogel et al. 2008).

Other vertical coordinate systems have also been proposed. For instance, one early suggestion was to use surfaces of potential temperature as the vertical coordinate. This was successfully tested by Eliassen and Raustein (1968, 1970), using a simplified primitive equation model, and by Bleck (1973), using the potential vorticity equation. It was successfully extended to full three-dimensional

L. P. Røed, *Atmospheres and Oceans on Computers*,
Springer Textbooks in Earth Sciences, Geography and Environment,
https://doi.org/10.1007/978-3-319-93864-6_8

dynamic–thermodynamic atmospheric models by Shapiro (1974). In the ocean, Bleck and Smith (1990) explored so-called isopycnal models, in which surfaces of potential density are used as vertical coordinates, doing this in the context of primitive equation models. More recently, it has also become quite common to use hybrid coordinate models, in which the vertical coordinate changes from one system to another throughout the height, both in the atmosphere and in the ocean (e.g., Bleck 2002). Finally, it should be emphasized that the various vertical coordinate systems all have their advantages and disadvantages (cf. Griffies 2004, Chap. 6).

The presentation here begins by discussing how equations formulated in geopotential coordinates can be transformed to a new non-orthogonal vertical coordinate, say $s = s(x, y, z, t)$. We follow the derivation by Kasahara (1974). We then show how the governing equations of a hydrostatic non-Boussinesq fluid (Sect. 1.3) are affected by such a transformation. The chapter ends with an explicit example using the σ-coordinate transformation.

8.1 Transformation to a General Vertical Coordinate

In general, one coordinate system of independent variables, say (x, y, z, t), is transformed to another system, say (x', y', z', t'), by specifying how the independent variables in the transformed system depend on the independent variables of the original system. Only the vertical height coordinate z is replaced here. Accordingly, the transformation is defined by

$$x' = x\,, \quad y' = y\,, \quad z' = s = s(x, y, z, t)\,, \quad t' = t\,. \tag{8.1}$$

Note that only the geopotential height coordinate z has been replaced by a general vertical coordinate s. The horizontal coordinates are left unchanged, which makes the new system (x', y', s, t') non-orthogonal. To ensure that the transformation is unique, s has to be a monotonic function of the height z. Mathematically, this implies

$$\partial_z s \gtrless 0\,, \quad \partial_z s \neq 0\,, \tag{8.2}$$

that is, the gradient of s with respect to z does not change sign within a fluid column. This condition also ensures that the inverse transformation $z = z(x', y', s, t')$ exists.

Equation (8.1) leads immediately to

$$\partial_z x' = \partial_z y' = \partial_z t' = 0\,, \quad \partial_t x' = \partial_t y' = 0\,, \quad \partial_y x' = \partial_x y' = 0\,, \quad \partial_x t' = \partial_y t' = 0\,, \tag{8.3}$$

while

$$\partial_x x' = \partial_y y' = \partial_t t' = 1\,. \tag{8.4}$$

Similarly, it follows that

$$\partial_s x = \partial_s y = \partial_s t = 0\,, \quad \partial_{t'} x = \partial_{t'} y = 0\,, \quad \partial_{y'} x = \partial_{x'} y = 0\,, \quad \partial_{x'} t = \partial_{y'} t = 0\,, \tag{8.5}$$

while

$$\partial_{x'} x = \partial_{y'} y = \partial_{t'} t = 1\,. \tag{8.6}$$

Since s is monotonic with respect to z, $\partial_z s \neq 0$ and $\partial_s z \neq 0$. Moreover, if $s = z$, then $\partial_z s = \partial_s z = 1$.

Let $\psi = \psi(x, y, z, t) = \psi(x', y', s, t')$ denote any scalar. Then, the first property of the transformation is

$$\partial_z \psi = \partial_z s \partial_s \psi \ . \tag{8.7}$$

Differentiating ψ with respect to one of the independent variables in the new coordinate system, say t', leads to

$$\partial_{t'} \psi = \partial_t \psi \partial_{t'} t + \partial_x \psi \partial_{t'} x + \partial_y \psi \partial_{t'} y + \partial_z \psi \partial_{t'} z = \partial_t \psi + \partial_z s \partial_s \psi \partial_{t'} z \ , \tag{8.8}$$

where the last equals sign follows by using (8.3)–(8.7). Solving (8.8) for $\partial_t \psi$ then yields

$$\partial_t \psi = \partial_{t'} \psi - \partial_z s \partial_s \psi \partial_{t'} z \ . \tag{8.9}$$

Similarly,

$$\partial_x \psi = \partial_{x'} \psi - \partial_z s \partial_s \psi \partial_{x'} z \ , \quad \partial_y \psi = \partial_{y'} \psi - \partial_z s \partial_s \psi \partial_{y'} z \ . \tag{8.10}$$

Let the horizontal gradient of ψ in the new coordinate system be defined by

$$\nabla_s \psi = \mathbf{i} \partial_{x'} \psi + \mathbf{j} \partial_{y'} \psi \ . \tag{8.11}$$

Then, making use of (8.9) and (8.10), the horizontal gradient of any scalar transforms as

$$\nabla_{\mathrm{H}} \psi = \nabla_s \psi - \partial_z s \partial_s \psi \nabla_s z \ . \tag{8.12}$$

Similarly, it follows that the horizontal divergence of any vector, say $\mathbf{A}$, transforms as

$$\nabla_{\mathrm{H}} \cdot \mathbf{A} = \nabla_s \cdot \mathbf{A} - \partial_z s \partial_s \mathbf{A} \cdot \nabla_s z \ . \tag{8.13}$$

Note that all vectors project onto the horizontal geopotential surface. This is also true for the gradient (8.12). The metric term associated with the vertical gradient of the surface s in the geopotential coordinate system is therefore eliminated.

Since the individual derivative $\mathrm{D}/\mathrm{d}t$ is independent of the coordinate transformation, the individual derivative in the geopotential coordinate system is equal to the individual derivative in the new general non-orthogonal coordinate system. Hence,

$$\frac{\mathrm{D}\psi}{\mathrm{d}t} = \partial_t \psi + \mathbf{u} \cdot \nabla_{\mathrm{H}} \psi + w \partial_z \psi = \partial_{t'} \psi + \mathbf{u} \cdot \nabla_s \psi + \dot{s} \partial_s \psi \ , \tag{8.14}$$

where

$$\dot{s} = \frac{\mathrm{D}s}{\mathrm{d}t} = \partial_t s + \mathbf{u} \cdot \nabla_{\mathrm{H}} s + w \partial_z s \tag{8.15}$$

is the speed of the isosurfaces of s in the direction of the three-dimensional velocity $\mathbf{v} = \mathbf{u} + w\mathbf{k}$. Making use of (8.9)–(8.12) to replace the appropriate terms on the right-hand side of the first equality in (8.14), we obtain

$$\frac{\mathrm{D}\psi}{\mathrm{d}t} = \partial_{t'} \psi + \mathbf{u} \cdot \nabla_s \psi + (w - \partial_{t'} z - \mathbf{u} \cdot \nabla_s z) \partial_z s \partial_s \psi \ . \tag{8.16}$$

Equating this to the individual derivative expressed in the new coordinate system, that is, the expression on the right-hand side of the second equality in (8.14), we then have

$$\dot{s} = (w - \partial_{t'} z - \mathbf{u} \cdot \nabla_s z)\partial_z s \equiv \omega \partial_z s \ , \tag{8.17}$$

where the identity in (8.17) defines the velocity ω by

$$\omega = w - (\partial_{t'} z + \mathbf{u} \cdot \nabla_s z) \ . \tag{8.18}$$

As revealed by (8.18), the difference between ω and w is associated with the speed of the surface z in the transformed coordinate system. If $s = \rho$, then the kinematic boundary condition requires $w = \partial_{t'} z + \mathbf{u} \cdot \nabla_s z$, in which case $\omega = 0$. Since a material surface is a surface that consists of the same fluid particles at all times, that is, no particles are transported through the surface, this is as expected. Thus, ω may be interpreted as the speed of the fluid particles moving through the surface s. As a check, let $s = z$. Then (8.18) leads to $\omega = w$, as expected.

8.2 Transformation of the Governing Equations

To give insight into the above, the general transformation formulas of the previous section are applied to the equations governing the motion of a non-Boussinesq hydrostatic fluid.

Hydrostatic Equation

The hydrostatic equation is

$$\partial_z p + \rho g = 0 \ . \tag{8.19}$$

Using the transformation formulas of the previous section yields

$$\partial_s p + \rho g \partial_s z = 0 \ . \tag{8.20}$$

Solving it with respect to the metric factors $\partial_s z$ and $\partial_z s$ leads to

$$\partial_s z = -\frac{\partial_s p}{\rho g} \ , \quad \partial_z s = -\frac{\rho g}{\partial_s p} \ . \tag{8.21}$$

In particular, if $s = p$, then $\partial_s p = 1$, and (8.20) reduces to

$$\partial_p z = -\frac{1}{\rho g} \ . \tag{8.22}$$

Mass Conservation

The continuity equation is

$$\partial_t \rho + \nabla \cdot (\mathbf{v}\rho) = 0 \ , \tag{8.23}$$

which may be rewritten as

$$\frac{1}{\rho}\frac{\mathrm{D}\rho}{\mathrm{d}t} + \nabla_{\mathrm{H}} \cdot \mathbf{u} + \partial_z w = 0 \ . \tag{8.24}$$

Making use of the transformation formulas of Sect. 8.1 to replace each of the terms then leads to

$$\frac{1}{\rho}\frac{\mathrm{D}\rho}{\mathrm{d}t} + \nabla_{\mathrm{H}} \cdot \mathbf{u} + \partial_z w = -\rho\big(\partial_{t'}\alpha + \mathbf{u} \cdot \nabla_s \alpha + \dot{s}\partial_s \alpha\big) \tag{8.25}$$
$$+\partial_z s\big[\partial_{t'}(\partial_s z) + \nabla_s \cdot (\mathbf{u}\partial_s z) + \partial_s(\dot{s}\partial_s z)\big]\,,$$

where $\alpha = 1/\rho$. To arrive at this result, (8.18) is solved for w to replace $\partial_s w$. Using (8.21) to replace the metric term $\partial_s z$, (8.25) may be further developed to yield

$$\frac{1}{\rho}\frac{\mathrm{D}\rho}{\mathrm{d}t} + \nabla_{\mathrm{H}} \cdot \mathbf{u} + \partial_z w = (\partial_s p)^{-1}\big[\partial_{t'}(\partial_s p) + \nabla_s \cdot (\mathbf{u}\partial_s p) + \partial_s(\dot{s}\partial_s p)\big]\,. \tag{8.26}$$

Hence, the transformed continuity equation finally takes the form

$$\partial_{t'}(\partial_s p) + \nabla_s \cdot (\mathbf{u}\partial_s p) + \partial_s(\dot{s}\partial_s p) = 0\,. \tag{8.27}$$

Once again, if $s = p$, then (8.27) reduces to

$$\nabla_p \cdot \mathbf{u} = \partial_p(\rho g \omega)\,, \tag{8.28}$$

that is, a diagnostic equation, as claimed in the introduction to this chapter.

Tracer Equation

Applying a similar procedure to the tracer equation (1.13) leads immediately to

$$\partial_{t'} C + \mathbf{u} \cdot \nabla_s C + \dot{s}\partial_s C = F_C + S_C\,, \tag{8.29}$$

where the right-hand side represents the transformed fluxes and source terms.

Momentum Equation

By slightly rewriting the horizontal component of the momentum equation for a non-Boussinesq hydrostatic fluid, as in (1.12), it takes the form

$$\frac{\mathrm{D}\mathbf{u}}{\mathrm{d}t} + f\mathbf{k} \times \mathbf{u} = -\alpha\nabla_{\mathrm{H}} p + \alpha\partial_z \boldsymbol{\tau} + \nabla_{\mathrm{H}} \cdot \mathscr{F}_{\mathrm{M}}^{\mathrm{H}}\,, \tag{8.30}$$

where $\boldsymbol{\tau}$ is the vertical mixing or flux vector, sometimes referred to as the vertical shear stress. The transformation of this equation is a bit more complicated, so it is treated term by term.

Consider first the pressure term. For a non-Boussinesq fluid, this is

$$\alpha\nabla_{\mathrm{H}} p = \alpha\nabla_s p + g\nabla_s z = \nabla_s M - p\nabla_s \alpha\,, \tag{8.31}$$

where

$$M = \alpha p + gz \tag{8.32}$$

is the *Montgomery potential*. Choosing $s = \rho$, the last term in the second equality of (8.31) vanishes, since $\nabla_s \alpha = 0$ under these circumstances. Thus, the Montgomery potential becomes a true potential and is a streamfunction for the geostrophic velocity. For any other choice of s, however, the last term in (8.31) must be retained. The Montgomery potential appears because all vectors are projected onto the horizontal surface (with respect to gravity), even though all gradients are evaluated in the transformed x', y', s, t' system.

Next, consider the vertical shear stress term. Applying (8.7) and (8.21) yields

$$\alpha \partial_z \boldsymbol{\tau} = \alpha \partial_z s \partial_s \boldsymbol{\tau} = -g \frac{\partial_s \boldsymbol{\tau}}{\partial_s p} = \partial_p \boldsymbol{\tau} \ . \tag{8.33}$$

Using (8.16), the acceleration term takes the form

$$\frac{\mathbf{Du}}{\mathrm{d}t} = \partial_{t'} \mathbf{u} + \mathbf{u} \cdot \nabla_s \mathbf{u} + \dot{s} \partial_s \mathbf{u} \ . \tag{8.34}$$

The second term may be further developed to yield

$$\mathbf{u} \cdot \nabla_s \mathbf{u} = \nabla_s \left(\frac{1}{2} \mathbf{u}^2 \right) + \zeta \mathbf{k} \times \mathbf{u} \ , \tag{8.35}$$

where $\zeta = \mathbf{k} \cdot \nabla_s \times \mathbf{u}$ is the relative vorticity in the new coordinate system. The momentum equation then finally takes the form

$$\partial_{t'} \mathbf{u} + \nabla_s \left(\frac{1}{2} \mathbf{u}^2 \right) + (\zeta + f) \mathbf{k} \times \mathbf{u} + \dot{s} \partial_s \mathbf{u} = -\nabla_s M + p \nabla_s \alpha - g \partial_p \boldsymbol{\tau} + \nabla_s \cdot \mathscr{F}_{\mathrm{M}}^{\mathrm{H}} \ . \tag{8.36}$$

It may also be written in flux form. To this end, the second and third terms on the left-hand side of (8.36) are recombined using (8.35). Next, multiplying (8.36) by $\partial_s p$ and making use of the continuity equation in the form (8.27), we obtain

$$\begin{aligned} &\partial_{t'} (\mathbf{u} \partial_s p) + \nabla_s \cdot (\mathbf{u}\mathbf{u} \partial_s p) + f \mathbf{k} \times \mathbf{u} \partial_s p + \partial_s (\dot{s} \mathbf{u} \partial_s p) \\ &\qquad = -\partial_s p \left(\nabla_s M + p \nabla_s \alpha \right) - g \partial_s \boldsymbol{\tau} + \partial_s p \nabla_s \cdot \mathscr{F}_{\mathrm{M}}^{\mathrm{H}} \ . \end{aligned} \tag{8.37}$$

If $s = p$, then (8.37) becomes

$$\partial_{t'} \mathbf{u} + \nabla_p \cdot (\mathbf{u}\mathbf{u}) + f \mathbf{k} \times \mathbf{u} + \partial_p (\dot{p} \mathbf{u}) = -\nabla_p M + p \nabla_p \alpha - g \partial_p \boldsymbol{\tau} + \nabla_p \cdot \mathscr{F}_{\mathrm{M}}^{\mathrm{H}} \ . \tag{8.38}$$

As mentioned earlier when treating the diffusion problem (Chap. 4), this is a good point to emphasize that the mixing term or "diffusion" term is generally added to prevent the numerical model from blowing up. Hence, its exact transformation is of secondary importance in a numerical sense, and is not treated here.

8.3 Terrain-Following Coordinates

As an example, let $s = \sigma$, where σ is the so-called σ-coordinate defined by

$$\sigma = \frac{z - \eta}{D} \quad \Longrightarrow \quad z = \sigma D + \eta \ , \tag{8.39}$$

with $D = H + \eta$ the total depth and η the deviation of a reference surface away from its equilibrium height (or depth) $z = H$. Models using σ, or some modified form of it, are commonly referred to as *terrain-following coordinate models*. Among the ocean models, examples are versions of POM (Blumberg and Mellor 1987; Engedahl 1995) and various versions of ROMS (Shchepetkin and McWilliams 2005; Haidvogel et al. 2008). It is also fairly popular in some form or another in numerical weather prediction models (Phillips 1957; Kasahara 1974; Bourke 1974), and more recently in the ETA model (Black 1994) and the global atmospheric model NICAM (Satoh et al. 2008).

First, using (8.21), the metric factor $\partial_z s$ and $\partial_s z$ take the form

$$\partial_z s = \partial_z \sigma = \frac{1}{D} , \quad \partial_s z = \partial_\sigma z = D . \tag{8.40}$$

Thus, the hydrostatic equation is

$$\partial_\sigma p = -\rho g D . \tag{8.41}$$

Furthermore, (8.39) implies

$$\omega = w - \sigma \partial_{t'} D - \partial_{t'} \eta - \sigma \mathbf{u} \cdot \nabla_s D - \mathbf{u} \cdot \nabla_s \eta , \tag{8.42}$$

where ω is the "vertical" velocity through the σ surfaces. The mass conservation equation (8.27) then takes the form

$$\partial_{t'}(\rho D) + \nabla_\sigma \cdot (\rho D \mathbf{u}) + \partial_\sigma(\dot{\sigma} \rho D) = 0 . \tag{8.43}$$

Separating the effect of the density leads to

$$\frac{\mathrm{D}\rho}{\mathrm{d}t} + \frac{\rho}{D}\big[\partial_{t'} D + \nabla_\sigma \cdot (D\mathbf{u}) + \partial_\sigma(\dot{\sigma} D)\big] = 0 . \tag{8.44}$$

Equation (8.17) then implies $\dot{\sigma} = \omega \partial_z \sigma = \omega D^{-1}$. Furthermore, $\partial_{t'} D = \partial_{t'} \eta$. Substituting these expressions into (8.44) and invoking the Boussinesq approximation (1.16) ($\mathrm{D}\rho/\mathrm{d}t = 0$), the continuity equation for a Boussinesq fluid formulated in terrain-following coordinates takes the form

$$\partial_{t'} \eta + \nabla_\sigma \cdot (D\mathbf{u}) + \partial_\sigma \omega = 0 . \tag{8.45}$$

The remaining equations are transformed into σ-coordinates in a similar fashion. For instance, using (8.41), the flux form (8.37) of the momentum equation becomes

$$\begin{aligned} &\partial_{t'}(D\mathbf{u}) + \nabla_\sigma \cdot (D\mathbf{u}\mathbf{u}) + f\mathbf{k} \times D\mathbf{u} + \partial_\sigma(\omega \mathbf{u}) \\ &\qquad = -D\left(\nabla_\sigma M + p \nabla_\sigma \alpha\right) - \frac{1}{\rho_0} \partial_\sigma \boldsymbol{\tau} + D \nabla_\sigma \cdot \mathscr{F}_{\mathrm{M}}^{\mathrm{H}} . \end{aligned} \tag{8.46}$$

8.4 Summary and Remarks

In this chapter, the reader is shown how to transform the governing equation from an orthogonal Cartesian coordinate system to a non-orthogonal coordinate system in which the vertical geopotential coordinate z is replaced by a general vertical coordinate $s = s(x, y, z, t)$. This involves some mathematics, but it is hoped that the reader will take the time necessary to grasp the main idea, since most of today's models of the atmosphere and oceans exploit this kind of transformation.

After some introductory remarks, Sect. 8.1 shows in general terms how to transform from a Cartesian coordinate system to a new one. In Sect. 8.2, these "rules" are applied to the governing equation of a non-Boussinesq hydrostatic fluid to get insight into the way the equations transform. This more or less leads to the equations used in isopycnal models. Finally, Sect. 8.3 provides an example of how to transform the governing equations into a terrain-following σ-coordinate system. This transformed system of equations is fairly common in ocean models, such as POM and ROMS, but also to some extent in atmospheric models, like the ETA model.

References

Black TL (1994) The new NMC mesoscale Eta model: description and forecast examples. Weather Forecast 9(2):265–278. https://doi.org/10.1175/1520-0434(1994)009<0265:TNNMEM>2.0.CO;2

Bleck R (1973) Numerical forecasting experiments based on conservation of potential vorticity on isentropic surfaces. J Appl Meteorol 12:737–752

Bleck R (2002) An oceanic general circulation model framed in hybrid isopycnic-Cartesian coordinates. Ocean Model. 4(1):55–88

Bleck R, Smith L (1990) A wind-driven isopycnic coordinate model of the north and equatorial Atlantic ocean. 1. Model development and supporting experiments. J Geophys Res 95C:3273–3285

Blumberg A, Mellor G (1987) A description of a three-dimensional coastal ocean circulation model. Three-dimensional coastal ocean models. In: Heaps N (ed) Coastal and estuarine sciences, vol 4. American Geophysical Union, Washington, pp 1–16

Bourke W (1974) A multilevel spectral model. I. Formulation and hemispheric integrations. Mon Weather Rev 102:687–701

Charney JG, Phillips NA (1953) Numerical integration of the quasi-geostrophic equations for barotropic and simple baroclinic flows. J Meteorol 10:71–99

Eliassen A (1949) The quasi-static equations of motion with pressure as independent variable. Geofys Publ 17(3):44 pp

Eliassen A, Raustein E (1968) A numerical integration experiment with a model atmosphere based on isentropic coordinates. Meteorol Ann 5:45–63

Eliassen A, Raustein E (1970) A numerical integration experiment with a six-level atmospheric model with isentropic information surface. Meteorol Ann 5:429–449

Engedahl H (1995) Use of the flow relaxation scheme in a three-dimensional baroclinic ocean model with realistic topography. Tellus 47A:365–382

Griffies SM (2004) Fundamentals of ocean climate models. Princeton University Press, Princeton. ISBN 0-691-11892-2

Haidvogel DB, Arango H, Budgell PW, Cornuelle BD, Curchitser E, Lorenzo ED, Fennel K, Geyer WR, Hermann AJ, Lanerolle L, Levin J, McWilliams JC, Miller AJ, Moore AM, Powell TM, Shchepetkin AF, Sherwood CR, Signell RP, Warner JC, Wilkin J (2008) Ocean forecasting in terrain-following coordinates: formulation and skill assessment of the regional ocean modeling system. J Comput Phys 227(7):3595–3624. https://doi.org/10.1016/j.jcp.2007.06.016

Kasahara A (1974) Various vertical coordinate systems used for numerical weather prediction. Mon Weather Rev 102:509–522

Phillips NA (1957) A coordinate system having some special advantages for numerical forecasting. J Meteorol 14:184–185

Satoh M, Matsuno T, Tomita H, Miura H, Nasuno T, Iga S (2008) Nonhydrostatic icosahedral atmospheric model (NICAM) for global cloud resolving simulations. J Comput Phys 227:3486–3514

Shapiro MA (1974) The use of isentropic coordinates in the formulation of objective analysis and numerical prediction models. Atmosphere 12:10–17
Shchepetkin AF, McWilliams JC (2005) The regional ocean modeling system (ROMS): a split-explicit, free-surface, topography-following coordinate ocean model. Ocean Model 9:347–404
Sutcliffe RC (1947) A contribution to the problem of development. Q J R Meteorol Soc 73:370–383

Two-Dimensional Problems

9

This chapter investigates the effect of including more than one dimension in space. In particular, it discusses the impact on numerical stability and the stability criterion. Extension to three dimensions is then straightforward.

9.1 Diffusion Equation

We begin with the diffusion equation. In two space dimensions, it takes the form

$$\partial_t\theta = \nabla_{\rm H} \cdot (\mathscr{K}_{\rm H} \cdot \nabla_{\rm H}\theta) \ , \tag{9.1}$$

where $\mathscr{K}_{\rm H} = \kappa_{ij}\mathbf{ij}$ is a 2×2 matrix containing the diffusivity coefficients. Assuming that $\mathscr{K}_{\rm H}$ is a diagonal matrix with $\kappa_{11} = \kappa_{22} = \kappa$, we have

$$\partial_t\theta = \kappa(\partial_x^2\theta + \partial_y^2\theta) \ . \tag{9.2}$$

Equation (9.2) refers to two dimensions, so we use the notation $\theta(x_j, y_k, t^n) = \theta^n_{jk}$ (see Sect. 2.8). Since a forward-in-time centered-in-space (FTCS) scheme worked well in the one-dimensional case (Chap. 4), the same scheme is investigated here. Thus, the two second-order derivatives are replaced by

$$\left[\partial_x^2\theta\right]^n_{jk} = \frac{\theta^n_{j+1k} - 2\theta^n_{jk} + \theta^n_{j-1k}}{\Delta x^2} \ , \qquad \left[\partial_y^2\theta\right]^n_{jk} = \frac{\theta^n_{jk+1} - 2\theta^n_{jk} + \theta^n_{jk-1}}{\Delta y^2} \ , \tag{9.3}$$

while the first-order derivative with respect to to time is replaced by

$$[\partial_t\theta]^n_{jk} = \frac{\theta^{n+1}_{jk} - \theta^n_{jk}}{\Delta t} \ , \tag{9.4}$$

where Δx, Δy are the space increments along the x, y axis, respectively, and Δt is the time step. Thus, (9.2) takes the finite difference form

$$\theta^{n+1}_{jk} = \theta^n_{jk} + \kappa\frac{\Delta t}{\Delta x^2}\left(\theta^n_{j-1k} - 2\theta^n_{jk} + \theta^n_{j+1k}\right) + \kappa\frac{\Delta t}{\Delta y^2}\left(\theta^n_{jk-1} - 2\theta^n_{jk} + \theta^n_{jk+1}\right) \ . \tag{9.5}$$

L. P. Røed, *Atmospheres and Oceans on Computers*,
Springer Textbooks in Earth Sciences, Geography and Environment,
https://doi.org/10.1007/978-3-319-93864-6_9

To investigate the stability of the scheme (9.5), we use the von Neumann method. The first step is, therefore, to replace θ^n_{jk} by its individual Fourier components. Since waves may now propagate in a random direction in the horizontal x, y space, the Fourier component must include waves propagating in the x direction as well as in the y direction. Thus, the discrete Fourier component takes the form

$$\theta^n_{jk} = \Theta_n e^{i\alpha j\Delta x} e^{i\beta k\Delta y} , \tag{9.6}$$

where α and β are wavenumbers in the x and y directions, respectively. The next step is to insert the discrete Fourier component into (9.5). After dividing through by $e^{i\alpha j\Delta x} e^{i\beta k\Delta y}$, this leads to

$$\Theta_{n+1} = \left[1 + \kappa\frac{\Delta t}{\Delta x^2}\left(e^{-i\alpha\Delta x} - 2 + e^{i\alpha\Delta x}\right) + \kappa\frac{\Delta t}{\Delta y^2}\left(e^{-i\beta\Delta y} - 2 + e^{i\beta\Delta y}\right)\right]\Theta_n . \tag{9.7}$$

Defining the growth factor by $G = \Theta_{n+1}/\Theta_n$, we obtain

$$G = 1 - 2\kappa\frac{\Delta t}{\Delta x^2}(1 - \cos\alpha\Delta x) - 2\kappa\frac{\Delta t}{\Delta y^2}(1 - \cos\beta\Delta y) , \tag{9.8}$$

where use has been made of the formula $e^{i\phi} + e^{-i\phi} = 2\cos\phi$. Comparing this expression to (4.35), that is, the similar expression for the growth function derived in the one-dimensional case, the difference is an extra term caused by the two-dimensionality of (9.2). Applying the von Neumann criterion for numerical stability (4.32), viz., $|G| \le 1$, this condition becomes

$$-1 \le 1 - 2\kappa\frac{\Delta t}{\Delta x^2}(1 - \cos\alpha\Delta x) - 2\kappa\frac{\Delta t}{\Delta y^2}(1 - \cos\beta\Delta y) \le 1 . \tag{9.9}$$

As in the one-dimensional case, the right-hand inequality is trivially satisfied. The left-hand inequality is satisfied if

$$\kappa\frac{\Delta t}{\Delta x^2}(1 - \cos\alpha\Delta x) + \kappa\frac{\Delta t}{\Delta y^2}(1 - \cos\beta\Delta y) \le 1 . \tag{9.10}$$

Since the left-hand side of (9.10) is maximum when $\cos\alpha\Delta x = \cos\beta\Delta y = -1$, the stability criterion is satisfied for all possible values of the wavenumbers α and β if

$$\kappa\frac{\Delta t}{\Delta x^2} + \kappa\frac{\Delta t}{\Delta y^2} \le \frac{1}{2} . \tag{9.11}$$

In terms of the time step, this criterion stipulates that

$$\Delta t \le \frac{1}{2\kappa}\frac{\Delta x^2\Delta y^2}{\Delta x^2 + \Delta y^2} . \tag{9.12}$$

In the special case when the grid is regular (a square grid), that is, $\Delta x = \Delta y = \Delta s$, (9.12) yields

$$\Delta t \le \frac{\Delta s^2}{4\kappa} . \tag{9.13}$$

Comparing (9.13) with the one-dimensional case (4.38), the maximum allowed time step is reduced by a factor of two. Thus, the inclusion of one more dimension in space has made the criterion for numerical stability stricter. It turns out, as will be exemplified, that this is a general result applying to all problems extended to two or three spatial dimensions.

9.2 Advection Equation

The two-dimensional version of the advection equation is

$$\partial_t \theta + \mathbf{u} \cdot \nabla_H \theta = 0 \,. \tag{9.14}$$

Assuming the horizontal velocity component $\mathbf{u}$ to be constant in time and space, this implies that

$$\partial_t \theta + u_0 \partial_x \theta + v_0 \partial_y \theta = 0 \,, \tag{9.15}$$

where u_0, v_0 are the velocity components in the x, y direction, respectively. To solve (9.15) numerically, we use the well-known CTCS (leap-frog) scheme, accurate to second order, which worked for the one-dimensional case. Thus, the finite difference version of (9.15) takes the form

$$\frac{\theta_{jk}^{n+1} - \theta_{jk}^{n-1}}{2\Delta t} + u_0 \frac{\theta_{j+1k}^{n} - \theta_{j-1k}^{n}}{2\Delta x} + v_0 \frac{\theta_{jk+1}^{n} - \theta_{jk-1}^{n}}{2\Delta y} = 0 \,, \tag{9.16}$$

or

$$\theta_{jk}^{n+1} = \theta_{jk}^{n-1} + u_0 \frac{\Delta t}{\Delta x} \left(\theta_{j+1k}^{n} - \theta_{j-1k}^{n} \right) + v_0 \frac{\Delta t}{\Delta y} \left(\theta_{jk+1}^{n} - \theta_{jk-1}^{n} \right) . \tag{9.17}$$

Once again, the von Neumann method is used to investigate numerical stability. Performing the first step, θ_{ij}^n in (9.17) is replaced by its discrete Fourier component (9.6), leading to

$$\Theta_{n+1} = \Theta_{n-1} - 2\mathrm{i} \left(u_0 \frac{\Delta t}{\Delta x} \sin \alpha \Delta x + v_0 \frac{\Delta t}{\Delta y} \sin \beta \Delta y \right) \Theta_n \,. \tag{9.18}$$

Defining the growth factor by $G = \Theta_{n+1}/\Theta_n$, we obtain

$$G^2 + 2\mathrm{i}\lambda G - 1 = 0 \,, \tag{9.19}$$

which is quite similar to the one derived for the one-dimensional case, except that the factor

$$\lambda = u_0 \frac{\Delta t}{\Delta x} \sin \alpha \Delta x + v_0 \frac{\Delta t}{\Delta y} \sin \beta \Delta y \tag{9.20}$$

contains an extra term due to the two-dimensionality. The equation for the growth function, therefore, has the two complex solutions

$$G_{1,2} = \mathrm{i}\lambda \pm \sqrt{1 - \lambda^2} \,, \tag{9.21}$$

as long as the radical is positive. Since $|G_{1,2}| = 1$, the two-dimensional scheme is, therefore, neutrally stable under the condition that $\lambda^2 \leq 1$ (no energy dissipation). The only difference from the one-dimensional problem is the extra term in λ. Requiring the radical to be a positive-definite quantity leads to

$$\left| u_0 \frac{\Delta t}{\Delta x} \sin \alpha \Delta x + v_0 \frac{\Delta t}{\Delta y} \sin \beta \Delta y \right| \leq 1 \,. \tag{9.22}$$

This criterion must be satisfied for all possible choices of wavenumbers α and β. Since the maximum value $\sin \alpha \Delta x$ and $\sin \beta \Delta y$ may attain is one, the only way to ensure that (9.22) is satisfied for all possible choices of wavenumbers α and β is to require

$$|u_0|\frac{\Delta t}{\Delta x} + |v_0|\frac{\Delta t}{\Delta y} \leq 1 \,. \tag{9.23}$$

In terms of time steps, this entails

$$\Delta t \leq \frac{\Delta x \Delta y}{|u_0|\Delta y + |v_0|\Delta x} \,. \tag{9.24}$$

To compare with the one-dimensional case, let $\Delta x = \Delta y = \Delta s$ and $u_0 = v_0 = c_0$. Then (9.24) takes the form

$$\Delta t \leq \frac{\Delta s}{2|c_0|} \,. \tag{9.25}$$

Thus, as in the diffusion problem, the increase in dimensions from one to two leads to a stricter stability condition.

9.3 Shallow Water Equations

Analytic Solutions

The two-dimensional linear rotating shallow water equations are given by (6.13) and (6.14). Before making finite difference versions of them, it is first worth analyzing the various wave motions supported by the two-dimensional problem. To this end, it is convenient to restate the two-dimensional shallow water equations in scalar form:

$$\partial_t u + \overline{u}\partial_x u + \overline{v}\partial_y u - f v + \partial_x \phi = 0 \,, \tag{9.26}$$

$$\partial_t v + \overline{u}\partial_x v + \overline{v}\partial_y v + f u + \partial_y \phi = 0 \,, \tag{9.27}$$

$$\partial_t \phi + \overline{\phi}(\partial_x u + \partial_y v) = 0 \,. \tag{9.28}$$

To analyze possible wave motions, all variables are assumed to be a sum of two-dimensional waves of frequency ω. We thus seek solutions in the form

$$\mathbf{X} = \mathbf{X}_0 \mathrm{e}^{-\mathrm{i}\omega t} \mathrm{e}^{\mathrm{i}(\alpha x + \beta y)} \,, \tag{9.29}$$

where α and β are wavenumbers in the x- and y-directions, respectively, while

$$\mathbf{X} = \begin{bmatrix} u \\ v \\ \phi \end{bmatrix} \tag{9.30}$$

contains the dependent variables and

$$\mathbf{X}_0 = \begin{bmatrix} u_0 \\ v_0 \\ \phi_0 \end{bmatrix} \tag{9.31}$$

is the amplitude. It is assumed that the two wave numbers are real quantities. Thus, the frequency ω may be complex.

Inserting (9.29) into the linearized equations (9.26)–(9.28) leads to

$$(\alpha\overline{u} + \beta\overline{v} - \omega)u_0 - \mathrm{i}fv_0 + \alpha\phi_0 = 0\,, \tag{9.32}$$

$$-\mathrm{i}fu_0 + \mathrm{i}(\alpha\overline{u} + \beta\overline{v} - \omega)v_0 - \beta\phi_0 = 0\,, \tag{9.33}$$

$$\overline{\phi}\alpha u_0 + \overline{\phi}\beta v_0 + (\alpha\overline{u} + \beta\overline{v} - \omega)\phi_0 = 0\,, \tag{9.34}$$

which, in turn, may be formulated as the homogeneous linear equation

$$\mathscr{A}\cdot\mathbf{x} = 0\,, \tag{9.35}$$

where the tensor $\mathscr{A}$ is

$$\mathscr{A} = \begin{bmatrix} (\alpha\overline{u} + \beta\overline{v} - \omega) & -f & \mathrm{i}\alpha \\ f & \mathrm{i}(\alpha\overline{u} + \beta\overline{v} - \omega) & \mathrm{i}\beta \\ \mathrm{i}\alpha\overline{\phi} & \mathrm{i}\beta\overline{\phi} & \mathrm{i}(\alpha\overline{u} + \beta\overline{v} - \omega) \end{bmatrix}. \tag{9.36}$$

For nontrivial solutions to exist, the determinant of the tensor $\mathscr{A}$ must be zero, implying

$$(\alpha\overline{u} + \beta\overline{v} - \omega)\left[(\alpha\overline{u} + \beta\overline{v} - \omega)^2 - \overline{\phi}(\alpha^2 + \beta^2) - f^2\right] = 0\,. \tag{9.37}$$

As in the one-dimensional case, there are three solutions for the frequency ω, namely,

$$\omega_1 = \overline{u}\alpha + \overline{v}\beta\,, \tag{9.38}$$

$$\omega_2 = \overline{u}\alpha + \overline{v}\beta + \sqrt{c_0^2(\alpha^2 + \beta^2) + f^2}\,, \tag{9.39}$$

$$\omega_3 = \overline{u}\alpha + \overline{v}\beta - \sqrt{c_0^2(\alpha^2 + \beta^2) + f^2}\,, \tag{9.40}$$

where

$$c_0 = \sqrt{\overline{\phi}} = \sqrt{gH} \tag{9.41}$$

is the wave speed of gravity waves. The first solution is simply the geostrophic balance, as displayed in (1.40) with $\phi = gh$, i.e.,

$$\mathbf{u} = \frac{1}{f}\mathbf{k}\times\nabla_{\mathrm{H}}\phi\,. \tag{9.42}$$

This interpretation is derived by substituting ω_1 from (9.38) into (9.32) and (9.33), respectively, i.e.,

$$0 - fv + \mathrm{i}\alpha\phi = 0\,, \tag{9.43}$$

$$fu + 0 + \mathrm{i}\alpha\phi = 0\,, \tag{9.44}$$

which leads to

$$u = -\frac{1}{f}\mathrm{i}\beta\phi \quad\text{and}\quad v = \frac{1}{f}\mathrm{i}\alpha\phi \quad\Longrightarrow\quad u = -\frac{1}{f}\partial_y\phi \quad\text{and}\quad v = \frac{1}{f}\partial_x\phi\,. \tag{9.45}$$

The last implication follows by using the Fourier solution backwards and shows that the geostrophic balance (9.42) is recovered.

The two other solutions represented by

$$\pm\sqrt{\overline{\phi}(\alpha^2 + \beta^2) + f^2}$$

are two-dimensional inertia–gravity waves. The inertia part is associated with frequencies ω proportional to f, while gravity waves are associated with frequencies

$$\pm c_0\sqrt{(\alpha^2+\beta^2)}\,.$$

To construct the analytic solution to (9.26)–(9.28) for any given initial and boundary conditions, the solution is expanded in a two-dimensional Fourier series

$$\mathbf{X}=\sum_{\alpha=-\infty}^{\infty}\sum_{\beta=-\infty}^{\infty}\mathbf{X}_0(\alpha,\beta)\mathrm{e}^{\mathrm{i}(\alpha x+\beta y-\omega t)}\,, \tag{9.46}$$

where $\mathbf{X}_0(\alpha,\beta)$ contains the amplitudes of each Fourier component at the initial time.

Finite Difference Form

To solve (9.26)–(9.28) by numerical means, we use the CTCS (leap-frog) scheme. Hence, all derivatives are replaced by centered FDAs and we obtain

$$\frac{u_{jk}^{n+1}-u_{jk}^{n-1}}{2\Delta t}+\overline{u}\frac{u_{j+1k}^{n}-u_{j-1k}^{n}}{2\Delta x}+\overline{v}\frac{u_{jk+1}^{n}-u_{jk-1}^{n}}{2\Delta y}-fv_{jk}^{n}=-\frac{\phi_{j+1k}^{n}-\phi_{j-1k}^{n}}{2\Delta x}\,, \tag{9.47}$$

$$\frac{v_{jk}^{n+1}-v_{jk}^{n-1}}{2\Delta t}+\overline{u}\frac{v_{j+1k}^{n}-v_{j-1k}^{n}}{2\Delta x}+\overline{v}\frac{v_{jk+1}^{n}-v_{jk-1}^{n}}{2\Delta y}+fu_{jk}^{n}=-\frac{\phi_{jk+1}^{n}-\phi_{jk-1}^{n}}{2\Delta y}\,, \tag{9.48}$$

$$\frac{\phi_{jk}^{n+1}-\phi_{jk}^{n-1}}{2\Delta t}+c_0^2\frac{u_{j+1k}^{n}-u_{j-1k}^{n}}{2\Delta x}+c_0^2\frac{v_{jk+1}^{n}-v_{jk-1}^{n}}{2\Delta y}=0\,. \tag{9.49}$$

Building on the experience gained by studying the stability of the one-dimensional shallow water equations, it is the part of the solution that contains free waves that is of special interest. Accordingly, momentum and volume sources, if any, are neglected. In addition, any steady state solution upon which the waves may ride is also neglected ($\overline{u}=\overline{v}=0$). The above equations then take the form

$$u_{jk}^{n+1}-u_{jk}^{n-1}=2\Delta t f v_{jk}^{n}-\frac{\Delta t}{\Delta x}\left(\phi_{j+1k}^{n}-\phi_{j-1k}^{n}\right)\,, \tag{9.50}$$

$$v_{jk}^{n+1}-v_{jk}^{n-1}=-2\Delta t f u_{jk}^{n}-\frac{\Delta t}{\Delta y}\left(\phi_{jk+1}^{n}-\phi_{jk-1}^{n}\right)\,, \tag{9.51}$$

$$\phi_{jk}^{n+1}-\phi_{jk}^{n-1}=-c_0^2\frac{\Delta t}{\Delta x}\left(u_{j+1k}^{n}-u_{j-1k}^{n}\right)-c_0^2\frac{\Delta t}{\Delta y}\left(v_{jk+1}^{n}-v_{jk-1}^{n}\right)\,, \tag{9.52}$$

which are similar to the finite difference version of the *linear* one-dimensional shallow water equations (6.56) employing the CTCS scheme, except for the additional terms caused by two-dimensionality.

The question arises as to whether the stability condition changes compared to the one-dimensional case, and if so, whether it is stricter or more relaxed. We use the von Neumann method once again. As before, the starting point is to replace each of the three variables u_{jk}^{n}, v_{jk}^{n}, and ϕ_{jk}^{n} by a discrete Fourier component corresponding to (9.6) in (9.50)–(9.52). The discrete Fourier components take the form

$$\mathbf{X}_{jk}^{n} = \mathbf{X}_{n} \mathrm{e}^{\mathrm{i}(\alpha j \Delta x + \beta k \Delta y)} , \tag{9.53}$$

where

$$\mathbf{X}_{jk}^{n\ \mathrm{T}} = \left[u_{jk}^{n}, v_{jk}^{n}, \phi_{jk}^{n}\right]$$

contains the three dependent variables and $\mathbf{X}_n^{\mathrm{T}} = \left[U_n, V_n, \Phi_n\right]$ contains their corresponding amplitudes.[1] Insertion into (9.50)–(9.52) then leads to

$$U_{n+1} - U_{n-1} = 2f \Delta t V_n - 2\mathrm{i}\frac{\Delta t}{\Delta x} \sin \alpha \Delta x \Phi_n , \tag{9.54}$$

$$V_{n+1} - V_{n-1} = -2f \Delta t U_n - 2\mathrm{i}\frac{\Delta t}{\Delta y} \sin \beta \Delta y \Phi_n , \tag{9.55}$$

$$\Phi_{n+1} - \Phi_{n-1} = -2\mathrm{i}c_0^2\frac{\Delta t}{\Delta x} \sin \alpha \Delta x U_n - 2\mathrm{i}c_0^2\frac{\Delta t}{\Delta y} \sin \beta \Delta y V_n . \tag{9.56}$$

To find an equation for the growth factor, V_n and U_n are first eliminated from the system (9.54)–(9.56) to arrive at one equation containing Φ_n alone. To this end, n is replaced by $n+1$ in (9.54) and (9.55), whence

$$U_{n+2} - U_{n} = 2f \Delta t V_{n+1} - 2\mathrm{i}\frac{\Delta t}{\Delta x} \sin \alpha \Delta x \Phi_{n+1} , \tag{9.57}$$

$$V_{n+1} - V_{n-1} = -2f \Delta t U_{n+1} - 2\mathrm{i}\frac{\Delta t}{\Delta y} \sin \beta \Delta y \Phi_{n+1} , \tag{9.58}$$

and then n is replaced by $n-1$, with the result

$$U_{n} - U_{n-2} = 2f \Delta t V_{n-1} - 2\mathrm{i}\frac{\Delta t}{\Delta x} \sin \alpha \Delta x \Phi_{n-1} , \tag{9.59}$$

$$V_{n} - V_{n-2} = -2f \Delta t U_{n-1} - 2\mathrm{i}\frac{\Delta t}{\Delta y} \sin \beta \Delta y \Phi_{n-1} . \tag{9.60}$$

Subtracting (9.59) from (9.57) and (9.60) from (9.58) yields

$$U_{n+2} - 2U_n + U_{n-2} = 2f \Delta t (V_{n+1} - V_{n-1}) - 2\mathrm{i}\frac{\Delta t}{\Delta x} \sin \alpha \Delta x (\Phi_{n+1} - \Phi_{n-1}) , \tag{9.61}$$

$$V_{n+2} - 2V_n + V_{n-2} = -2f \Delta t (U_{n+1} - U_{n-1}) - 2\mathrm{i}\frac{\Delta t}{\Delta y} \sin \beta \Delta y (\Phi_{n+1} - \Phi_{n-1}) . \tag{9.62}$$

Using the expression for $V_{n+1} - V_{n-1}$ from (9.55) in (9.61) and the expression for $U_{n+1} - U_{n-1}$ from (9.54) in (9.62), we obtain

$$\begin{aligned} U_{n+2} - 2(1 - 2f^2 \Delta t^2) U_n + U_{n-2} = &- 4\mathrm{i} f \Phi_n \frac{\Delta t^2}{\Delta y} \sin \beta \Delta y \\ &- 2\mathrm{i}(\Phi_{n+1} - \Phi_{n-1}) \frac{\Delta t}{\Delta x} \sin \alpha \Delta x , \end{aligned} \tag{9.63}$$

[1] $\mathbf{A}^{\mathrm{T}}$ denotes the transpose of of the vector $\mathbf{A}$.

$$V_{n+2} - 2(1-2f^2\Delta t^2)V_n + V_{n-2} = 4\mathrm{i}f\Phi_n\frac{\Delta t^2}{\Delta x}\sin\alpha\Delta x - 2\mathrm{i}(\Phi_{n+1}-\Phi_{n-1})\frac{\Delta t}{\Delta y}\sin\beta\Delta y\,. \tag{9.64}$$

Furthermore, replacing n by $n+2$ in (9.56), and then n by $n-2$ leads to

$$\Phi_{n+3} - \Phi_{n+1} = -2\mathrm{i}c_0^2\frac{\Delta t}{\Delta x}\sin\alpha\Delta x U_{n+2}\,, \tag{9.65}$$

$$\Phi_{n-1} - \Phi_{n-3} = -2\mathrm{i}c_0^2\frac{\Delta t}{\Delta x}\sin\alpha\Delta x U_{n-2}\,. \tag{9.66}$$

Multiplying (9.56) by $2(1-2f^2\Delta t^2)$ and subtracting it from the sum of (9.65) and (9.66) leads finally to

$$\Phi_{n+3} - (1+2\lambda)\Phi_{n+1} + (1+2\lambda)\Phi_{n-1} - \Phi_{n-3} = 0\,, \tag{9.67}$$

where use has been made of (9.63) and (9.64), and where

$$\lambda = 1 - 2f^2\Delta t^2 - 2\left(c_0\frac{\Delta t}{\Delta x}\right)^2\sin^2\alpha\Delta x - 2\left(c_0\frac{\Delta t}{\Delta y}\right)^2\sin^2\beta\Delta y\,. \tag{9.68}$$

The last step in von Neumann's stability analysis is to defining a growth factor. Defining it by $G = \Phi_{n+2}/\Phi_n$, we obtain the third-order equation

$$G^3 - (1+2\lambda)G^2 + (1+2\lambda)G - 1 = 0\,, \tag{9.69}$$

which takes the form

$$(G-1)(G^2 - 2\lambda G + 1) = 0\,. \tag{9.70}$$

There are, therefore, three solutions:

$$G_1 = 1\,, \quad G_{2,3} = \lambda \pm \mathrm{i}\sqrt{1-\lambda^2}\,. \tag{9.71}$$

Once again, requiring G_2 and G_3 to be complex, the radical has to be a positive-definite quantity. Under these circumstances, the magnitude of all three possible growth functions equals one. As expected, the scheme is, therefore, neutrally and conditionally stable (energy-conserving) under the condition

$$\lambda^2 \le 1\,. \tag{9.72}$$

This is satisfied as long as

$$-1 \le 1 - 2f^2\Delta t^2 - 2\left(c_0\frac{\Delta t}{\Delta x}\right)^2\sin^2\alpha\Delta x - 2\left(c_0\frac{\Delta t}{\Delta y}\right)^2\sin^2\beta\Delta y \le 1\,. \tag{9.73}$$

The right-hand inequality is always satisfied, while the left-hand inequality requires

$$\left(c_0\frac{\Delta t}{\Delta x}\right)^2\sin^2\alpha\Delta x + \left(c_0\frac{\Delta t}{\Delta y}\right)^2\sin^2\beta\Delta y \le 1 - f^2\Delta t^2\,. \tag{9.74}$$

This should be satisfied for all possible wavenumbers α and β. Since the maximum value that the two sine functions can reach is one, the criterion for stability is

$$c_0^2 \Delta t^2 \left(\frac{1}{L_R^2} + \frac{1}{\Delta x^2} + \frac{1}{\Delta y^2} \right) \leq 1 , \tag{9.75}$$

where $L_R = c_0/f$ is the Rossby deformation radius. In terms of the time step, the criterion is

$$\Delta t \leq \frac{\Delta x}{c_0 \sqrt{1 + (\Delta x/\Delta y)^2 + (\Delta x/L_R)^2}} . \tag{9.76}$$

If in addition the Rossby deformation radius is required to be well resolved by the grid, which is the case in most applications, then $\Delta x/L_R \ll 1$ and the last term in the radical may be neglected. For all practical purposes, the stability condition is then

$$\Delta t < \frac{\Delta x}{c_0 \sqrt{1 + (\Delta x/\Delta y)^2}} . \tag{9.77}$$

Finally, letting $\Delta x = \Delta y = \Delta s$ in (9.77), we obtain the condition

$$\Delta t < \frac{\Delta s}{2c_0} \sqrt{2} . \tag{9.78}$$

Comparing this to (6.68), the corresponding conditions for the one-dimensional case, the present condition is indeed stricter by a factor of $\sqrt{2}/2$. As mentioned at the end of Sect. 9.1, this is common when the dimensionality is increased.

9.4 Summary and Remarks

This chapter aims to show the reader that it is easy to extend from the one-dimensional problems considered hitherto to two- and three-dimensional problems. Following the general organization of this book, we discussed diffusion, advection, and shallow water equations separately. The focus was on how to construct the finite difference version of the equations, and on possible implications regarding the stability criterion.

Section 9.1 began with a description of how to extend the diffusion equation to two spatial dimensions, which turned out to be straightforward. This was also true for the possible effects on the stability criterion. Once again, von Neumann's stability analysis was straightforward, leading to a stricter condition for stability.

In Sect. 9.2, it was shown that the same was true for the advection equation, and again the investigation of the stability criterion led to a stricter time step constraint. Regarding the shallow water equations treated in Sect. 9.3, the extension of the finite difference equations to an additional independent variable was similarly straightforward. It was a little more complicated to deduce the stability criterion. The result was a stricter condition. Thus, the conclusion was that the extension to more than one independent variable leads to a stricter stability condition for all three equations.

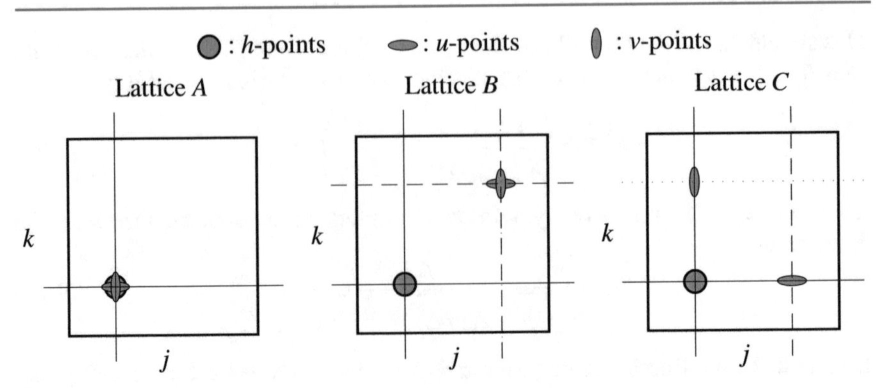

Fig. 9.1 Cell structure of lattice A (non-staggered), lattice B, and lattice C of Mesinger and Arakawa (1976) in two space dimensions. *Filled circles* are associated with h-points, *horizontal ellipses* with u, and *vertical ellipses* with v-points. Only one cell is drawn for each lattice. The cells are counted using the counters j, k. The distance between two adjacent h-points (and hence also u- and v-points) is Δx in the x direction and Δy in the y direction. *Solid straight lines* correspond to the h-grid, *dashed straight lines* to the u-grid, and *dotted lines* to the v-grid. Notice that the three grids overlap in lattice A, and that the u and v grids overlap in lattice B

Finally, regarding the shallow water equations, staggered grids were mentioned in Sect. 6.5. When we increase the number of independent variables to two, this means that the grid can also be staggered in the second spatial direction. In the pioneering work of Mesinger and Arakawa (1976), the non-staggered A-grid and the staggered B- and C-grids were as illustrated in Fig. 9.1.

There is one computer problem at the end of this chapter which the reader is strongly urged to do. It concerns the way equatorially and coastally trapped Kelvin waves can explain how an event occurring on the eastern shores of Brazil can manifest itself on the shores of northwest Africa, a kind of oceanic telecommunication. As such, it is close to a realistic computer problem, illustrating Einstein's mantra of making things as simple as possible, but no simpler.

9.5 Exercises

1. Consider the linear two-dimensional nonrotating shallow water equations in the form

$$\begin{aligned} \partial_t u &= -g\partial_x h \, , \\ \partial_t v &= -g\partial_y h \, , \\ \partial_t h &= -H_0\partial_x u - H_0\partial_y v \, . \end{aligned} \tag{9.79}$$

Show using the leap-frog scheme (Sect. 9.3) that the finite difference version of (9.79) on lattice C of Mesinger and Arakawa (1976) (Fig. 9.1) takes the form

$$
\begin{aligned}
u_{jk}^{n+1} &= u_{jk}^{n} - 2g\frac{\Delta t}{\Delta x}\left(h_{j+1k}^{n} - h_{jk}^{n}\right) , \\
v_{jk}^{n+1} &= v_{jk}^{n} - 2g\frac{\Delta t}{\Delta y}\left(h_{jk+1}^{n} - h_{jk}^{n}\right) , \\
h_{jk}^{n+1} &= h_{jk}^{n} - 2g\frac{\Delta t}{\Delta x}\left(u_{jk}^{n} - u_{j-1k}^{n}\right) - 2g\frac{\Delta t}{\Delta y}\left(v_{jk}^{n} - v_{jk-1}^{n}\right) .
\end{aligned}
\tag{9.80}
$$

2. Using the von Neumann stability analysis, show that the stability condition of (9.80) is

$$\Delta t \leq \frac{1}{2c_0}\sqrt{\frac{\Delta x^2 \Delta y^2}{\Delta x^2 + \Delta y^2}} . \tag{9.81}$$

3. Show that, if $\Delta x = \Delta y = \Delta s$, then (9.81) reduces to

$$\Delta t \leq \frac{\Delta s}{4c_0}\sqrt{2} . \tag{9.82}$$

Compare (9.82) and (9.78), where the latter is the stability condition for the finite difference version of the linear two-dimensional rotating shallow water equations on lattice A. Is it stricter or more relaxed with regard to the time step Δt, and if so why?

9.6 Computer Problems

9.6.1 Upwelling in the Bay of Guinea

In their pioneering work, Adamec and O'Brien (1978) demonstrated how remote wind events in the western equatorial Atlantic could be the source of an observed coastal upwelling event in the Bay of Guinea, Africa. Prior to their work, the prevailing explanation of coastal upwelling phenomena was that they were forced by local winds. Thus, the observed upwelling in the Bay of Guinea was rather puzzling, and remain unexplained at the time.

Their starting point was that the Russian research vessel R/V Passat, located on the equator about 10° W, observed a significant and unexpected drop in the sea surface temperature *before* the seasonal upwelling in the Bay of Guinea was about to start. They also noted that the time difference between the two observations was just about right for an internal Kelvin wave to travel from the location of the research vessel to the Bay of Guinea. They, therefore, hypothesized that the observed upwelling event in the Bay of Guinea was remotely forced by an equatorial trapped upwelling Kelvin wave generated by a pressure disturbance in the western part of the equatorial Atlantic. Once created, it would travel eastward across the Atlantic basin toward the African coast, where it would transform into two coastal internal Kelvin waves, one propagating northward and the other propagating southward. After some time, the northward propagating one would hit the Bay of Guinea, causing an upwelling event as the wave passed by. To support their hypothesis, Adamec and O'Brien (1978) performed an experiment using a fairly simple linear numerical ocean model.

The objective of this computer problem is to reproduce the results of the numerical experiments conducted by Adamec and O'Brien (1978). What they did was to simulate equatorial and coastal Kelvin waves using a linear reduced gravity model. A reduced gravity model is based on the shallow water equations but the gravitational acceleration is replaced by the reduced gravitational acceleration. Thus, the governing equations are

$$\partial_t \mathbf{u} + \beta y \mathbf{k} \times \mathbf{u} + g' \nabla_{\mathrm{H}} h = \frac{\boldsymbol{\tau}}{\rho H} + A \nabla_{\mathrm{H}}^2 \mathbf{u} \,, \tag{9.83}$$

$$\partial_t h + H \nabla_{\mathrm{H}} \cdot \mathbf{u} = 0 \,, \tag{9.84}$$

where $g' = \Delta\rho/\rho$ is the reduced gravity, $\mathbf{u}$ is the horizontal velocity component, h is the thickness of the upper layer, β is the change in the Earth's rotation with latitude (the β-plane approximation), $\boldsymbol{\tau} = \tau_{\mathrm{s}}^x \mathbf{i} + \tau_{\mathrm{s}}^y \mathbf{j}$ is the wind stress acting on the sea surface, H is the equilibrium depth of the upper layer of the ocean, $\nabla_{\mathrm{H}} = \mathbf{i}\partial_x + \mathbf{j}\partial_y$ is the two-dimensional del operator, and A is the eddy viscosity (diffusion) coefficient. Due to the β plane approximation, the x-axis points eastward along the equator, while the y-axis points northwards.

To perform the simulation of Adamec and O'Brien (1978), we use the same setup here to solve (9.83) and (9.84). Initially, the ocean is at rest and in equilibrium ($\mathbf{u} = 0$, $h = H$). The computational domain is an idealized rendition of the Atlantic basin (Fig. 9.2). The basin is 5000 km long and stretches 1500 km to the north and south of the equator. The African continent is represented by a rectangular box 2000 km long and 1000 km wide, protruding into the basin from the northeast corner. Motion is forced by applying a wind stress of the form

$$\tau_{\mathrm{s}}^x = -\tau_0 \begin{cases} 1 \,, & 0 < x \le a \,, \\ \dfrac{b-x}{b-a} \,, & a < x \le b \,, \end{cases} \qquad \tau_{\mathrm{s}}^y = 0 \,, \tag{9.85}$$

as shown in Fig. 9.3. Here τ_0 is its strength and a its extension eastward from the western boundary, before its strength falls off linearly to zero at $x = b$. The wind

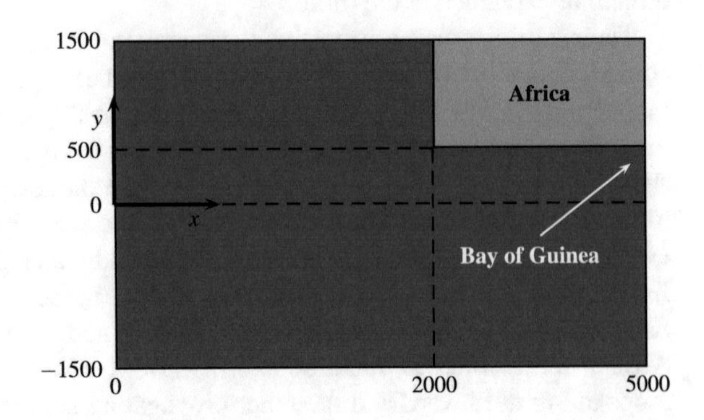

Fig. 9.2 Model domain for which (9.83) and (9.84) are to be solved. Numbers along axes are in kilometers

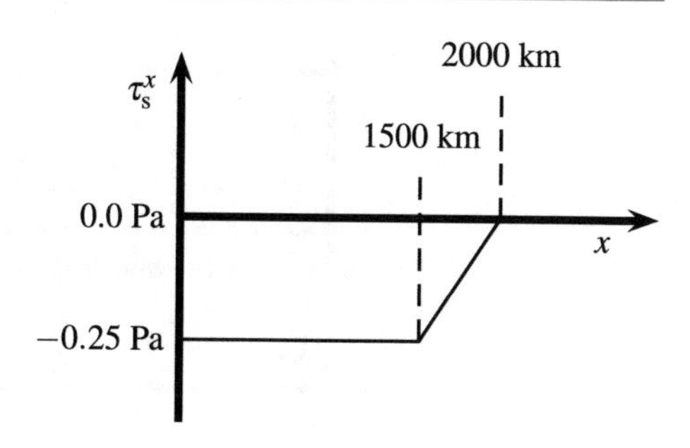

Fig. 9.3 Distribution of the zonal wind stress component in the zonal direction

forcing, therefore, only acts in the western part of the basin. It is directed westward (from east towards west), so that upwelling takes place at the equator. There is no wind stress acting in the north–south direction, so $\tau_s^y = 0$. Moreover, τ_s^x does not vary with the latitude, but depends on the longitude, in such a way that there is no wind forcing east of $x = b$.

Moreover, the mesh is the staggered Arakawa C-grid of size $\Delta x = \Delta y = 25$ km (see Fig. 9.4), and the density difference between the upper and lower layers is $\Delta\rho = 2$ kg/m^3. The time step is set to maximum 1/8 of a day. Furthermore, set $\beta = 10^{-11}$ m^{-1}s^{-1}, $H = 50$ m, $\tau_0 = 0.025$ Pa, $a = 1500$ km, $b = 2000$ km, the eddy viscosity $A = 10^2$ m^2/s, and let the simulation span at least 120 days.

The finite difference version of (9.83) and (9.84) used by Adamec and O'Brien (1978) is the leap-frog scheme (CTCS). To avoid instabilities due to the diffusive term, the Dufort–Frankel scheme is employed for these terms. A no-slip condition

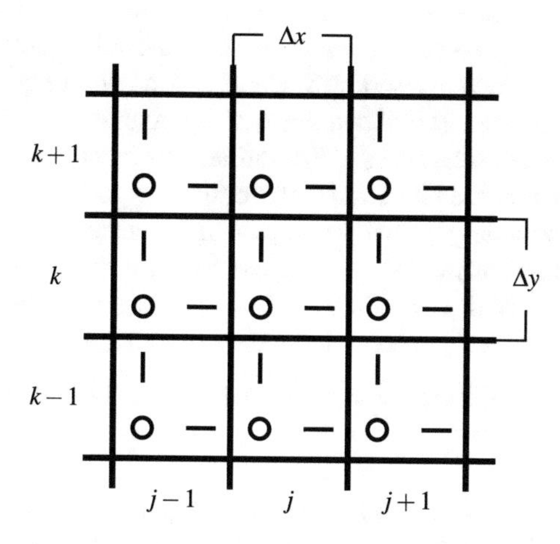

Fig. 9.4 Organization of grid cells and grid points corresponding to lattice C of Mesinger and Arakawa (1976), used to solve (9.83) and (9.84) by numerical means. The grid increments are Δx and Δy in the x and y directions. There is a total of $(J+1) \times (K+1)$ grid cells along the x- and y-axes, counted by the dummy indices j, k. The coordinates of the grid points are $x_j = (j-1)\Delta x$ and $y_k = (k-1)\Delta y$. *Circles* correspond to h-points, *horizontal dashes* to u-points, and *vertical dashes* to v-points

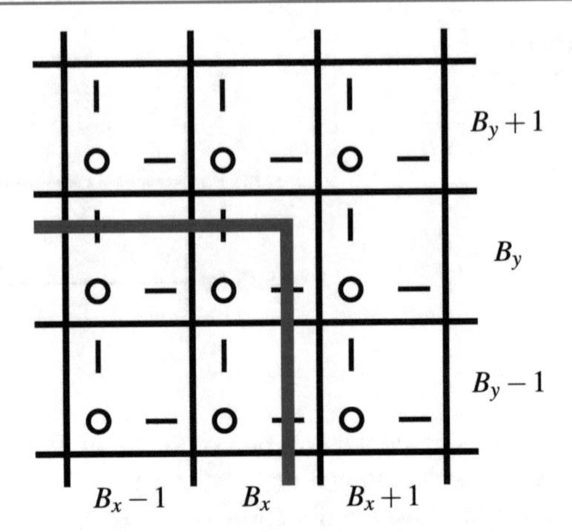

Fig. 9.5 Cells required to account for the no-slip boundary conditions at the walls. The walls are drawn as *solid blue lines*, corresponding to the right-hand corner of the Bay of Guinea (Fig. 9.2). The three cells along the x-axis are numbered $B_x - 1$, B_x, and $B_x + 1$, while the three cells along the y-axis are numbered $B_y - 1$, B_y, and $B_y + 1$. Otherwise the notation is as in Fig. 4.1. Note that the five cells $(B_x + 1, B_y - 1)$, $(B_x + 1, B_y)$, $(B_x + 1, B_y + 1)$, $(B_x, B_y + 1)$, and $(B_x - 1, B_y + 1)$ are on land, and hence lie outside the land–sea boundary. To account for the no-slip boundary condition of zero velocity at the walls, the along-wall velocity component is mirrored across the walls. Thus, for the velocity points shown, the condition of zero along-wall velocity (no slip) is satisfied if $v_{B_x+1B_y-1} = -v_{B_xB_y-1}$ and $u_{B_x-1B_y+1} = -u_{B_x-1B_y}$

is enforced as boundary condition along the solid boundaries of the domain, so that both velocity components are zero there. To make $v = 0$ along the eastern (and western) boundary, so-called mirror points are used outside the boundary, as sketched in Fig. 9.5. Thus, by setting v at the mirror point equal to v just inside the boundary, the linearly interpolated value at the boundary itself becomes zero as required. A similar approach is used at the northern and southern boundaries to make u zero there. Notice that the physical (real) boundaries go through the points where the normal to the boundary velocity component is located. Thus, the eastern and western boundaries go through u-points, while the northern and southern boundaries go through v-points. With this configuration, the coastlines fall exactly halfway between the points where the velocity component along the boundaries is computed.

A. Construct the finite difference version of (9.83) and (9.84) using the required scheme, as discussed above.

B. Why is it permissible to use the diffusive, dissipative, and inconsistent Dufort–Frankel scheme for the diffusion terms? What are the advantages? Explain how you would construct a scheme which is forward in time and centered in space, but still stable.

C. Construct a graph displaying contour lines of the thickness deviation $\Delta h = H - h$ after 10, 50, and 100 days and with a contour interval of 5 m. Furthermore, construct a similar graph showing the velocities of the upper layer at day 50.

D. Next construct a graph showing the thickness deviation in a Hovmöller diagram. Let the vertical axis be the time axis, ranging from 0 to 120 days, and let the horizontal axis be 6000 km long. Let the first part run along the equator from $x = 2500$ km to $x = 5000$ km, the next 500 km along the eastern boundary north of the equator, the next 2000 km along the southern coast of Africa, and the last 1000 km along the northeastern boundary (West Africa).

E. Use the Hovmöller diagram to estimate the phase speed of the Kelvin wave and compare it to the analytic one entering the governing equations. Discuss the result.

F. Solve the nonlinear version of (9.83) and (9.84), that is,

$$\partial_t \mathbf{u} + \mathbf{u} \cdot \nabla_\mathrm{H} \mathbf{u} + \beta y \mathbf{k} \times \mathbf{u} + g' \nabla_\mathrm{H} h = \frac{\boldsymbol{\tau}}{\rho H} + A \nabla_H^2 \mathbf{u} \,, \tag{9.86}$$

$$\partial_t h + \nabla_\mathrm{H} \cdot (h \mathbf{u}) = 0 \,, \tag{9.87}$$

and plot the results in a similar manner as for the linear case. Discuss differences and similarities.

G. Finally, construct an animation (movie) displaying the linear as well as the non-linear results.

References

Adamec D, O'Brien JJ (1978) The seasonal upwelling in the Gulf of Guinea due to remote forcing. J Phys Oceanogr 8:1050–1060

Mesinger F, Arakawa A (1976) Numerical methods used in atmospheric models. GARP publication series, vol 17. World Meteorological Organization, Geneva, 64 pp

Advanced Topics

10

The purpose of this chapter is to use the knowledge acquired in the previous chapters to learn about some slightly more advanced topics. For instance, we sketch ways to construct schemes of higher order accuracy, and ways to solve problems when advection and diffusion are equally important. Furthermore, we consider ways to treat nonlinearities numerically, and ask whether they harbor implications for instability. Since two-way nesting is becoming more and more popular, we also say a few words about smoothing and filtering, and give a detailed presentation of two-way nesting itself. Since the spectral method mentioned in the preface is rather common in global atmospheric models, the chapter ends with a brief description of a one-dimensional application of this method.

10.1 Higher Order Advection Schemes

As discussed in Sect. 2.5, schemes with higher order accuracy may be constructed using Taylor series expansions. Here, we present a few examples to give an idea of how to do this. The focus is on the advection scheme (Sect. 5.1), but the method is applicable to any PDE. For completeness, the one-dimensional advection equation is repeated here:

$$\partial_t \theta + u_0 \partial_x \theta = 0 \,, \tag{10.1}$$

where u_0 is a constant advection velocity. Recall that to arrive at a convergent system, the scheme must be stable and the finite difference approximations (FDAs) of the derivatives in (10.1) must be constructed from Taylor series (Sect. 2.5). Again for completeness, expanding $\theta^n_{j\pm1}$ into a Taylor series yields

$$\theta^n_{j+1} = \theta^n_j + \partial_x \theta|^n_j \Delta x + \frac{1}{2}\partial_x^2 \theta|^n_j \Delta x^2 + \frac{1}{6}\partial_x^3 \theta|^n_j \Delta x^3 + \sum_{m=4}^{\infty} \frac{1}{m!}\partial_x^m \theta|^n_j \Delta x^m \,, \tag{10.2}$$

L. P. Røed, *Atmospheres and Oceans on Computers*,
Springer Textbooks in Earth Sciences, Geography and Environment,
https://doi.org/10.1007/978-3-319-93864-6_10

$$\theta^n_{j-1} = \theta^n_j - \partial_x\theta|^n_j\Delta x + \frac{1}{2}\partial^2_x\theta|^n_j\Delta x^2 - \frac{1}{6}\partial^3_x\theta|^n_j\Delta x^3 + \sum_{m=4}^{\infty}\frac{(-1)^m}{m!}\partial^m_x\theta|^n_j\Delta x^m . \tag{10.3}$$

Likewise, and for later convenience, it is also possible to expand $\theta^n_{j\pm2}$ into Taylor series:

$$\theta^n_{j+2} = \theta^n_j + 2\partial_x\theta|^n_j\Delta x + 2\partial^2_x\theta|^n_j\Delta x^2 + \frac{4}{3}\partial^3_x\theta|^n_j\Delta x^3 + \sum_{m=4}^{\infty}\frac{(2)^m}{m!}\partial^m_x\theta|^n_j\Delta x^m , \tag{10.4}$$

$$\theta^n_{j-2} = \theta^n_j - 2\partial_x\theta|^n_j\Delta x + 2\partial^2_x\theta|^n_j\Delta x^2 - \frac{4}{3}\partial^3_x\theta|^n_j\Delta x^3 + \sum_{m=4}^{\infty}\frac{(-2)^m}{m!}\partial^m_x\theta|^n_j\Delta x^m . \tag{10.5}$$

Thus, in general

$$\theta^n_{j\pm l} = \theta^n_j + \sum_{m=1}^{\infty}\frac{(\pm l)^m}{m!}\partial^m_x\theta|^n_j\Delta x^m , \tag{10.6}$$

where $l = 1, 2, 3, \ldots$.

Fourth-Order Leap-Frog Scheme

Recall from Sect. 2.5 that a second-order-accurate FDA of $\partial_x\theta$ is constructed by first subtracting (10.3) from (10.2). Solving the resulting equation for $\partial_x\theta|^n_j$ yields

$$\partial_x\theta|^n_j = \frac{\theta^n_{j+1} - \theta^n_{j-1}}{2\Delta x} - \frac{1}{6}\partial^3_x\theta|^n_j\Delta x^2 + \mathcal{O}(\Delta x^4) . \tag{10.7}$$

Furthermore, truncating by neglecting terms of order $\mathcal{O}(\Delta x^2)$ and higher leads to the FDA

$$[\partial_x\theta]^n_j = \frac{\theta^n_{j+1} - \theta^n_{j-1}}{2\Delta x} . \tag{10.8}$$

This is the FDA used in Sect. 5.3 to construct the second-order-accurate leap-frog scheme. As mentioned at the end of Sect. 2.5, there are also other possibilities for constructing a second-order-accurate FDA of $\partial_x\theta$. For instance, subtracting (10.5) from (10.4) and solving for $\partial_x\theta|^n_j$, we obtain

$$\partial_x\theta|^n_j = \frac{\theta^n_{j+2} - \theta^n_{j-2}}{4\Delta x} - \frac{2}{3}\partial^3_x\theta|\Delta x^2 + \mathcal{O}(\Delta x^4) . \tag{10.9}$$

Once again, truncating by neglecting all terms on the right-hand side of order $\mathcal{O}(\Delta x^2)$ and higher leads to the FDA

$$[\partial_x\theta]^n_j = \frac{\theta^n_{j+2} - \theta^n_{j-2}}{4\Delta x} , \tag{10.10}$$

which is an equally valid centered second-order-accurate FDA of $\partial_x\theta$. Moreover, since both (10.8) and (10.10) tend to the continuous derivative $\partial_x\theta$ in the limit $\Delta x \to 0$, they are both numerically consistent FDAs of $\partial_x\theta$.

To arrive at a fourth-order-accurate FDA, we have to get rid of the second-order terms. To this end, we may use (10.7) and (10.9). The trick is to first multiply (10.7) by an as yet unknown coefficient a, and similarly multiply (10.9) by an unknown coefficient b. Adding the results, we have

$$(a+b)\partial_x\theta|_j^n = a\frac{\theta_{j+1}^n - \theta_{j-1}^n}{2\Delta x} + b\frac{\theta_{j+2}^n - \theta_{j-2}^n}{4\Delta x} - \frac{1}{6}(a+4b)\partial_x^3\theta|_j^n\Delta x^2 + \mathscr{O}(\Delta x^4) . \tag{10.11}$$

Thus, if $a+4b=0$, the second-order term on the right-hand side vanishes. Moreover, demanding $a+b=1$ leads to $a=4/3$ and $b=-1/3$. Under these circumstances, (10.11) takes the form

$$\partial_x\theta|_j^n = \frac{4}{3}\frac{\theta_{j+1}^n - \theta_{j-1}^n}{2\Delta x} - \frac{1}{3}\frac{\theta_{j+2}^n - \theta_{j-2}^n}{4\Delta x} + \mathscr{O}(\Delta x^4) . \tag{10.12}$$

Consequently, neglecting terms of $\mathscr{O}(\Delta x^4)$ or higher leads to

$$[\partial_x\theta]_j^n = \frac{4}{3}\frac{\theta_{j+1}^n - \theta_{j-1}^n}{2\Delta x} - \frac{1}{3}\frac{\theta_{j+2}^n - \theta_{j-2}^n}{4\Delta x} , \tag{10.13}$$

which is a fourth-order-accurate FDA of $\partial_x\theta$. A scheme for the advection equation (10.1) that is second-order accurate in time and fourth-order accurate in space then takes the form

$$\theta_j^{n+1} = \theta_j^{n-1} - \frac{4}{3}u_0\frac{\Delta t}{\Delta x}\left[\theta_{j+1}^n - \theta_{j-1}^n - \frac{1}{8}(\theta_{j+2}^n - \theta_{j-2}^n)\right] . \tag{10.14}$$

Since the fourth-order scheme is based on a Taylor series, it is consistent. It is also expected to be conditionally stable, but under what conditions? Will it be more restrictive or more tolerant? As usual, the von Neumann method is used to investigate the stability. Inserting the Fourier component $\theta_j^n = \Theta_n e^{i\alpha j\Delta x}$ into (10.14), we obtain

$$\Theta_{n+1} = \Theta_{n-1} - i\frac{8}{3}\,\text{sgn}(u_0)C\left(\sin\alpha\Delta x - \frac{1}{8}\sin 2\alpha\Delta x\right)\Theta_n , \tag{10.15}$$

where $C = |u_0|\Delta t/\Delta x$ is the Courant number. If we now recall the trigonometric formula $\sin 2\phi = 2\sin\phi\cos\phi$ and define the growth factor by $G = \Theta_{n+1}/\Theta_n$, we have

$$G^2 + 2i\lambda G - 1 = 0 , \tag{10.16}$$

where

$$\lambda = \frac{1}{3}\,\text{sgn}(u_0)C\sin\alpha\Delta x(4 - \cos\alpha\Delta x) . \tag{10.17}$$

Hence, there are two solutions

$$G_{1,2} = -i\lambda \pm \sqrt{1-\lambda^2} . \tag{10.18}$$

As expected, the scheme that is fourth-order accurate in space is neutrally and conditionally stable ($|G_{1,2}| = 1$) under the condition that the radical in (10.18) is a positive-definite quantity. The stability condition is , therefore,

$$|\lambda| \le 1 \quad\Longrightarrow\quad C|\sin\alpha\Delta x|(4 - \cos\alpha\Delta x) \le 3 . \tag{10.19}$$

Since the maximum value of $(4 - \cos\alpha\Delta x)$ is five and the maximum value of $|\sin\alpha\Delta x|$ is one, the inequality is valid for all possible wavenumbers if

$$C \leq \frac{3}{5} \quad \text{or} \quad \Delta t \leq \frac{3\Delta x}{5|u_0|} , \tag{10.20}$$

which is indeed a stricter condition than the one obtained for the second-order-accurate leap-frog scheme ($C \leq 1$, or $\Delta t \leq \Delta x/|u_0|$).

As shown in Sect. 5.5, the leap-frog scheme is numerically dispersive. The question, therefore, arises as to whether the use of a higher order scheme has an effect on the numerical dispersion. To investigate this, the starting point is to decompose θ_j^n into its Fourier components in time and space:

$$\theta_j^n = \Theta_0 e^{i\alpha(j\Delta x - c^* n\Delta t)} , \tag{10.21}$$

where c^* is the numerical phase speed. Substitution of (10.21) into (10.14) and solving for c^* leads to the dispersion relation

$$c^* = \frac{1}{\alpha\Delta t} \arcsin\left[u_0\alpha\Delta t\left(\frac{4}{3}\frac{\sin\alpha\Delta x}{\alpha\Delta x} - \frac{1}{3}\frac{\sin 2\alpha\Delta x}{2\alpha\Delta x}\right)\right] . \tag{10.22}$$

To leading order in $\alpha\Delta x$, the dispersion relation is, therefore,[1]

$$c^* \approx u_0\left[1 - \frac{4}{5!}(\alpha\Delta x)^4\right] . \tag{10.23}$$

This may be compared with the dispersion relation (5.36) associated with the second-order-accurate leap-frog scheme. To leading order in $\alpha\Delta x$, the latter is

$$c^*_{\text{leap-frog}} \approx u_0\left[1 - \frac{1}{3!}(\alpha\Delta x)^2\right] . \tag{10.24}$$

Thus, the fourth-order scheme is actually less dispersive, as long as $0 \leq \alpha\Delta x \leq \pi$.

Finally, recall that the second-order leap-frog scheme contained a computational mode (Sect. 5.7). Note that the growth factor for the fourth-order scheme (10.18) has exactly the same two solutions as the second-order leap-frog scheme (5.25). The only difference is the expression for λ, which for the fourth-order scheme is as in (10.16). Thus, the fourth-order leap-frog scheme also contains a computational mode. To conclude, the fourth-order scheme is superior to the second-order scheme with regard to accuracy and dispersivity, but this has no effect on the computational mode, even for well resolved wavelengths.

Sixth-Order Leap-Frog Scheme

The process of constructing higher order finite difference approximations may be continued. Invoking the Taylor series (10.6), and adding the series for $l = 1$, 2, and 3 multiplied by a, b, and c, respectively, results in a sixth-order-accurate centered-in-space scheme that takes the form

$$\frac{\theta_j^{n+1} - \theta_j^{n-1}}{2\Delta t} + \frac{3}{2}u_0\left(\frac{\theta_{j+1}^n - \theta_{j-1}^n}{2\Delta x} - \frac{2}{5}\frac{\theta_{j+2}^n - \theta_{j-2}^n}{4\Delta x} + \frac{1}{15}\frac{\theta_{j+3}^n - \theta_{j-3}^n}{6\Delta x}\right) = 0 . \tag{10.25}$$

[1]Note that $\sin z/z = 1 - z^2/6 + \cdots$, while $\arcsin z = 1 + z^2/6 + \cdots$, for $|z| < 1$.

The associated numerical dispersion relation is

$$c^* = \frac{1}{\alpha\Delta t}\arcsin\left[\frac{4}{3}u_0\alpha\Delta t\left(\frac{\sin\alpha\Delta x}{\alpha\Delta x} - \frac{9}{20}\frac{\sin 2\alpha\Delta x}{2\alpha\Delta x} + \frac{3}{40}\frac{\sin 3\alpha\Delta x}{3\alpha\Delta x}\right)\right], \tag{10.26}$$

which yields to leading order

$$c^* \approx u_0\left[1 - \frac{36}{7!}(\alpha\Delta x)^6\right]. \tag{10.27}$$

Comparing (10.27) with (10.23) and (10.24) shows that the sixth-order scheme is superior to the fourth-order scheme, and so on.

Second-Order Upstream Scheme
It is also of interest to construct higher order schemes based on the upstream scheme. Once again, the starting point is the Taylor series. For instance, solving (10.3) for $\partial_x\theta|_j^n$ leads to

$$\partial_x\theta|_j^n = \frac{\theta_j^n - \theta_{j-1}^n}{\Delta x} - \frac{1}{2}\partial_x^2\theta|_j^n\Delta x + \mathcal{O}(\Delta x^2). \tag{10.28}$$

Similarly, solving (10.5) for $\partial_x\theta|_j^n$ leads to

$$\partial_x\theta|_j^n = \frac{\theta_j^n - \theta_{j-2}^n}{2\Delta x} - \partial_x^2\theta|_j^n\Delta x + \mathcal{O}(\Delta x^2). \tag{10.29}$$

Multiplying (10.28) by a and (10.29) by b, then adding the results, we obtain

$$(a+b)\partial_x\theta|_j^n = a\frac{\theta_j^n - \theta_{j-1}^n}{\Delta x} + b\frac{\theta_j^n - \theta_{j-2}^n}{2\Delta x} + \frac{1}{2}(a+2b)\partial_x^2\theta|_j^n\Delta x + \mathcal{O}(\Delta x^2). \tag{10.30}$$

Choosing $a + b = 1$ and $a + 2b = 0$, that is, $a = 2$ and $b = -1$, and neglecting terms of order $\mathcal{O}(\Delta x^2)$ or higher, leads to the second-order-accurate FDA

$$[\partial_x\theta]_j^n = \frac{1}{2\Delta x}(3\theta_j^n - 4\theta_{j-1}^n + \theta_{j-2}^n). \tag{10.31}$$

Hence, assuming a positive advection velocity, an upstream scheme that is one-sided and second order in space and first order in time is

$$\theta_j^{n+1} = \theta_j^n - u_0\frac{\Delta t}{2\Delta x}(3\theta_j^n - 4\theta_{j-1}^n + \theta_{j-2}^n). \tag{10.32}$$

Third-Order Upstream Schemes
A third-order upstream scheme may also be constructed. The starting point is again to expand θ_{j-1}^n, θ_{j-2}^n, and θ_{j-3}^n in terms of Taylor series in which terms of third order are retained. Solving each of them for $\partial_x\theta_j^n$ yields

$$\partial_x\theta|_j^n = \frac{\theta_j^n - \theta_{j-1}^n}{\Delta x} + \frac{1}{2}\partial_x^2\theta|_j^n\Delta x - \frac{1}{6}\partial_x^3\theta|_j^n\Delta x^2 + \mathcal{O}(\Delta x^3), \tag{10.33}$$

$$\partial_x\theta|_j^n = \frac{\theta_j^n - \theta_{j-2}^n}{2\Delta x} + \partial_x^2\theta|_j^n\Delta x - \frac{2}{3}\partial_x^3\theta|_j^n\Delta x^2 + \mathcal{O}(\Delta x^3), \tag{10.34}$$

$$\partial_x\theta|_j^n = \frac{\theta_j^n - \theta_{j-3}^n}{3\Delta x} + \frac{3}{2}\partial_x^2\theta|_j^n\Delta x - \frac{3}{2}\partial_x^3\theta|_j^n\Delta x^2 + \mathcal{O}(\Delta x^3). \tag{10.35}$$

Multiplying (10.33) by a, (10.34) by b, and (10.35) by c, and adding them together, we obtain

$$(a+b+c)\partial_x\theta|_j^n = a\frac{\theta_j^n-\theta_{j-1}^n}{\Delta x}+b\frac{\theta_j^n-\theta_{j-2}^n}{2\Delta x}+c\frac{\theta_j^n-\theta_{j-3}^n}{3\Delta x}$$
$$+\frac{1}{2}(a+2b+3c)\,\partial_x^2\theta|_j^n\Delta x-\frac{1}{6}(a+4b+9c)\,\partial_x^3\theta|_j^n\Delta x^2$$
$$+\mathscr{O}(\Delta x^3)\,. \quad (10.36)$$

Moreover, choosing $a = 3$, $b = -3$, and $c = 1$ the coefficients in front of the first and second-order terms vanish and $a + b + c = 1$. Thus, by neglecting terms of $\mathscr{O}(\Delta x^3)$ and higher, a third-order-accurate FDA of $\partial_x\theta|_j^n$ takes the form

$$[\partial_x\theta]_j^n = \frac{1}{6\Delta x}\left(11\theta_j^n - 18\theta_{j-1}^n + 9\theta_{j-2}^n - 2\theta_{j-3}^n\right)\,, \quad (10.37)$$

in which case the third-order upstream scheme for the advection equation is

$$\theta_j^{n+1} = \theta_j^n - \frac{1}{6}\,\mathrm{sgn}(u_0)C\left(11\theta_j^n - 18\theta_{j-1}^n + 9\theta_{j-2}^n - 2\theta_{j-3}^n\right)\,, \quad (10.38)$$

where $C = |u_0|\Delta t/\Delta x$ is the Courant number. Alternatively, one may use θ_{j+1}^n, θ_{j-1}^n, and θ_{j-2}^n, which results in the third-order scheme

$$\theta_j^{n+1} = \theta_j^n - \frac{1}{6}\,\mathrm{sgn}(u_0)C(2\theta_{j+1}^n + 3\theta_j^n - 6\theta_{j-1}^n + \theta_{j-2}^n)\,. \quad (10.39)$$

It should be mentioned that complications can arise from the higher order spatial treatments exemplified above. As already stated, the stability condition tends to be more restrictive, as shown for instance by (10.20). A more serious complication, however, is associated with the boundaries. Since points further and further away from the x_j-point are needed when constructing the higher order schemes, additional boundary conditions are required. The one-dimensional advection equation (10.1) allows only one boundary condition to be specified in x, while the higher order schemes (10.14), (10.25), (10.31), and (10.39) all require specification of more than one. One way to avoid this problem is to use a lower order scheme close to the boundary, but then the accuracy is reduced there. Finally, regarding the upstream (or upwind) scheme, the higher order schemes are less diffusive than the lowest order scheme.

10.2 Combined Advection–Diffusion

As stated in Chap. 4, a centered-in-time centered-in-space scheme is unstable for the diffusion equation, and according to Chap. 5, a forward-in-time centered-in-space scheme (Euler scheme) is unstable for the advection equation. On the other hand, as discussed in Chap. 3, most problems encountered regarding evolution of tracers in the atmosphere and oceans contain both advection and diffusion in one and the same equation. The question is, therefore, what scheme should be employed when solving equations which are a combination of the two processes, that is, when solving the so-called advection–diffusion equation?

For simplicity, a one-dimensional advection equation is used as an example. Using the common parameterization for the diffusive and advective fluxes, the advection–diffusion equation (3.1) takes the form

$$\partial_t \theta + u_0 \partial_x \theta = \kappa \partial_x^2 \theta \ , \tag{10.40}$$

where both the advection velocity u_0 and the diffusivity κ are constants. To obtain a stable scheme the diffusive part must be treated forward in time, while the advective part must be treated centered in time.

Consider, for instance, the scheme

$$\frac{\theta_j^{n+1} - \theta_j^{n-1}}{2\Delta t} = -u_0 \frac{\theta_{j+1}^n - \theta_{j-1}^n}{2\Delta x} + \kappa \frac{\theta_{j+1}^{n-1} - 2\theta_j^{n-1} + \theta_{j-1}^{n-1}}{\Delta x^2} \ , \tag{10.41}$$

in which a centered-in-space FDA is used for both the diffusion term and the advection term. Setting κ to zero (no diffusion term), the scheme reduces to the neutrally and conditionally stable second-order leap-frog scheme, which is stable under the condition

$$C \le 1 \ , \quad \text{or} \quad \Delta t \le \frac{\Delta x}{|u_0|} \ . \tag{10.42}$$

If u_0 is zero (no advection term), the scheme is also stable under the condition

$$\frac{\kappa \Delta t}{\Delta x^2} \le 1/4 \ , \quad \text{or} \quad \Delta t \le \frac{\Delta x^2}{4\kappa} \ . \tag{10.43}$$

This looks like a stricter condition by a factor of 1/2 than the one presented in (4.38). However, to ensure that the diffusive part is forward in time, it is evaluated at the time level $n-1$. The forward time step is therefore $2\Delta t$, and thus a factor of 1/4 replaces the factor of 1/2 in (4.38). In contrast, the advective part is evaluated at time step n, so it is centered in time with a time step of Δt.

In the general case, where $u_0 \neq 0$ and $\kappa \neq 0$, we expect a condition that is a combination of the two conditions above. Once again, we use von Neumann's method to investigate the stability. Hence, a single discrete Fourier component is inserted into (10.41) to get an equation for the growth factor. The result is[2]

$$G^2 + 2\mathrm{i}\lambda G - \lambda_2 = 0 \ , \tag{10.44}$$

where

$$\lambda = \mathrm{sgn}(u_0) C \sin \alpha \Delta x \ , \quad \lambda_2 = 1 - 4K(1 - \cos \alpha \Delta x) \ , \tag{10.45}$$

are two real numbers, $C = |u_0|\Delta t/\Delta x$ is the Courant number, and $K = \kappa \Delta t/\Delta x^2$. Since $1 - \cos \alpha \Delta x \ge 0$, it follows that $\lambda_2 \le 1$. In accordance with (10.44), the growth factor has two solutions given by

$$G_{1,2} = -\mathrm{i}\lambda \pm \sqrt{\lambda_2 - \lambda^2} \ . \tag{10.46}$$

[2]The algebra is left to the reader, in Exercise 10.8 at the end of this chapter.

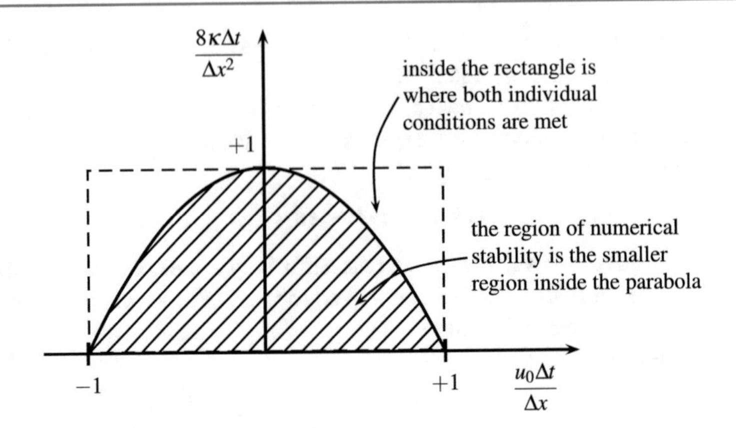

Fig. 10.1 Region of stability for the scheme (10.41) constructed to solve the combined advection–diffusion equations numerically. This is the area inside the parabola (*hatched area*). The area inside the rectangle is where both the advection and the diffusion are stable individually. Notice that a more restrictive stability condition is obtained for both the advection equation and the diffusion equation, when they are combined

To ensure that the two complex solutions have a real part, the radical is required to be positive, that is, $\lambda^2 \leq \lambda_2$, which is equivalent to

$$C^2 \sin^2 \alpha \Delta x + 4K(1 - \cos \alpha \Delta x) \leq 1 . \tag{10.47}$$

Since this must be satisfied for all possible wavenumbers, a sufficient condition for stability of the combined advection–diffusion scheme (10.41) is

$$C^2 + 8K \leq 1 , \quad \text{or} \quad \frac{(u_0 \Delta t)^2 + 8\kappa \Delta t}{\Delta x^2} \leq 1 . \tag{10.48}$$

Note that when $u_0 = 0$ or $\kappa = 0$, the stability conditions for the individual advective and diffusive schemes are recovered. Imposing both conditions is not sufficient; the conditions have to be combined as above. This leads to the somewhat surprising result that adding explicit diffusion in the advection equation reduces the maximum time step allowed for advection. In fact, (10.48) says that adding diffusion results in a more restrictive condition, and this is visualized in Fig. 10.1. For most cases in oceanography and meteorology, this is not a serious problem since we commonly have

$$K \ll C^2 . \tag{10.49}$$

As mentioned earlier (Sect. 4.9), it is common to add a diffusion term to avoid non-linear problems becoming numerically unstable through so-called nonlinear instabilities. This will be discussed in the next section (Sect. 10.3). When used for this purpose, the diffusion term is not part of the physics, but is included to prevent the numerical solution from blowing up. Under these circumstances, we may use the Dufort–Frankel scheme (Sect. 4.9) to approximate the diffusion term, even though it makes the scheme inconsistent. This is acceptable as long as the remaining terms in the equations that govern the motion are treated using consistent schemes. An alternative to the above scheme (10.41) is, therefore,

$$\theta_j^{n+1} = \theta_j^{n-1} - \frac{u_0 \Delta t}{\Delta x}(\theta_{j+1}^n - \theta_{j-1}^n) + 2K(\theta_{j+1}^n - \theta_j^{n+1} - \theta_j^{n-1} + \theta_{j-1}^n) \,, \quad (10.50)$$

in which a consistent conditionally stable scheme for advection is combined with an unconditionally stable inconsistent scheme for diffusion. Performing the von Neumann stability analysis, we obtain

$$(1 + 2K)G^2 - 2\lambda G - (1 - 2K) = 0 \,, \quad (10.51)$$

where G is the growth factor and

$$\lambda = 2K \cos \alpha \Delta x - \mathrm{i} \frac{u_0 \Delta t}{\Delta x} \sin \alpha \Delta x \,. \quad (10.52)$$

This leads to the two solutions

$$G_{1,2} = \frac{1}{1 + 2K}\left(\lambda \pm \sqrt{4K^2 + \lambda^2 - 1}\right) \,. \quad (10.53)$$

It can be shown that, for the one-dimensional case, it is sufficient to satisfy the CFL condition $C \leq 1$. In the more general case, for instance for a two-dimensional case, a more stringent condition has to be imposed (Cushman-Roisin 1984).

Many authors (e.g., Clancy 1981) suggest using the unstable forward-in-time centered-in-space (FTCS) scheme when combining advection and diffusion. The approximation to (10.40) becomes

$$\theta_j^{n+1} = \theta_j^n - \frac{u_0 \Delta t}{2\Delta x}(\theta_{j+1}^n - \theta_{j-1}^n) + \frac{\kappa \Delta t}{\Delta x^2}(\theta_{j+1}^n - 2\theta_j^n + \theta_{j-1}^n) \,. \quad (10.54)$$

The amplification or growth factor then obeys the equation

$$G = 1 - \mathrm{i} \frac{u_0 \Delta t}{2\Delta x} \sin \alpha \Delta x - 2\frac{\kappa \Delta t}{\Delta x^2}(1 - \cos \alpha \Delta x) \,. \quad (10.55)$$

As shown by Clancy (1981), the scheme is stable provided that the two conditions

$$\frac{\kappa \Delta t}{\Delta x^2} \leq \frac{1}{2} \,, \qquad \frac{|u_0| \Delta t}{\kappa} \leq 1 \,, \quad (10.56)$$

are both satisfied at the same time. Despite the enthusiasm of several authors, the present author would not recommend using the FTCS scheme. The more conservative schemes (10.41) and (10.50) are advocated here.

10.3 Nonlinear Instabilities

Toward the end of Sect. 3.2, we noted that every numerical solution to a nonlinear problem of hyperbolic nature, in which friction is neglected, eventually becomes numerically unstable. This is independent of the time step chosen, and is associated with the energy cascade toward smaller and smaller scales, which is characteristic of nonlinear problems. It is not, therefore, sufficient to satisfy the linear CFL criterion, for instance, when solving the nonlinear advection equation (3.16).

To satisfy oneself that this is indeed true, it suffices to solve a simple nonlinear advection problem like (3.16). Sooner or later, disturbances to wavelengths in the range $2\Delta x$ to $4\Delta x$ crop up. These disturbances start out small in amplitude, but then grow. At some stage in the calculation, the solution falls short of satisfying the linear CFL condition and the solution blows up, that is, becomes linearly numerically unstable. The solution is then useless. It is common to credit (Phillips 1959) as the first to demonstrate this phenomenon by analytic means. Richtmyer (1963) provided another example, which is reproduced below, while (Robert et al. 1970) generalized the previous example.

Before going into the details, we note the following points:

- All well-behaved functions may be expanded in terms of a discrete set of waves or exponentials.
- In a linear system, waves of different wavelengths exist independently of each other, and without interacting.
- In a nonlinear system, the latter is no longer true. Waves of different wavenumbers will interact and generate waves with new periods and wavelengths.
- In a finite grid of size Δx, the resolution is band-limited in wavenumber space, that is, only a finite number of discrete waves can exist.

The first point is well known. It simply says that all well-behaved functions $\Psi(x)$ within a domain $x \in [-L, L]$ may be expanded in a Fourier series, i.e.,

$$\Psi(x) = a_0 + \sum_{m=1}^{\infty} a_m \sin(\alpha_m x) + b_m \cos(\alpha_m x) , \tag{10.57}$$

where $\alpha_m = m\pi/L$ is the discrete wavenumber, a_m and b_m are the amplitude or energy associated with the wavenumber α_m, and a_0 is the mean or average value of Ψ for $x \in [-L, L]$. The second point tells us that if the system is linear there is no exchange of energy between the individual waves. Thus, two wave trains of different amplitude, wavelength, and direction will pass through each other without change.

The third point emphasizes the fact that it is the nonlinearity that causes exchange to happen. To illustrate this, consider a solution, say wind or current,

$$u(x, t) = \sum_{n} u_n(t) \sin(\alpha_n x) . \tag{10.58}$$

Then nonlinear products like $u_1 u_2 \sin\alpha_1 x \sin\alpha_2 x$ will give rise to terms having wavenumbers which are the sum of and difference of the two original wavenumbers, e.g.,

$$\sin(\alpha_1 x)\sin(\alpha_2 x) = \frac{1}{2}\big[\cos(\alpha_1 - \alpha_2)x - \cos(\alpha_1 + \alpha_2)x\big] . \tag{10.59}$$

Thus, in a nonlinear situation, the two wave trains will be different after passing through one another, that is, they will experience a change either in wavelength, as illustrated by (10.59), or in amplitude, or in direction.

The fourth and last point says that when formulating the function $\Psi(x)$ as a sum of discrete waves on a grid of size Δx, the wavenumber space is band-limited. The shortest wave that can possibly be resolved is then $2\Delta x$. Thus, the wavenumber space is limited to wavenumbers $\pi/L \leq \alpha_m \leq \pi/\Delta x$. For a nonlinear problem in which

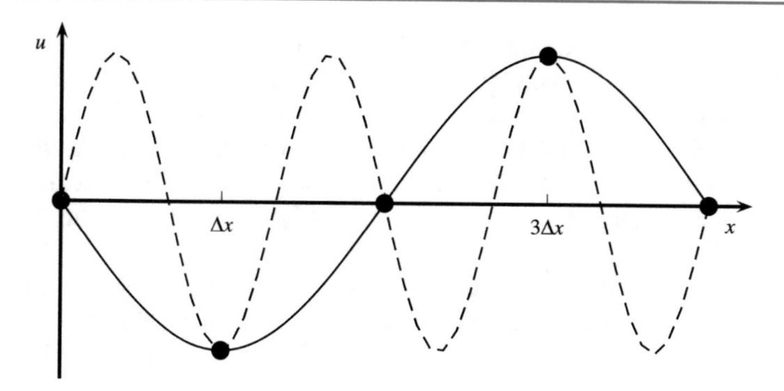

Fig. 10.2 Two waves with wavelength $4\Delta x$ (*solid curve*) and $4\Delta x/3$ (*dashed curve*) in a grid of grid size Δx. Note that the grid cannot distinguish between the unresolved wave of wavelength $4\Delta x/3$ and the resolved wave of wavelength $4\Delta x$. The energy contained in the unresolved wave will, therefore, be folded into the low wavenumber space represented by the $4\Delta x$ wave

the various waves interact to produce waves of wavenumber $\alpha > \pi/\Delta x$, i.e., waves with wavelengths shorter than $2\Delta x$, they are unresolved by the grid. Unfortunately, the energy of these unresolved waves is folded into the resolved waves, implying that the energy of the resolved waves, and in particular those of short wavelengths, increases artificially. In fact, as shown in Fig. 10.2, a wave of wavelength $4\Delta x/3$ is indistinguishable from a wave of wavelength $4\Delta x$. If $\alpha < \pi/2\Delta x$ are called low wavenumbers and $\pi/\Delta x < \alpha < \pi/2\Delta x$ classify as high wavenumbers, then waves with wavelengths between $2\Delta x$ and $4\Delta x$ correspond to the shortest waves that are resolved by our grid of size Δx.

Points three and four above make us expect a priori that, even though all the initial energy is in low wavenumbers (long waves), nonlinear interactions will eventually provide energy (or variance) at high wavenumbers (short waves). This is easily checked by investigating

$$\partial_t u + u\partial_x u = 0 , \tag{10.60}$$

as an example problem.[3] The difference between (10.60) and the earlier advection equation, for instance (5.3), is the appearance of the nonlinear term $u\partial_x u$. Suppose a solution has been found at time level n. Moreover, suppose that this solution is $u^n(x) = u_0 \sin \alpha x$, that is, a monochromatic wave of wavelength $2\pi/\alpha$, where α is the wavenumber. Using a scheme that is centered in time, (10.60) leads to

$$u^{n+1} - u^{n-1} = -u_0^2 \sin \alpha x \partial_x (\sin \alpha x) = -u_0^2 \sin \alpha x \cos \alpha x = -\frac{1}{2} u_0^2 \sin 2\alpha x , \tag{10.61}$$

which shows that the solution at the next time level is a wave of wavelength $2\pi/2\alpha$. This is half the original wavelength at time level n, indicating a cascade of energy

[3]Note that (6.128) is the acceleration term in the momentum equation for a one-dimensional problem. Hence, nonlinearity is ubiquitous in all realistic atmospheric and oceanographic models.

to higher wavenumbers (shorter waves). If this is allowed to continue, all the energy originally contained at low wavenumbers (long waves) will end up at high wavenumbers (short waves) that are unresolved by the grid. Due to the folding of the energy contained in the unresolved waves, energy will accumulate in the shortest wave resolved by our grid. After a sufficiently long time, the numerical model, therefore, blows up due to ordinary numerical linear instability, even though the linear problem is numerically stable.

To inspect the nonlinear instability in more detail, we use the example from Richtmyer (1963). The starting point is to assume that the problem has variance at infinite wavelengths (zero wavenumber), $4\Delta x$, and $2\Delta x$, and that the model problem is the simple nonlinear advection equation in one dimension (10.60). As in Richtmyer (1963), the numerical scheme employed is the classic leap-frog (CTCS) scheme:

$$u_j^{n+1} = u_j^{n-1} - \frac{\lambda}{2}\left[(u_{j+1}^n)^2 - (u_{j-1}^n)^2\right] , \tag{10.62}$$

where $\lambda = \Delta t/\Delta x$. Note that, as in the von Neumann analysis (Sect. 4.4), a local analysis is performed, and hence boundary conditions are of no concern. Making use of (10.57) with $\Psi = u$ and $a_0 = V$, an exact formal solution to (10.62) is

$$u_j^n = C_n \cos\frac{\pi j}{2} + S_n \sin\frac{\pi j}{2} + U_n \cos(\pi j) + V . \tag{10.63}$$

The amplitudes C_n and S_n are then associated with a wave of wavelength $4\Delta x$, U_n with a wave of wavelength $2\Delta x$, and V with the wave of infinite wavelength. Substituting (10.63) into (10.62) and using the formula

$$(u_{j+1}^n)^2 - (u_{j-1}^n)^2 = (u_{j+1}^n - u_{j-1}^n)(u_{j+1}^n + u_{j-1}^n) ,$$

we obtain the following relationships among the amplitudes:

$$\begin{aligned} C_{n+1} - C_{n-1} &= 2\lambda S_n(U_n - V) , \\ S_{n+1} - S_{n-1} &= 2\lambda C_n(U_n + V) , \\ U_{n+1} &= U_{n-1} . \end{aligned} \tag{10.64}$$

The last equation says that U_n takes on different values for the odd and even time steps, say A for odd time steps ($n = 1, 3, 5, \ldots$) and B for even time steps ($n = 2, 4, 6, \ldots$), that is, $U_{2m} = A$ and $U_{2m-1} = B$ for $m = 1, 2, 3, \ldots$. Eliminating S_n from the first equation in (10.64), we have

$$C_{n+2} - 2C_n + C_{n-2} = 4\lambda^2(A + V)(B - V)C_n . \tag{10.65}$$

The question then arises—is this solution stable in the von Neumann sense? As in the simple linear case, using the von Neumann method, we define a growth factor associated with the $4\Delta x$ wave. For the high wavenumber wave of wavelength $4\Delta x$, the growth factor has to be less than or equal to one to be stable. Defining the growth factor by $G \equiv C_{n+2}/C_n$ and substituting this into (10.65), it takes the form

$$G^2 - 2\gamma G + 1 = 0 , \tag{10.66}$$

where

$$\gamma = 1 + 2\lambda^2(A + V)(B - V) \tag{10.67}$$

is a real number. The roots of (10.66) are

$$G_{1,2} = \gamma \pm \mathrm{i}\sqrt{1-\gamma^2}\,. \tag{10.68}$$

As long as the radical is real ($\gamma^2 \leq 1$),

$$|G_{1,2}| = \sqrt{\gamma^2 + 1 - \gamma^2} \equiv 1\,, \tag{10.69}$$

and the $4\Delta x$ wave is neutrally stable.

The condition $\gamma^2 \leq 1$ is equivalent to

$$-1 \leq \gamma \leq 1\,, \tag{10.70}$$

and it is satisfied if the amplitude of the $2\Delta x$ wave is such that $|A| < V$ and/or $|B| < V$. This is violated when the amplitude of the $2\Delta x$ wave is large in comparison with the energy contained in the longer waves (low wavenumbers). Under these circumstances, the $4\Delta x$ wave will grow exponentially and the scheme will be unstable.

10.4 Smoothing and Filtering

Nonlinearity is the heart and soul of the dynamics of atmospheres and ocean. As shown in the previous section, when nonlinearity is present it leads to nonlinear instability. To control this instability, it is important to damp out the smallest space scales. As discussed in Sect. 6.7, this may be done by adding explicit eddy viscosity or momentum diffusion (Sects. 3.2 and 4.9). Focus here is on another method, namely, filtering techniques. Such techniques also control spurious growth of short waves due to numerical errors and computational instabilities that would otherwise obscure a good forecast. In fact, even if a catastrophic instability does not occur, it is still appealing to remove the noise in the shortest wavenumber band for aesthetic reasons, in particular with regard to the final product.

The simplest form of smoothing is to apply a one-dimensional three-point operator or filter, often referred to as the Shapiro filter (Shapiro 1970, 1975). This is defined by

$$\overline{u}_j^n = (1-\mu)u_j^n + \frac{1}{2}\mu\left(u_{j+1}^n + u_{j-1}^n\right)\,, \tag{10.71}$$

where μ is a constant. If the solution is a monochromatic wave, say of the form $u_j^n = U_n \mathrm{e}^{\mathrm{i}\alpha(j\Delta x - cn\Delta t)}$, then the filtered solution is

$$\overline{u}_j^n = R u_j^n\,, \tag{10.72}$$

where

$$R = 1 - \mu(1-\cos\alpha\Delta x) = 1 - 2\mu\sin^2\frac{\alpha\Delta x}{2} \tag{10.73}$$

is the *response function* associated with the filter. Provided that $\mu \leq 1/2$, the response function is $0 \leq R \leq 1$. Thus, the filtering does not affect the sign, the wavelength, or the phase speed. However, for waves with $R < 1$, the amplitude of that particular wave is damped. Regarding the shortest wave resolved, i.e., $\alpha = \pi/\Delta x$, the response function is

$$R = 1 - 2\mu \, . \tag{10.74}$$

Choosing $\mu = 1/2$ then leads to $R = 0$, implying that the shortest resolved wave (wavelength $2\Delta x$) is completely removed by that particular filter.

Furthermore, observe that the filter may be rewritten to yield

$$\overline{u}_j^n = u_j^n + \frac{1}{2}\mu \left(u_{j+1}^n - 2u_j^n + u_{j-1}^n \right) \, . \tag{10.75}$$

The last term on the right-hand side of (10.75) is recognizable as the FDA of the second-order derivative in space, with a truncation error of $\mathcal{O}(\Delta x^2)$. Hence, making use of Taylor series, (10.75) takes the form

$$\overline{u}_j^n = u_j^n + \frac{1}{2}\mu \Delta x^2 \left[\partial_x^2 u \right]_j^n \, . \tag{10.76}$$

The filter, therefore, acts similarly to diffusion, with a mixing coefficient given by $\kappa = \mu \Delta x^2/2$. For further details, the reader is referred to Chap. 11-8, p. 392 ff Haltiner and Williams (1980).

10.5 Two-Way Nesting

Nesting techniques generally consist of a local high resolution grid (child grid or just child) embedded in a coarse resolution grid (parent) that provides the boundary conditions for the child grid (see Fig. 7.2 in Sect. 7.2). If information is transferred only from the parent to the child, with no feedback, this is referred to as *one-way nesting* (Sects. 7.5 and 7.6). If, in addition, there is a transfer of information from the child to the parent, it is referred to as *two-way nesting*. A general review of two-way nesting algorithms may be found in Debreu and Blayo (2008) and Debreu et al. (2012), along with applications focusing on upscaling impact (Biastoch et al. 2008), fine-scale dynamics (Marchesiello et al. 2011), and topographic refinement (Sannino et al. 2009).

The starting point is that the governing equations of the underlying mathematical model to be solved is the same for the child and the parent, e.g.,

$$\partial_t q = \mathcal{L}[q] \, , \tag{10.77}$$

in which q represents an array containing the variables and $\mathcal{L}$ is a spatial operator. The model considered may, for instance, be one of the three equations treated in Chaps. 4–6. For instance, $\mathcal{L} = \kappa \partial_x^2$ if the model is the diffusion equation (4.1), while $\mathcal{L} = -u_0 \partial_x$ if the governing equation is the advection equation (5.3). Finally, if the model is the nonlinear rotating shallow water equations (6.128)–(6.130), then

$$q = \begin{bmatrix} u \\ v \\ h \end{bmatrix}, \qquad \mathscr{L} = \begin{bmatrix} -u\partial_x & f & -g\partial_x \\ -f & -u\partial_x & 0 \\ -h\partial_x & 0 & -u\partial_x \end{bmatrix}. \tag{10.78}$$

The next step is to discretize (10.77) by choosing FDAs for $\partial_t q$ and the various derivatives entering the $\mathscr{L}$ operator. In principle, a different choice of numerical schemes and parameterizations may be adopted in the child, but this would complicate the issue of interface continuity already posed by the grid refinement itself. For pedagogical reasons, we focus here on problems where the same discretization is used for the child and parent. In general, the discretized version of (10.77) is therefore

$$[\partial_t q]_\mathrm{p} = \mathscr{L}_\mathrm{p}[q_\mathrm{p}], \qquad [\partial_t q]_\mathrm{c} = \mathscr{L}_\mathrm{c}[q_\mathrm{c}], \tag{10.79}$$

where the subscripts p and c denote the parent and child domains, respectively. We stress that $\mathscr{L}_\mathrm{p}$ and $\mathscr{L}_\mathrm{c}$ are the same discretizations of the same continuous operator $\mathscr{L}$, but at different resolutions. The problem is to solve (10.79) within the parent and child domains (Fig. 7.2) so that the parent domain solution impacts the child domain solution, while at the same time the child domain solution is allowed to impact the parent domain solution.

Furthermore, we assume that the discretization used to solve (10.79) results in an explicit scheme, and denote the time step and space increments by $\Delta t_{\mathrm{p,c}}$ and $\Delta x_{\mathrm{p,c}}$, respectively. Due to the refinement of the grid employed for the child domain, both the time step and space increments are smaller. Let i_r denote the spatial refinement factor, implying $\Delta x_\mathrm{p} = i_\mathrm{r}\Delta x_\mathrm{c}$. For convenience, let the time refinement factor be i_r as well, so that $\Delta t_\mathrm{p} = i_\mathrm{r}\Delta t_\mathrm{c}$. For reasons of interface continuity, the refinement factor cannot be too large. A number between 1 and 5 is common (Debreu et al. 2012). Moreover, if (10.77) is the advection equation or the shallow water equations, then the parent and child discretizations obey the same stability criterion, namely $C \le 1$, where C is the Courant number.

With this in mind, the steps needed to achieve two-way nesting are as follows:

1. Run the parent forward from time level n to time level $n+1$ for all interior parent grid points, e.g.,

$$q_\mathrm{p}^{n+1} = q_\mathrm{p}^n + \Delta t_\mathrm{p}\mathscr{L}_\mathrm{p}^n[q], \quad \mathbf{r}_\mathrm{p} \in \Omega, \tag{10.80}$$

where $\mathbf{r}_\mathrm{p}$ is a vector containing all the grid points *inside* the domain specified here by Ω, and update the solution at its outer boundary by taking into account the boundary conditions of the parent domain. Notice that any time discretization may be used to solve (10.80), as long as an appropriate spatial discretization is used to render the scheme stable and consistent.

2. Update the child at the interface Γ by interpolating parent values using a time interpolator $\mathscr{P}$:

$$q_\mathrm{c}^{n+m/i_\mathrm{r}}|_\Gamma = \mathscr{P}[q_\mathrm{p}^n, q_\mathrm{p}^{n+1}], \quad m = 1, 2, \ldots, i_\mathrm{r}. \tag{10.81}$$

3. Run the child forward for all interior child grid points, e.g.,

$$q_\mathrm{c}^{n+m/i_\mathrm{r}} = q_\mathrm{c}^{n+(m-1)/i_\mathrm{r}} + \Delta t_\mathrm{c}\mathscr{L}_\mathrm{c}^{n+(m-1)/i_\mathrm{r}}[q], \quad \mathbf{r}_\mathrm{c} \in \omega, \quad m = 1, 2, \ldots, i_\mathrm{r}. \tag{10.82}$$

4. Finally, update the parent within the child domain by filtering the child at their common grid points by

$$q_{\mathrm{p}}^{n+1} = \mathscr{R}[q_{\mathrm{c}}^{n+1}]_{\mathrm{p}}\,, \quad \mathbf{r}_{\mathrm{c}} \in \omega\,, \tag{10.83}$$

where $\mathscr{R}$ is the filter (sometimes referred to as the restriction operator).

Steps 1–3 are the same as those one would perform when applying one-way nesting, except that the OBC (Chap. 7) is replaced by a simple interpolation scheme (Step 2), which forces q_{c}^{n+1} to equal q_{p}^{n+1} on the interface Γ. The innovation is the fourth step, in which the child is allowed to update the parent solution. This is done by replacing the parent domain solution obtained through Step 1 by a filtered child domain solution at the common grid points within the child domain, whence the subscript p on the right-hand side of (10.83). Thus, when the parent is run forward to the next time level, it uses the corrected solutions within the child domain. It is emphasized that the parent is only corrected in the interior of the child domain ω. No corrections are made at the interface Γ of the two domains, or outside the child domain. Nevertheless, since the parent solution is changed within the child, the child solution impacts the parent even outside the child domain.

The rationale behind applying the filter $\mathscr{R}$ in Step 4 is to filter out those smaller scale variations in the child that are more or less unresolved by the parent, but at the same time to include as much as possible of those scales that are resolved by the parent. The filter may be a simple average, e.g.,

$$[q_{\mathrm{p}}]_{j_{\mathrm{p}}}^{n+1} = \frac{1}{3}\left([q_{\mathrm{c}}]_{j_{\mathrm{c}}+1}^{n+1} + [q_{\mathrm{c}}]_{j_{\mathrm{c}}}^{n+1} + [q_{\mathrm{c}}]_{j_{\mathrm{c}}-1}^{n+1}\right)\,, \tag{10.84}$$

or a Shapiro filter (Sect. 10.4),

$$[q_{\mathrm{p}}]_{j_{\mathrm{p}}}^{n+1} = [q_{\mathrm{c}}]_{j}^{n+1} + \frac{1}{2}\mu\left([q_{\mathrm{c}}]_{j_{\mathrm{c}}+1}^{n+1} - 2[q_{\mathrm{c}}]_{j_{\mathrm{c}}}^{n+1} + [q_{\mathrm{c}}]_{j_{\mathrm{c}}-1}^{n+1}\right)\,, \tag{10.85}$$

which becomes a simple 1-2-1 filter when $\mu = 1/2$. It may also be a more sophisticated filter, for instance, a full weighted filter

$$[q_{\mathrm{p}}]_{j_{\mathrm{p}}}^{n+1} = \frac{1}{9}\left([q_{\mathrm{c}}]_{j_{\mathrm{c}}+2}^{n+1} + 2[q_{\mathrm{c}}]_{j_{\mathrm{c}}+1}^{n+1} + 3[q_{\mathrm{c}}]_{j_{\mathrm{c}}}^{n+1} + 2[q_{\mathrm{c}}]_{j_{\mathrm{c}}-1}^{n+1} + [q_{\mathrm{c}}]_{j_{\mathrm{c}}-2}^{n+1}\right)\,. \tag{10.86}$$

In (10.84)–(10.86), j_{p} and j_{c} are the grid point counters for the parent and child, respectively. It is assumed in these filter formulas that the midpoint j_{c} is a grid point which the child and the parent have in common. Like the Shapiro filter, these filters have different response functions. Nevertheless, they all damp the amplitudes of small scale variations in the child solution before applying them to the parent.

A Simple Example: Advection of a Bell Function

As an example consider the one-dimensional advection equation (5.3). Let $q = \psi$, where ψ is the concentration of a solute in percent, $\mathscr{L}[q] = u_0\partial_x\psi$, and u_0 is the advection speed (assumed positive). Assume that the initial concentration is bell-shaped, i.e.,

$$\psi(x, 0) = \psi_0 \mathrm{e}^{-(x/\sigma)^2}\,, \tag{10.87}$$

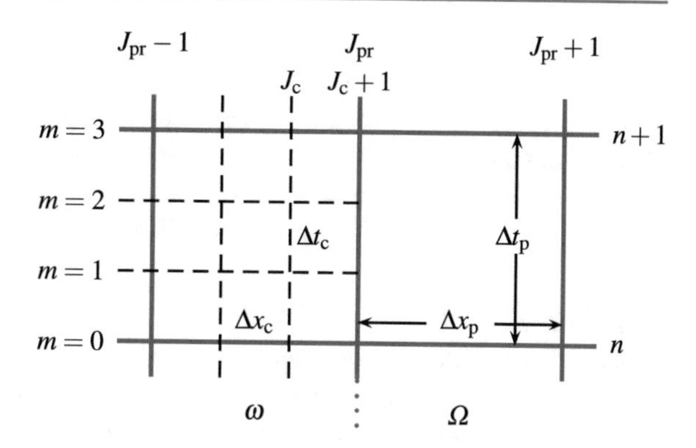

Fig. 10.3 Close-up of the parent (*solid lines*) and child (*dashed*) grid point locations in time and space in the vicinity of a right-hand interface. The refinement factor is $i_r = 3$

where σ is a measure of the width of the bell and ψ_0 the initial (100%) maximum concentration (Fig. 10.4).

Let the parent span the domain $-L < x_p < L$, where L is the halfwidth of the domain, and let $j_p = 1(1)J_p + 1$ and $j_c = 1(1)J_c + 1$ be the counters for the grid points of the parent and child domains, respectively. The locations of the common grid points at the left-hand ($j_c = 1$) and right-hand ($j_c = J_c + 1$) interfaces between the parent and child domains are denoted by $j_p = J_{pl}$ and $j_p = J_{pr}$, respectively (Fig. 10.3). Furthermore, let the child grid occupy a portion of the parent domain, that is, $-aL < x_c < bL$, where a and b are positive-definite constants less than or equal to one. Thus,

$$J_{pl} = 1 + \frac{(1-a)L}{\Delta x_p} \quad \text{and} \quad J_{pr} = 1 + \frac{(1+b)L}{\Delta x_p} . \tag{10.88}$$

If $a = b = 1$, then $J_{pl} = 1$ and $J_{pr} = J_p + 1$, in which case the child grid covers the entire parent domain.[4]

Let $\hat{\psi}_{j_p}, \tilde{\psi}_{j_c}$ denote ψ at parent and child grid points, respectively. Employing the convergent, stable, and consistent upstream scheme to discretize (5.3), the first step (Step 1) in the two-way nesting procedure leads to

$$\hat{\psi}^{n+1}_{j_p} = \hat{\psi}^{n}_{j_p} - C(\hat{\psi}^{n}_{j_p} - \hat{\psi}^{n}_{j_p-1}) , \quad j_p = 2(1)J_p + 1 , \tag{10.89}$$

where $\hat{\psi}_{j_p}$ is used to denote ψ at the parent grid points and $C = u_0 \Delta t_p / \Delta x_p$ is the Courant number. Imposing cyclic boundary conditions at the left-hand and right-hand boundaries of the domain, we obtain

$$\hat{\psi}^{n+1}_{1} = \hat{\psi}^{n+1}_{J_p+1} . \tag{10.90}$$

The next step (Step 2) updates the child at the boundaries for $m = 1, 2, \ldots, i_r$. Using a two-point linear interpolator yields

$$\tilde{\psi}^{n+m/i_r}_{1} = \left(1 - \frac{m}{i_r}\right) \hat{\psi}^{n}_{J_{pl}} + \frac{m}{i_r} \hat{\psi}^{n+1}_{J_{pl}} , \tag{10.91}$$

[4]Recall that $2L = (J_p + 1)\Delta x_p$ in accordance with (2.49), and hence $J_{pr} = J_p + 1$ when $b = 1$.

$$\tilde{\psi}_{J_c+1}^{n+m/i_r} = \left(1 - \frac{m}{i_r}\right)\hat{\psi}_{J_{pr}}^{n} + \frac{m}{i_r}\hat{\psi}_{J_{pr}}^{n+1} . \quad (10.92)$$

The third step (Step 3) is to solve (5.3) numerically on the child grid. Note that the Courant number is exactly the same, since $\Delta t_p/\Delta x_p = \Delta t_c/\Delta x_c$. Using the same scheme as for the parent grid, we have

$$\tilde{\psi}_{j_c}^{n+m/i_r} = \tilde{\psi}_{j_c}^{n+(m-1)/i_r} - C\left[\tilde{\psi}_{j_c}^{n+(m-1)/i_r} - \tilde{\psi}_{j_c-1}^{n+(m-1)/i_r}\right], \quad j_c = 2(1)J_c , \quad (10.93)$$

for $m = 1, 2, \ldots, i_r$.

The last step (Step 4) is to update the concentration using the child solution at the parent grid points in the interior of the child grid at those locations where the child and the parent grid points coincide. To this end, the Shapiro filter (10.85) may be used with $\mu = 1/2$ (a 1-2-1 filter). The result is

$$\hat{\psi}_{j_p}^{n+1} = \frac{1}{4}\left(\tilde{\psi}_{j_c^*+1}^{n+1} + 2\tilde{\psi}_{j_c^*}^{n+1} + \tilde{\psi}_{j_c^*-1}^{n+1}\right), \quad j_p = J_{pl} + 1(1)J_{pr} - 1 , \quad (10.94)$$

where j_c^* denotes a child grid point that coincides with the parent grid point. The solution is displayed in Fig. 10.4, where $L = 50$ km, $\sigma = L/5$ and the refinement factor is set to $i_r = 5$.

Recall that the upstream scheme is numerically diffusive with a diffusion coefficient given by (5.115)

$$\kappa_p^* = \frac{1}{2}(1 - C)|u_0|\Delta x_p , \qquad \kappa_c^* = \frac{1}{2}(1 - C)|u_0|\Delta x_c , \quad (10.95)$$

respectively, implying that $\kappa_p^* = i_r\kappa_c^*$. The child solution is, therefore, expected to be less diffusive than the parent solution if a refinement is taken into account ($i_r > 1$). This is illustrated by the top right-hand panel of Fig. 10.4 displaying the solutions to (10.89) and (10.93) for the full parent domain. No nesting is applied and $a = b = 1$ in (10.88). Under these circumstances, the parent and child solutions are stand-alone solutions, and a cyclic boundary condition (10.90) is applied at the boundaries $x = \pm L$. Neither of the two solutions is perfect compared to the analytic solution (shown by the dashed curve), but the child solution is clearly less diffusive.

Two questions arise. First, what is the impact on the child solution of applying a one-way nesting procedure as described in Chap. 7? And second, what is the impact on the parent and child solutions of applying a two-way nesting procedure as described here? The answer to these questions is illustrated in the two lower panels of Fig. 10.4. In deriving the solutions depicted, the above two-way nesting procedure is used, embedding the child domain in the parent domain by letting $a = b = 1/2$. If we apply one-way nesting, as illustrated in the lower left-hand panel of Fig. 10.4 (bottom left), nothing happens to the parent solution, while the child solution is severely degraded compared to the control shown top right. This is to be expected, since one-way nesting has no impact on the parent solution, either outside or inside the child domain. Applying two-way nesting changes this picture, as can be seen from the lower right-hand panel of Fig. 10.4. Both the parent and the child solutions are improved compared to the one-way nesting. Moreover, the parent solution is improved outside as well as inside the child grid domain. Although the child solution is still degraded compared to the no-nesting case, it is nonetheless improved compared to the one-way nesting case.

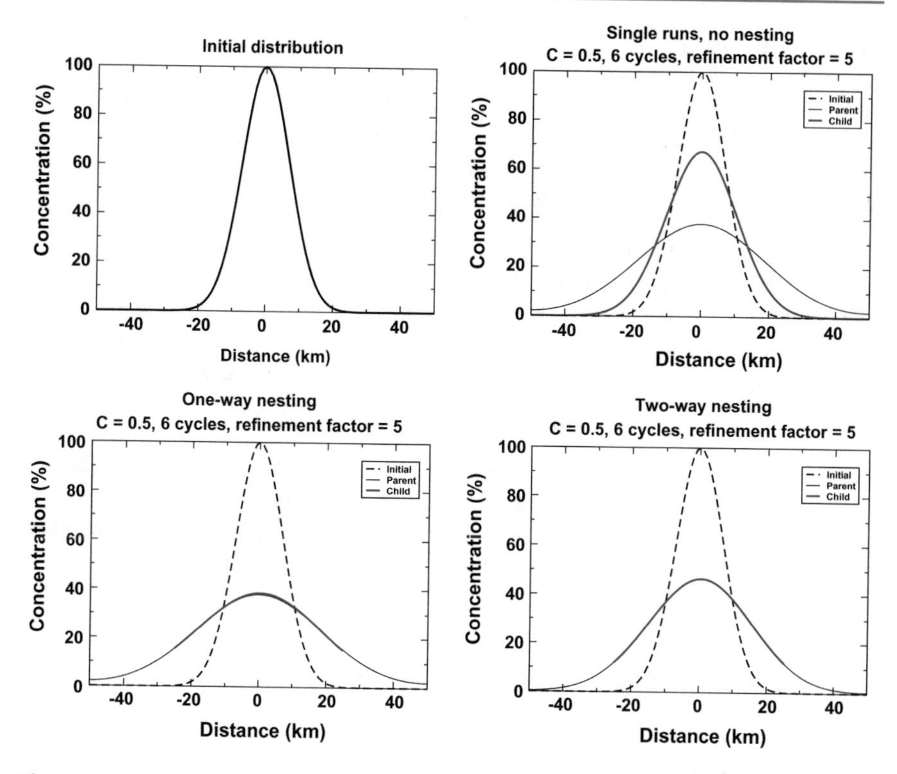

Fig. 10.4 *Top left*: Initial distribution. The other panels show the distributions after six cycles with no, one-way, and two-way nesting, from *top right*, through *bottom left* to *bottom right*. In the *top right*, the child grid covers the entire domain of the parent grid [$a = b = 1$ in (10.88)], and a cyclic boundary condition is applied on the boundaries. Interfaces between the parent and child domains are at ± 25 km. In the *top right*, *bottom left*, and *bottom right* panels, the *blue line* is the child solution while the *solid black line* is the parent solution. The *dashed line* is the analytic solution after six cycles and corresponds to the initial distribution shown *top left*

10.6 Spectral Method

When applying the various finite difference approximations to the advection equation, only grid point values of the dependent variables were considered. No assumptions were made regarding the behavior of the variables in-between grid points, other than assuming that they were well-behaved functions. And under this assumption, an alternative to the finite difference approach is to expand the dependent variables in terms of finite series of orthogonal functions (Sect. 2.9). The problem is then reduced to solving a set of ordinary differential equations determining the time evolution of the expansion coefficients. The spatial "resolution", as well as the computational burden, is then determined by the number of waves in the finite series.

This approach is known as the *spectral method*. It is particularly suitable for global (atmospheric) models. The dependent variables are then zonally cyclic functions and hence easy to expand. The method is, therefore, commonly applied in modern global atmospheric models, as in the global model used at the European Centre for Medium-Range Weather Forecasts (ECMWF). The method can also be used for limited-area models, but it is then a bit more cumbersome. Since oceanographic models have to deal with continental land boundaries, it is also cumbersome to apply the spectral method to global oceanographic models.

Application to the One-Dimensional Linear Advection Equation

To illustrate the method, consider the one-dimensional advection equation on the globe, that is, along a latitude. The natural boundary condition is then the periodic or cyclic boundary condition (Sect. 2.4). The starting point is the one-dimensional advection equation (Sect. 5.2), which we restate here for completeness:

$$\partial_t \phi = -u_0 \partial_x \phi \ , \quad \text{for} \ \ x \in [0, L] \ , \quad t > 0 \ . \tag{10.96}$$

Here, L is the length of the circumference at a particular latitude. Recall that (10.96) is just a special case of the general equation (2.66). Hence, the linear operator is $\mathscr{H} = -u_0 \partial_x$. Since (10.96) is to be solved along a latitude, it is convenient to transform to a coordinate system in which

$$2\pi x = \xi L \ , \quad \text{or} \ \ \xi = \frac{2\pi x}{L} \ , \tag{10.97}$$

where $\xi \in \langle 0, 2\pi \rangle$ is the new (dimensionless) zonal coordinate. Since

$$\partial_x \phi = \partial_\xi \phi \partial_x \xi \ , \tag{10.98}$$

(10.96) takes the form

$$\partial_t \phi = -\gamma \partial_\xi \phi \ , \tag{10.99}$$

where

$$\gamma = \frac{2\pi u_0}{L} \tag{10.100}$$

is the angular velocity (in units s^{-1}). The cyclic boundary condition is written

$$\phi(\xi, t) = \phi(\xi + 2\pi m) \ , \quad m = 1, 2, 3, \dots \ , \tag{10.101}$$

where m describes the number of times a signal has traveled around the world at that latitude. Furthemore, let the initial condition be described by the well-behaved function $f(\xi)$. Hence,

$$\phi(\xi, 0) = f(\xi) \ . \tag{10.102}$$

As outlined at the beginning of Sect. 5.2, the analytic or true solution to (10.99) is therefore

$$\phi = f(\xi - \gamma t) \ . \tag{10.103}$$

In order to solve (10.99) using expansions in terms of orthogonal functions, we must first choose a suitable set of expansion functions. Since trigonometric functions or complex exponentials (sine and cosine functions) are eigenfunctions of the differential operator $\mathscr{H} = -u_0\partial_x$, the obvious choice in our case is to use complex exponentials. In general, this leads to

$$\phi = \sum_{m=-\infty}^{\infty} \phi_m(t)\mathrm{e}^{\mathrm{i}\alpha_m\xi} . \tag{10.104}$$

Solving (10.99) using numerical methods implies that the wavenumber space is band-limited. Hence, a truncated version of (10.104) must be used, which takes the form

$$\phi(\xi, t) = \sum_{m=-M}^{M} \phi_m(t)\mathrm{e}^{\mathrm{i}\alpha_m\xi} , \tag{10.105}$$

where M gives the maximum wavenumber α_M, corresponding to the shortest wavelength resolved by the choice of grid size (here $2\Delta\xi$). Furthermore, since $\phi^*_{-m} = \phi_m$, we need only consider waves associated with $0 \le m \le M$. Thus, the expansion that needs to be considered is

$$\phi(\xi, t) = \sum_{m=0}^{M} \phi_m(t)\mathrm{e}^{\mathrm{i}\alpha_m\xi} . \tag{10.106}$$

Substituting (10.106) into (10.99), we obtain

$$\sum_{m=0}^{M} (\partial_t\phi_m + \mathrm{i}\alpha_m\gamma\phi_m) = 0 \implies \partial_t\phi_m = -\mathrm{i}\alpha_m\gamma\phi_m , \quad m = 0(1)M , \tag{10.107}$$

leading to $M+1$ equations for the expansion coefficients ϕ_m. In this particular case, (10.107) can be integrated exactly for each wavenumber α_m separately, whence

$$\phi_m(t) = \phi^0_m\mathrm{e}^{-\mathrm{i}\alpha_m\gamma t} , \quad m = 0(1)M , \tag{10.108}$$

where ϕ^0_m is the initial condition associated with the wavenumber α_m. Expanding the well-behaved function $f(\xi)$ in terms of the complex exponentials, our chosen orthogonal functions, we have

$$f(\xi) = \sum_{m=0}^{M} a_m\mathrm{e}^{\mathrm{i}\alpha_m\xi} = \sum_{m=0}^{M} \phi^0_m\mathrm{e}^{\mathrm{i}\alpha_m\xi} , \tag{10.109}$$

and hence $\phi^0_m = a_m$. The complete solution to (10.99) is, therefore,

$$\phi(\xi, t) = \sum_{m=0}^{M} a_m\mathrm{e}^{\mathrm{i}\alpha_m(\xi-\gamma t)} , \tag{10.110}$$

which is exactly the same as the true solution (10.103) when the latter is expanded in complex exponentials. Hence, there is no dispersion due to the "space" discretization, in stark contrast to the finite difference solution (Sect. 5.5). This is to be expected, since the space derivatives are computed analytically when we apply the spectral method, while they are approximated by finite differences when we apply the finite difference method.

Recall that the orthogonality property of the expansion functions, in our case $e^{i\alpha_m \xi}$, is

$$\int_0^{2\pi} e^{i\alpha_m \xi} e^{-i\alpha_n \xi} d\xi = A_m \delta_{mn} , \tag{10.111}$$

where A_m are the normalization factors. Multiplying (10.106) by the complex conjugate of the expansion functions and integrating in space yields

$$\int_0^{2\pi} \phi(\xi, t) e^{-i\alpha_n \xi} d\xi = \sum_{m=0}^{M} \phi_m(t) \left(\int_0^{2\pi} e^{i\alpha_m \xi} e^{-i\alpha_n \xi} d\xi \right) = A_n \phi_n(t) . \tag{10.112}$$

Thus, since m and n are dummy counters,

$$\phi_m(t) = A_m^{-1} \int_0^{2\pi} \phi(\xi, t) e^{-i\alpha_m \xi} d\xi . \tag{10.113}$$

In passing, we note that the integral in (10.113) is the so-called direct Fourier transform (Sect. 2.11). The normalization coefficients are determined from the initial condition. Letting $t = 0$ in (10.113), and recalling that $\phi_m(0) = a_m$, we obtain

$$A_m = a_m^{-1} \int_0^{2\pi} \phi(\xi, 0) e^{-i\alpha_m \xi} d\xi . \tag{10.114}$$

In practice, rather than being a continuous function in space, $\phi(\xi, t)$ is only known at the grid points $\xi_j = j\Delta\xi$, $j = 0(1)J$, where $\xi_0 = 0$ and $\xi_J = 2\pi$. Under these circumstances, the truncated Fourier series of ϕ in (10.106) represents an interpolating function which exactly fits the values of ϕ at the $J + 1$ grid points. Hence, in practice (10.113) is written as a discrete direct Fourier transform, which takes the form

$$\phi_m(t) = A'^{-1}_m \sum_{j=1}^{J} \phi(\xi_j, t) e^{-i\alpha_m \xi_j} , \tag{10.115}$$

where the normalization coefficients are found by discretization of (10.114):

$$A'_m = a_m^{-1} \sum_{j=1}^{J} \phi(\xi_j, 0) e^{-i\alpha_m \xi_j} . \tag{10.116}$$

The corresponding discrete inverse Fourier transform is then

$$\phi(\xi_j, t) = \sum_{m=-M}^{M} \phi_m(t) e^{i\alpha_m \xi_j} . \tag{10.117}$$

Both (10.115) and (10.117) can be computed with the Fast Fourier Transform (FFT) algorithm. It can be shown that, starting from the set $\phi_m(t)$, going to the set $\phi(\xi_j, t)$ with $j = 0(1)J$, and returning to the set $\phi_m(t)$, the original values are exactly recovered, provided that the number of grid points J is such that $J > 2M + 1$. Recall that M is the number of waves used to compute the direct Fourier transform in (10.117). In addition, the points ξ_j must be equally spaced, that is, $\Delta\xi$ must be constant.

It remains to find the expansion coefficients $\phi_m(t)$ at an arbitrary time, given their initial values ϕ_m^0. This is done using the finite difference method, for instance applying a centered-in-time scheme to (10.107), viz.,

$$\phi_m^{n+1} = \phi_m^{n-1} - 2\mathrm{i}\alpha_m \gamma \Delta t \phi_m^n , \quad m = 0(1)M , \tag{10.118}$$

for each wavenumber α_m. The stability condition, using von Neumann's method once again, is

$$|\alpha_m \gamma \Delta t| \leq 1 , \quad \forall m . \tag{10.119}$$

Since the maximum wavenumber is $\alpha_M = L/2\Delta x$, and recalling that $\gamma = 2\pi u_0/L$, it follows that the stability condition in terms of the Courant number $C = u_0 \Delta t/\Delta x$ is

$$C \leq \frac{1}{\pi} . \tag{10.120}$$

This condition is actually stricter than the one derived using the finite difference approximation. Nonetheless, the spectral method and the derived scheme have the advantage that they are nearly nondispersive, and that the dispersiveness is very small even for the shortest waves.

10.7 Summary and Remarks

Section 10.1 outlined the procedure for constructing higher order finite difference schemes. This procedure was exemplified in an application to the advection equation. Properties such as stability, numerical dispersion, and computational modes were discussed. In general, it was found that applying higher order schemes lessened the dispersion and led to stricter stability conditions, but had no effect on the computational mode. A more serious implication is that all higher order schemes make use of information farther and farther away from the grid point j, where the solution is to be computed at the new time level. For instance, the centered sixth-order-accurate leap-frog scheme (10.25) makes use of information from grid points $j \pm 3$, and $j \pm 2$ in addition to $j \pm 1$ at time level n. Thus, close to any natural physical boundary, we require information from points outside the computational domain and its boundaries. This information is not available, so ad hoc methods have to be found to specify this information. Consequently, the correctness of the solution may or may not be seriously jeopardized, since such an approach is equivalent to specifying more boundary conditions than required by the underlying continuous problem.

Section 10.2 raised the question as to what scheme should be used when advection and diffusion are equally important. As stated in Chap. 4, the scheme used to solve the diffusion equation must be forward-in-time to be numerically stable. In contrast, and as stated in Chap. 5, the scheme used to solve the advection equation requires a centered-in-time scheme. At least two options were discussed. One was to use a centered-in-time and centered-in-space scheme, but to evaluate the diffusion part at time level $n - 1$ rather than at time level n. In this way, the diffusion part was forward-in-time with a time step of $2\Delta t$. This had implications for the stability. As it turned out, the stability criterion became stricter when the two processes were of equal importance.

In Sect. 10.3, we discussed nonlinear instabilities and ways to avoid them. When nonlinear terms are present, which they must be in any credible model of the atmosphere and/or ocean, energy is continuously cascaded down to shorter and shorter waves. In the real world, this cascade goes all the way down to the molecular scale and hence is lost to internal energy. In a numerical model, the shortest wave resolved is the wave of wavelength equal to the Nüquist wavelength or $2\Delta x$. If the cascade to even smaller wavelengths is not properly treated, the energy contained in the unresolved waves will accumulate by folding into the highest wavenumber (shortest wavelength) band resolved, and sooner or later the model will blow up. We discussed methods for handling the cascade, such as adding artificial diffusion.

One such method was investigated further in Sect. 10.4, in which we presented various spatial filters like the Shapiro filter. In many ways, this section was a prelude to Sect. 10.5, which presented details of the two-way nesting technique. This technique is the most recent one suggested to avoid accumulation of properties like precipitation, vorticity, etc., which accumulate in the vicinity of interfaces between a coarse mesh model domain and a finer mesh model domain when the latter is wholly or partly embedded in the former (Debreu and Blayo 2008; Debreu et al. 2012). The idea is to let the coarser mesh model, or parent, influence the solution within the fine mesh model domain, or child domain, but at the same time let the child impact the solution in the parent domain. This effectively inhibits any such accumulation at the interface between the two. Additionally, since the child is usually closer to the true solution due to its finer mesh, it also helps the parent to be a better approximation, even outside the child domain. Once again, the advection equation was used to exemplify the technique.

Finally, in Sect. 10.6, we touched upon a method that differs somewhat from the finite difference approach. This method is the spectral method, which uses the fact that all well-behaved functions may be expanded in an infinite series of orthogonal functions. This opens up the possibility of having an extremely accurate interpolator between the grid points. Once again, the one-dimensional advection equation was used to illustrate the method. The main advantage of the spectral method is that it is almost nondispersive. However, the stability criterion is somewhat stricter.

10.8 Exercises

1. Using the procedure outlined in Sect. 10.1 and assuming that the advection velocity $u_0 \geq 0$, show that a third-order upstream scheme for the advection equation is given by (10.39).
2. Use von Neumann's method (Chap. 4.4) to show that the expression (10.44) is indeed the correct expression for the growth factor when using the scheme in (10.41).
3. Show that the condition (10.48) is a sufficient condition for numerical stability.

10.9 Computer Problems

10.9.1 Combined Advection–Diffusion

Here, we consider the equation

$$\partial_t u + A\partial_x u - \partial_x(\kappa\partial_x u) = 0\,, \quad \text{for } \forall x \in \langle 0, L\rangle\,, \quad t > 0\,, \tag{10.121}$$

where $u = u(x,t)$ is the dependent variable, x, t are the independent variables, and A and κ are constants. Moreover, the boundary conditions are

$$u(x,0) = f_0(x)\,, \quad \forall x \in \langle 0, L\rangle\,, \tag{10.122}$$

$$u(0,t) = g_0(t)\,, \quad \forall t > 0\,. \tag{10.123}$$

A. What physics do the various terms represent? Let $\kappa = 0$ and $A \neq 0$, and discuss whether (10.121) is a hyperbolic, parabolic, or elliptic equation under these circumstances. If $\kappa \neq 0$ and $A = 0$, is (10.121) then hyperbolic, parabolic, or elliptic?

B. Construct a scheme to solve (10.121) numerically, which renders your scheme explicit, stable, and consistent. Explain how you would analyze your scheme to satisfy yourself that it is stable and consistent.

C. Change the scheme to make it implicit. How would you solve it?

D. Next, consider the nonlinear version of (10.121), namely

$$\partial_t u + u\partial_x u - \kappa\partial_x^2 u = 0\,. \tag{10.124}$$

Solve (10.124) numerically using the method of characteristics within a domain $x \in 0, L$, where the boundaries at $x = 0, L$ are open.

E. Set up a CTCS scheme that employs Dufort–Frankel for the diffusion term, and solve it. Explain why this scheme is inconsistent, and explain under what conditions this scheme could be used even though it is inconsistent. Compare the results with the solution obtained in (D).

References

Biastoch A, Böning C, Lutjeharms J (2008) Agulhas leakage dynamics affects decadal variability in Atlantic overturning circulation. Nature 456:489–492

Clancy RM (1981) On wind-driven quasi-geostrophic water movement at fast ice edges. Mon Weather Rev 109:1807–1809

Cushman-Roisin B (1984) Analytic, linear stability criteria for the leap-frog, Dufort-Frankel method. J Comput Phys 53:227–239

Debreu L, Blayo E (2008) Two-way embedding algorithms: a review. Ocean Dyn 58:415–428. https://doi.org/10.1007/s10236-008-0150-9

Debreu L, Marchesiello P, Penven P, Cambon G (2012) Two-way nesting in split-explicit ocean models: algorithms, implementation and validation. Ocean Model 49–50:1–21
Haltiner GJ, Williams RT (1980) Numerical prediction and dynamic meteorology, 2nd edn. Wiley, New York, 477 pp
Marchesiello P, Capet X, Menkes C, Kennan SC (2011) Submesoscale dynamics in tropical instability waves. Ocean Model 39:31–46. https://doi.org/10.1016/j.ocemod.2011.04.011
Phillips NA (1959) An example of non-linear computational instability. In: Bolin B (ed) The atmosphere and the sea in motion. Rockefeller Institute Press, New York, pp 501–504
Richtmyer RD (1963) A survey of difference methods for non-steady fluid dynamics. Technical Note 63-2, National Center for Atmospheric Research (NCAR)
Robert AJ, Shuman FG, Gerrity JP (1970) On partial difference equations in mathematical physics. Mon Weather Rev 98. https://doi.org/10.1175/1520-0493(1970)0982.3.CO;2
Sannino G, Herrmann M, Carillo A, Rupolo V, Ruggiero V, Artale V, Heimbach P (2009) An eddy-permitting model of the Mediterranean Sea with a two-way grid refinement at the Strait of Gibraltar. Ocean Model 30:56–72
Shapiro R (1970) Smoothing, filtering, and boundary effects. Rev Geophys Space Phys 8:359–387
Shapiro R (1975) Linear filtering. Math Comput 29:1094–1097

11 Quality Assurance Procedures

The aim here is to summarize a set of sound procedures for establishing what is referred to below as a "good" model. The text is based on earlier reports by the author on the subject, in particular, McClimans et al. (1992) and Røed (1993). For more extensive reading on the subject, the reader is referred to the in-depth analysis documented in the GESAMP report (GESAMP 1991), or the review (Lynch and Davies 1995).

Many of today's engineering models are formulated by mathematical equations leading to a numerical model that can be solved reliably with almost "canned" routines that require little understanding on the part of the user (referred to as "expert systems for nonexperts"). However, the atmospheric and oceanographic weather prediction models and climate models available today are not yet among them. They are prime examples of complex numerical models involving coupling of intricate physical, and sometimes chemical and biological model modules. Inherently, most complex models require a minimum of expertise to be transferred with the model, so that only in exceptional circumstances is it possible to turn a complex model developed by one group over to another. The reason is simply that all complex model systems have their inherent limitations that demand an understanding of the underlying processes, and in the case of a numerical model also the numerical techniques used to solve them. The simulation is never perfect. Different models and methods preserve different features of the original problem, implying that one needs to understand what is important for the purpose at hand. Nevertheless, in meteorology and oceanography such almost "canned" systems are publicly available for downloading on the web. Regarding models of the atmosphere, the Weather Research and Forecasting Model (WRF, https://www.mmm.ucar.edu/weather-research-and-forecasting-model) is a prime example, while the same is true for the Regional Ocean Modeling System (ROMS, http://www.myroms.org/) for the ocean.

In this light, any numerical model is at best an approximate representation of the real world. Hence, its predictions are inherently uncertain. This uncertainty results

L. P. Røed, *Atmospheres and Oceans on Computers*,
Springer Textbooks in Earth Sciences, Geography and Environment,
https://doi.org/10.1007/978-3-319-93864-6_11

from both a lack of knowledge of the full set of equations, and an inability to solve them. Therefore, approximations have to be made that involve the use of parameterizations of the processes in space and time. Uncertainty also arises from errors in observational data used to derive input and parameter values, that is, the initial state of the model and the boundary conditions. In addition, there may be problems with the accuracy of the computer code, and the method and techniques used to solve the discretized numerical analogue of the original continuous mathematical model. All these factors need attention when determining the accuracy of model predictions.

As an introduction, Sect. 11.1 highlights one of the common problems in solving a mathematical model by numerical techniques, using the advection equation as an example. This is the problem of the parameterization of unresolved scales. This paves the way for Sect. 11.2, which gives a general description of what is meant by a *good* model. Moreover, the terms *tuned* model, *transportable* model, and *robust* model are defined. Section 11.3 describes what are sometimes referred to as quality assurance or model validation procedures. These involve a three-step process in which the first step is to check or *verify* that the finite difference equations are solved correctly (referred to as model verification). The next step is to perform a *sensitivity analysis* to uncover the model's response to changes in the input data, parameter values, and parameterizations. The sensitivity analysis also serves to uncover the predictive skills of the model. The third and final step is to investigate the agreement between the model's predictions and observation, a task referred to as *model validation*. We also discuss *model calibration*, that is, the tuning of parameter values to make the model output fit a given data set. We end with a summary and some final remarks (Sect. 11.4).

11.1 Sub-grid Scale Parameterizations and Spectral Cutoffs

An exact numerical solution of the governing equations for the atmosphere and ocean as outlined in Chap. 1 is impossible, mainly due to processes that are not resolved by the grid. Thus, there exist spectral cutoffs for processes on scales smaller than the grid resolution, i.e., sub-grid scale processes (Sect. 10.3 and Griffies 2004, Chap. 7, p. 154). The effect of these sub-grid scale processes on the resolved scales must then be parameterized, i.e., it must be given an approximate mathematical formulation. This is commonly in the form of simplified formulas involving the specification of one or several parameters that may or may not be functions of the resolved scales. We can illustrate the need for such parameterizations by considering the advection–diffusion equation and simplifying by including only one space dimension.

Let $C = C(x, t)$ denote any property of the fluid, e.g., potential temperature or concentration of a particular contaminant, at location x and time t. Then, mass conservation requires

$$\partial_t C + \partial_x (uC) = 0 , \tag{11.1}$$

where $u = u(x,t)$ is the speed along the x-axis by which the property C is advected[1] (or propagated). The first term on the left-hand side of (11.1) then represents the time rate of change of the potential temperature or the contaminant in question, while the second term is the divergence of the advective flux $F_{\text{adv}} = uC$.

In practice, it is impossible to describe such a flow field, as u changes rapidly in both space and time. Hence, we must invoke an ensemble average or a space–time averaging process, the latter taken over a certain length scale T in time and/or L in space. This separates the current into an average or mean current, $\overline{u}$, and a random component, u', such that $u = \overline{u} + u'$, where $\overline{u'} = 0$. Here, $\overline{u}$ may be thought of as the mean flow over a certain time period and u' as the motion deviating from the mean, so that $\overline{u} + u'$ makes up the instantaneous flow at any time or location.

If the same separation is used for the concentration, it follows from (11.1) that

$$\partial_t(\overline{C} + C') + \partial_x\left[(\overline{u} + u')(\overline{C} + C')\right] = 0\,. \tag{11.2}$$

Averaging (11.2) over the chosen averaging period (or length), and noting that terms like u', C', $u'\overline{C}$, and $\overline{u}C'$ average out, leads to

$$\partial_t\overline{C} + \partial_x(\overline{u}\overline{C}) = \partial_x\left(\overline{u'C'}\right)\,. \tag{11.3}$$

The left-hand side of (11.3) is very similar to (11.1), except for the nonzero term on the right-hand side. As such it is an advection equation for a concentration $\overline{C}$ with a speed $\overline{u}$, where the right-hand side is a source/sink term. With reference to a numerical model, $\overline{C}$ and $\overline{u}$ may be thought of as being the solution resolved by the grid, while C' and u' represent the sub-grid scale motion. The term on the right-hand side of (11.3) then represents the way sub-grid scale processes influence the resolved concentration $\overline{C}$. To solve (11.3) with respect to $\overline{u}$ and $\overline{C}$, the right-hand side must somehow be expressed in terms of the average or the resolved quantities, i.e., it must be parameterized. Concerning the advection–diffusion equation, the parameterization is often in terms of a diffusive process (Sect. 10.3), i.e.,

$$F_{\text{diff}} = -\overline{u'C'} = -\kappa\partial_x C\,, \tag{11.4}$$

where F_{diff} is the diffusive flux.[2] The parameter or coefficient κ is called the eddy diffusivity or dispersion coefficient, and is in general a function of the motion $\overline{u}$, and hence time and space. As discussed below (Sect. 11.3), it is important to test the sensitivity of a model to these parameterizations and their parameters. This may provide insight into the fitness of the parameterization and may help to build confidence in the model and its predictive capability. Model prognoses that are highly sensitive to a particular parameterization or to the value given to a particular parameter should be treated with caution. Similar problems occur when parameterizing, e.g., biological and chemical processes. For instance, details of turbulent motion in atmospheres and oceans are poorly understood and must be parameterized.

---

[1] In a three-dimensional problem, the speed becomes a current, that is, a vector $\mathbf{v}$, and (11.1) becomes $\partial_t C + \nabla\cdot(\mathbf{v}C) = 0$, where ∇ is the three-dimensional gradient vector.

[2] In a three-dimensional problem, the diffusive flux becomes a vector $\mathbf{F}_{\text{diff}} = -\mathscr{K}\cdot\nabla C$, where $\mathscr{K}$ is a tensor. The right-hand side of (11.3) is then written $\nabla\cdot\mathbf{F}_{\text{diff}}$.

11.2 Good Models

Tuned, Transportable, and Robust Models

In theory, a numerical model with refined descriptions of the many processes involved and the interactions between them, and which includes complex and sophisticated parameterizations, should provide more accurate results and be more applicable to different situations and/or geographical areas than a model invoking simpler and coarser descriptions and parameterizations. This philosophy reflects a conviction that more detailed formulations provide a better description of the processes than simpler ones. In practice, this concept breaks down in many cases for the following reasons:

- it is sometimes questionable whether a "true" description exists for all relevant processes (e.g., turbulent mixing),
- too many processes are included that have to be parameterized, or
- our knowledge and understanding of the unresolved processes upon which the parameterizations are based is poor.

Under these circumstances, a complex model is of little value, since no more fundamental knowledge is being incorporated into the model, only parameterizations of poorly understood processes. This is visualized in Fig. 11.1.

Before constructing a "good" model a set of criteria has to be selected in order to make the necessary choices, that is, which processes and which interactions between

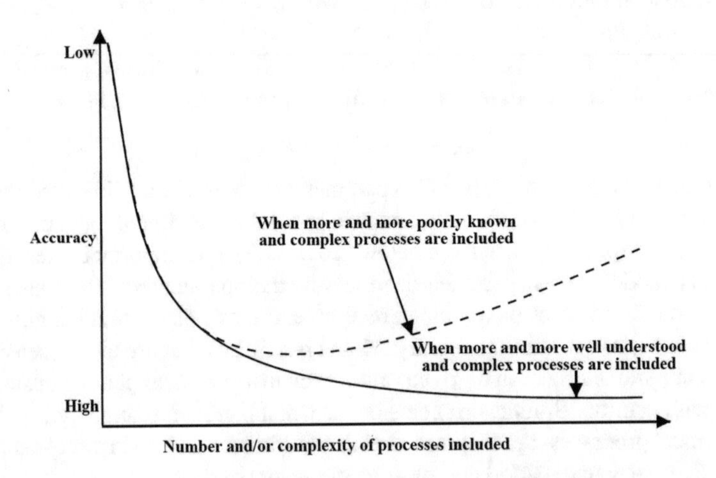

Fig. 11.1 Effect on accuracy of including more and more sophisticated parameterizations of processes. Decreasing accuracy is shown along the *vertical axis*, while increasing number and/or complexity of processes are shown along the *horizontal axis*. The accuracy first increases, then decreases (*dashed curve*) when more and more poorly known processes are included. This is in contrast to the case when the processes included are well understood (*solid curve*)

processes should be included. It is convenient in this respect to introduce the terms tuned model, transportable model, and robust model:

- A *tuned* model is one in which the parameterizations and parameter selection have been adjusted to reproduce, as accurately as possible, a given data set in a specific region for a specific time interval. In general, the more adjustable parameters there are in a model, the more difficult it will be to tune it, the more data will be required, and the more site-specific the model will become.
- A model is *transportable* if the parameterizations within the model are sufficiently comprehensive and representative of all relevant processes to ensure that, once calibrated and validated in one geographical area, the model can be used in any area containing the same generic processes. This does not imply that the specific parameter values that have been chosen for one area should remain invariant when the model is transported. However, a transportable model should yield similar levels of accuracy in a different geographical area, once it has been properly calibrated.
- A model is *robust* when it can provide similar degrees of accuracy over a wide range of variations in the forcing functions. For instance, storm surge models are normally validated against observations during major storm events. Nevertheless, they are expected to give the same level of accuracy for even more extreme events (e.g., the hundred year storm) for which no direct validation is usually possible.

The Concept of a Good Model

With the definitions given in Sect. 11.2 in mind, one may think of a good model as one which:

- retains a conceptual representation of the processes known to be important,
- uses parameterizations consistent with our understanding of those processes,
- does not use parameterizations that are so complex that the model needs to be highly tuned,
- can reproduce and/or predict phenomena over a wide range of geographical locations (i.e., it is transportable), and
- can reproduce and/or predict phenomena over a wide range of differing conditions (i.e., it is robust).

These criteria differ somewhat from the commonly accepted definitions of a good model, which may be drawn from many examples in the literature. There, a good model is simply one which accurately reproduces a given set of field observations. However, if this has been accomplished by tuning the model, so that it essentially fits a narrow set of conditions, it is no longer good since it may not accurately predict different events in the same area or the same kind of events in another geographical area, or it may not provide any insight into the nature of the underlying processes.

11.3 Quality Assurance Procedures

To ensure that a particularly model fulfills the requirements of a good model, it should be possible somehow to assess its "quality". Ideally, there should exist documentation that reports results from procedures or steps that have been followed to test the model's behavior with regard to the above definitions of robustness, transportability, and tuning. The idea is that it should be possible for a third party to assess the procedures adopted to ensure the quality of a particular model, and to evaluate the results of these procedures. Collectively, such procedures, including their documentation, are commonly referred to as quality assurance procedures. They are also sometimes pragmatically referred to as model validation (Dee 1995). This reflects the fact that the ultimate goal is that the model should be able to reliably simulate what happens in nature.

In the following, we describe three steps which, taken together, allow us to construct useful quality assurance procedures. These are referred to loosely as follows:

1. model verification,
2. sensitivity analysis,
3. model validation.

Other discussions can be found in the review by Lynch and Davies (1995). Besides the steps reviewed here, model–model comparison exercises can also help to elucidate a model's quality. Examples of the latter can be found in Hackett and Røed (1994), Røed et al. (1995), and Hackett et al. (1995).

Model Verification

Model verification is a particularly important first step in model development, and in quality assurance procedures. The aim is to ensure that there are no errors in the computer coding or in the method of numerical solution. The first part involves rigorously checking the computer program (or code) by comparing the coding with the numerical algorithm chosen to solve the model's governing equations.[3] Obviously, this should be done at an early stage in the development of a model. More often than not, it is very time-consuming, especially for complex numerical models which may include many tens of thousands of lines of coding, written by a diverse group of scientists and programmers.

The second part of model verification is to ensure that there are no errors in the chosen method of numerical solution. Ideally, this can be done by comparing the numerical solution with analytical solutions, or at least other accurate numerical solutions. However, for most problems of a certain complexity, no analytical solutions exist. This is simply because most processes of interest are nonlinear. Nevertheless, it is usually possible to test separate parts of the entire model against either analytical solutions or accurately known numerical solutions. It is only in this way that we

[3]The governing equations are the mathematical formulation of the physical and/or biochemical processes that are to be simulated. This is usually a set of coupled partial differential equations.

may establish confidence in the accuracy of the model. In this context, accuracy is the extent to which the numerical solution approximates the true solution of the governing equations.

Additional complications arise because once the mathematical model is formulated, the methods used to solve the governing equations are usually wholly numerical. It is, therefore, a major task to ensure that the numerical methods employed are sufficiently rigorous to ensure that the solution is accurate under a range of conditions. This can be particularly difficult since most numerical solutions are prone to errors in regions where the predicted contaminant or any other variable in the model exhibits strong gradients, sometimes referred to as fronts, across which the variable in question experiences a large or abrupt change. In some cases, the occurrence of such a front can be anticipated, e.g., high-velocity gradients in shear boundary layers and high concentration gradients close to the source of a contaminant discharge. However, in many cases, the occurrence of fronts or frontal structures cannot be anticipated. Fronts may also evolve in time, e.g., a frontal structure may form within a certain time span (frontogenesis) and may break down due to strong outflow or wind events and/or diffusion, but may later reestablish under low outflow or wind conditions.

It is reasonable, therefore, to state that the idea of a numerical model of a certain complexity that aims to give the correct solution for all times and at all locations (an "all singing, all dancing" model) is still in its infancy. It is also fair to state that the development of numerical techniques and advanced supercomputing, which can provide an accurate solution of the coupled partial differential equations representing the processes in a numerical model, is also currently in its infancy although growing fast.

Although the numerical model, once verified, is deemed a proper representation of the mathematical formulation, the mathematical model itself may still be only a gross approximation of the real system. Thus, model verification is only one step toward the ultimate goal of establishing confidence in the validity of the results that the numerical model can produce when used to answer certain specific management questions. Regrettably, this important stage in model development is often omitted in the construction and application of numerical models, and more often than not, the task is not taken seriously enough (Hannay et al. 2009).

Sensitivity Analysis

The next step in determining whether a chosen model accurately reproduces conditions in the "real world" is a model sensitivity study. The aim here is to establish the predictive power of the model. In essence, there are two distinct components in a sensitivity analysis. The first deals with sensitivity to input data and conditions, and the second involves sensitivity to the chosen parameter values and to the parameterizations themselves.

When evaluating the sensitivity of a model's output to variations in the input data, it is accepted that the model equations will contain certain terms that have been approximated, and also that intricate processes will have been reduced to simple parameterizations. The range of variations in the input data may be determined from knowledge of the variations and estimated errors in the observed data. Such an exercise, often referred to as an uncertainty analysis, is particularly revealing, both in terms of establishing the sensitivity of the model and in identifying crucial field observations. If the sensitivity study reveals that the model output is crucially dependent on the precision and accuracy of certain measurements, then an effort must be made to reduce the error in these measurements. If, for example, the model shows that the representation of processes and boundary conditions at one geographical location has a larger effect upon the model output than this representation at other locations, then an observational program can be designed to sample more intensely in that critical area.

The second component of the sensitivity analysis involves the various assumptions made in developing the model. The main difficulty here relates to the problem of parameterizing small-scale processes, e.g., turbulent mixing processes in hydrodynamical and biogeochemical models, which cannot be resolved explicitly within the numerical model (Sect. 11.1). Consider, for instance, the parameterization of turbulent mixing processes. Physically, these processes are associated with turbulent motions in the fluid. The mechanisms producing this turbulence and its intensity provide a prime example of a poorly understood process. However, they are clearly related to larger scale physical phenomena. In this context, surface roughness determines near-surface turbulence, and larger scale obstructions, e.g., mountains, steep slopes, and ridges, cause hydraulic jumps associated with vigorous turbulent shocks downstream. The representation of such processes in the physical compartment of the model is particularly difficult. They can sometimes be parameterized by a single coefficient, e.g., a diffusion coefficient as exemplified in Sect. 11.1, or they can be represented in a hydrodynamical model by a complex system of turbulence energy equations. In any true sensitivity study, a range of parameterizations of these mixing processes must be considered. If such a sensitivity study shows that the contaminant distribution is sensitive to the mixing formulation (which is normally the case), then confidence limits can be placed upon the model on the basis of the accepted range of parameterizations of the mixing process. In the unlikely event that a sensitivity analysis reveals that the model is insensitive to the formulation of mixing, then only the simplest formulation of this process is required. However, in reality, the main problem arising in a sensitivity analysis of an "all-encompassing model" model is that, in certain circumstances, the results may be insensitive to one part of the model. In other circumstances, the formulation of this same part of the model may be critical. Such a finding obviously leads to the conclusion that, in practice, a range of models is required.

A conservative approach when developing numerical models is, therefore, that each model should only incorporate those processes that are essential for providing accurate answers to the specific management questions raised. In most applications, this approach to modeling is to be preferred. In contrast, an "all singing, all dancing" model, designed to cover every conceivable situation, is rarely constructed, because of the requirements for immense computer power, a large body of supporting field data, and the problems imposed in conducting a comprehensive sensitivity analysis. The latter is particularly relevant, since the task of carrying out such an analysis is virtually open-ended, potentially requiring huge resources, so it is rarely undertaken.

In fact, it is often a relief to find out that a model does not have to be perfect, all-encompassing, or complicated to be useful.

When a numerical model has been verified and a proper sensitivity analysis has been carried out, some confidence in the model can definitely be claimed. It is then sure that the model code is accurate, that the numerical methods are sound for the problem at hand, and that the numerical solution is able to reproduce known solutions, albeit in a reduced, simplified, and idealized context. A properly conducted (and documented) sensitivity study further increases our confidence in the predictive power of the model and helps us to understand which parameters and parameterizations must be known with a high degree of certainty and which need not. However, we still cannot be sure that the model will be able to reproduce a given observational data set for the correct reasons, so one final step remains necessary.

Model Validation

The ultimate test of the usefulness of a numerical model is its predictive power. At first sight, the ability of a model to reproduce a given data set would appear to be a good guide to its predictive capability. However, some care must be exercised in reaching this conclusion. If a sensitivity analysis has revealed that the model is sensitive to variations in a poorly known parameter, and a good fit between model output and observations is achieved by adjusting this parameter, then the model may legitimately be regarded as a well-tuned or at least a highly calibrated model. Ideally, such a model should be able to produce similarly accurate results under similar conditions elsewhere. In practice, however, a tuned model is probably neither transportable nor robust in the sense defined in Sect. 11.2. A potential user of the model should then be alerted to this, and the conditions under which the model may be applied with some confidence should be clearly stated, or at least be part of the documentation accompanying the model.

The user must be aware of how extensively a model has been validated before it can be used. If the model is able to reproduce various observational data obtained under a broad range of physical and biogeochemical conditions without adjusting its parameterizations, the model may be regarded as transportable. It may, therefore, be applied with confidence over a wide range of situations.

In general, data are required for both model operation and model validation. Data used for model operation include initial conditions, boundary conditions, source terms, and other forcing functions. One of the most difficult aspects of modeling is to provide a suitable description of conditions at an open boundary (Chap. 7). In particular, data for specifying conditions of the hydrodynamics and contaminant fluxes between the far field and the open ocean beyond are not always available. In the absence of appropriate data, the only recourse is to use simple assumptions, such as diffusion into an infinite field or periodic flow conditions at the boundary, in order to keep the model operational.

Ideally, model validation is achieved when the model output compares favorably with data sets independent of those used during model calibration, that is, those used to tune the parameters of the model. In the case of a complex deterministic model, this could be an overwhelmingly difficult task. In theory, the predictions of the model should be compared at all appropriate levels with different data obtained from real systems. However, this is rarely done in practice. In some cases, all or part of the calibration data is reused in the validation process. Such a partial validation, using data from the same site and/or under similar conditions, is called model confirmation.

In the final analysis, it is, therefore, crucial that independent data sets from many different regimes should be used to establish the credibility of the model through rigorous statistical tests. For instance, the standard deviation between modeled and measured values gives a quantitative measure of how well the model is working. Very few validations provide this quantitative measure or metric. Obviously, if the observational data are very limited, then all will be used in the calibration stage and model validation, and even a model confirmation, will not be possible until a wider range of data sets become available.

It is also important to realize that a good validation performance does not necessarily guarantee that the model will predict future conditions accurately. Some uncertainties will always remain in the model coefficients, the model variables, and the model structure itself. Therefore, models should be subjected to post-audits in which their predictions are tested again and again under changed conditions. The purpose of this validation stage is to check whether the model reproduces the expected changes. Except for atmospheric models, it is only in exceptional cases that a post-audit is feasible, and hence it is rarely carried out. Only recently has there been any activity in this phase of validation.

Model users must also be aware of the quality and relevance of observational data. Even the best of models cannot make reasonable and accurate predictions if these predictions are based on imprecise or inaccurate input data. Although the adage "garbage in, garbage out" has become common modeling jargon, it nonetheless provides an important cautionary note for potential model users. In many cases, the underlying cause of such a situation is that data used for model development were originally collected for a purpose other than modeling. If data collection programs are more closely linked with modeling studies, then the constraints imposed by the lack of suitable data can be substantially reduced. The bottom line is that there is always an acute need for high quality and relevant data sets for model calibrations and validations.

One of the most difficult problems is proving that the model is robust (Sect. 11.2), i.e., that it can predict extreme conditions with some confidence. Since most validation data are collected under normal conditions, they are of little value in establishing confidence in the model's output under extreme circumstances. A sensitivity analysis is then probably the most appropriate way to determine the value of the model under such circumstances. If corrected parameterizations of the various processes are included in the model, and if we can be confident in our knowledge of these processes, then the model should be reasonably accurate under extreme conditions. Fortunately, extreme conditions are seldom a problem for most quality management issues.

11.4 Summary and Remarks

This chapter suggests ways to develop criteria that can be used for objective assessment of the quality of numerical models. We discuss three steps that we consider necessary to establish confidence in models designed to answer specific management questions about problems involving hydrodynamic and biochemical processes. The guidelines proposed are referred to as quality assurance procedures, although quality assurance is rarely undertaken to its full extent, mainly due to lack of data. Nevertheless, these guidelines are put forward as the backbone for developing a set of objective criteria to evaluate the usefulness of numerical models of atmospheres and oceans.

The first step in the quality assurance procedures is model verification. It involves checking the numerical code and the methods of numerical solution used to solve the underlying mathematical formulation of the model. The next step is to perform a model sensitivity analysis, whose aim is to establish which parameters and/or parameterizations must be accurately known and under what conditions. Sometimes these parameters are tuned to make the model match a given observational data set. However, if this involves tuning of critical parameters, the predictive ability of the model will be poor in the sense that it will probably fail when used under different conditions. The final step is to perform a model validation. This aims to measure the extent to which the model output compares with observational data. Here, it is important that the data set used should consist of data collected under different conditions than the data used to calibrate the model, i.e., the data used to determine the model parameters and parameterizations. If this is not the case, the exercise will not be a true model validation, but will be classified as a model confirmation.

We also introduce the concept of a good model. This is one which can be used in any area containing the same generic processes (transportable model) and one which can provide a similar degree of accuracy under a wide range of conditions (robust model). It should be emphasized that this does not imply that all models have to be good models to be useful. Highly tuned models may also be useful under certain conditions. However, these underlying conditions should be clearly stated and transparent to potential users.

Finally, the above description is general in nature, because there are so many processes and interactions between processes that must be formulated and parameterized to construct a useful numerical model for even the simplest specific management questions. There is therefore a general feeling that the routine development and application of an "all singing, all dancing" model to answer specific problems would be too expensive and too complicated. And this is indeed why there exist a plethora of models from the simplest box type models to the most expensive three-dimensional ones.

It is thus likely, and probably perfectly reasonable, that there should exist a wide range of numerical models. Fortunately, many of the existing models use standard formulations and parameterizations that are well proven and/or widely accepted by the international community. The important message here is that anyone who offers a model as a tool to answer a specific management question or problem for a potential user should also provide documentation about the quality of the their model, i.e., they should provide documentation about the quality assurance procedures that have been followed, so that the potential user can assess the quality and suitability of the model in an objective way. The bottom line is therefore that any numerical model must come with quality assurance documentation in the sense described above. This is the only way that such models could be applied with any confidence.

References

Dee DP (1995) A pragmatic approach to model validation. In: Lynch D, Davies A (eds) Quantitative skill assessment for coastal ocean models. Coastal and estuarine studies, vol 47. American Geophysical Union, Washington, pp 1–13

GESAMP (1991) (IMO/FAO/UNESCO/WMO/WHO/IAEA/UN/UNEP Joint Group of Experts on the Scientific Aspects of Marine Pollution), Coastal Modelling, Technical report. International Atomic Energy Agency (IAEA), GESAMP Reports and Studies 43, 192 pp

Griffies SM (2004) Fundamentals of ocean climate models. Princeton University Press, Princeton. ISBN 0-691-11892-2

Hackett B, Røed LP (1994) Numerical modeling of the Halten Bank area: a validation study. Tellus 46A:113–133

Hackett B, Røed LP, Gjevik B, Martinsen EA, Eide LI (1995) A review of the metocean modeling project (MOMOP). Part 2: model validation study. In: Lynch DR, Davies AM (eds) Quantitative skill assessment for coastal ocean models. Coastal and estuarine studies, vol 47. American Geophysical Union, Washington, pp 307–327

Hannay JE, MacLeod C, Singer J, Langtangen HP, Pfahl D, Wilson G (2009) How do scientists develop and use scientific software? In: SECSE '09: proceedings of the 2009 ICSE workshop on software engineering for computational science and engineering. IEEE Computer Society, Washington, pp 1–8. https://doi.org/10.1109/SECSE.2009.5069155

Lynch DR, Davies AM (1995) Quantitative skill assessment for coastal ocean models. Coastal and estuarine studies, vol 47. American Geophysical Union, Washington

McClimans TA, Røed LP, Thendrup A (1992) Fjord water quality/ecological modelling. State of the art and needs, Technical report. Royal Norwegian Council for Scientific and Industrial Research, Programme on Marine Pollution, 54 p. Plus appendices and attachments. ISBN 82-7224-335-0

Røed LP (1993) Models as management tools for environmental problems in water, SFT-Rapport 93:14, Statens forurensningstilsyn - SFT, Box 8100 Dep, 0032 Oslo, Norway (in Norwegian), 38 pp. Plus attachments. ISBN 82-7655-130-0

Røed LP, Hackett B, Gjevik B, Eide LI (1995) A review of the metocean modeling project (MOMOP). Part 1: model comparison study. In: Lynch DR, Davies AM (eds)Quantitative skill assessment for coastal ocean models. Coastal and Estuarine Studies, vol 47. American Geophysical Union, Washington, pp 285–305

Appendix A Introduction to Fortran 2003 via Examples

A

by Gunnar Wollan[1] and Lars Petter Røed

The following gives a brief insight into the Fortran 2003 programming language via specific examples. The reader will learn how to solve a computational problem, and how to handle reading and writing of data to files. For more details, in particular regarding Fortran 90/95, the reader may download Fortran texts from the net. We particularly recommend the site: www.nsc.liu.se/~boein/f90/, which contains versions in both English and Swedish.

A.1 Why Use Fortran?

First, the obvious question: Why should I learn to program in Fortran? In the field of meteorology and oceanography, you will probably come across atmospheric models like WRF and CAM, and ocean models like ROMS, NEMO, or similar. Depending on your project, you will sometimes have to make changes or additions to an already existing model written in Fortran. This task will be decidedly easier if you acquire some knowledge of programming in Fortran. Moreover, valuable time may be lost if you have to acquire that knowledge later on, when you need to make changes to the model.

In the last 15–20 years or so, Fortran has been looked upon as an old-fashioned and poorly structured programming language by researchers and students alike in the field of informatics. The reason is that earlier versions of Fortran lacked most of the features found in modern programming languages like C++, Java, etc. The lack of object orientation has been the main drawback with Fortran. But this is no longer true. Fortran 2003 and Fortran 2008 have all modern features, including object-oriented programming (OOP).

[1]Former scientific programmer at the Department of Geosciences, University of Oslo.

L. P. Røed, *Atmospheres and Oceans on Computers*,
Springer Textbooks in Earth Sciences, Geography and Environment,
https://doi.org/10.1007/978-3-319-93864-6

The most important reason why Fortran is still favored as a programming language within the meteorology and oceanography community is the execution speed of the compiled program. In number crunching speed, Fortran is much faster than C and C++. Tests show that an optimized Fortran program may in some cases run up to 30 percent faster than the equivalent C or C++ program. Thus, for large and complex programs and codes with a runtime of weeks, even a small increase in speed will significantly reduce the overall time it takes to solve a problem. This is important in the field of meteorology and oceanography, since speed is everything when producing a forecast. In addition, laboratory experiments and fieldwork are sometimes costly to perform. Computer simulations are less costly and are therefore becoming increasingly important as an addition to laboratory and fieldwork. Even then speed is a factor.

A.2 Historical Background

Fortran is indeed an old programming language. As early as 1954, John W. Backus[2] and his team at IBM began developing the scientific programming language Fortran. It was first introduced in 1957 for a limited set of computer architectures. After a short time, the language had spread to other architectures. Since then, it has been the most widely used programming language for solving numerical problems in the natural sciences in general and in atmosphere and ocean science in particular.

The name Fortran is derived from Formula Translation and, as already mentioned, is still the language of choice for fast numerical computations. In 1959, a new version, Fortran II, was introduced. This version was more advanced, and among the new features it could use complex numbers and split a program into subroutines. In the years to follow, Fortran was further developed to become a programming language that was fairly easy to understand and well adapted to solve numerical problems.

In 1962, a new version, Fortran IV, emerged. Among its new features was the ability to read and write direct access files. In addition, it introduced a new data type called `LOGICAL`. This was a Boolean data type with two states *true* or *false*. At the end of the 1970s, Fortran 77 was introduced. This version contained better loop and test structures. In 1992, Fortran 90 was formally introduced as an ANSI/ISO standard, followed shortly afterwards by Fortran 95, a minor extension of Fortran 90. These versions turned Fortran into a modern programming language and included many of the features expected of a modern programming language. Finally, Fortran 2003 was released, incorporating even OOP with type extension and inheritance, polymorphism, dynamic type allocation, and type-bound procedures.

[2]John Warner Backus (1924–2007) was an American computer scientist. He directed the team that invented the first widely used high-level programming language (FORTRAN) and was the inventor of the Backus–Naur form, a widely used notation to define formal language syntax. He also did research in function-level programming and helped to popularize it. The IEEE awarded Backus the W.W. McDowell Award in 1967 for the development of FORTRAN. He received the National Medal of Science in 1975 (Source: Wikipedia).

A.3 Fortran Syntax

The starting point is that Fortran, like all programming languages, has its own syntax. To start programming in Fortran, a knowledge of its syntax is therefore required. In Fortran, as well as other programming languages, the code is separated into a part declaring variables through declaration statements, and a part carrying out instructions for manipulating the contents of the variables. An important difference between the earlier Fortran 77 and Fortran 90/95/2003 is the way in which each separate line of code is written.

In Fortran 77, it was written in fixed form, where each line of code was divided into 80 columns, and each column had its own meaning. This division had a historical background. In the 1960s and part of the 1970s, the standard medium for data input was punched cards like the one shown in Fig. A.1. The cards were divided into 80 columns and it was therefore natural to set the length of each line of code to 80 characters. Table A.1 gives an overview of the subdivision of the line of code. Note that Fortran 77 is a subset of Fortran 2003. Thus, all programs written in Fortran 77 can be compiled using a Fortran 2003 compiler.

In addition to the fixed code format from Fortran 77, Fortran 2003 also supports free format coding. This means that the separation into columns is no longer necessary and the program code can be written in a more structured way. This makes it more readable and far easier to maintain. Today, *the free format is the default* setting for Fortran 90/95 and Fortran 2003 compilers.

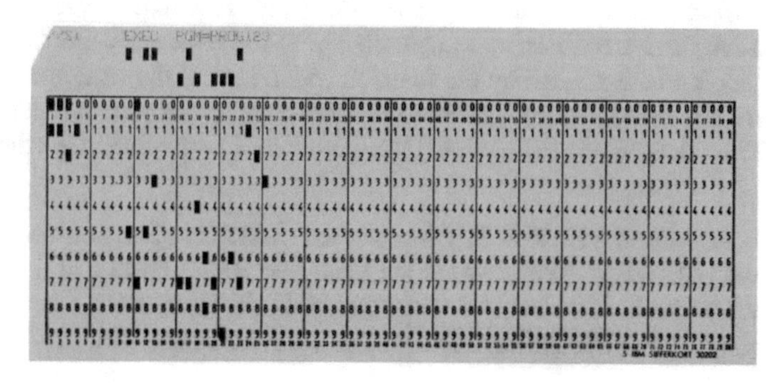

Fig. A.1 The punched card

Table A.1 Fortran 77 (F77) fixed format

Column	Meaning
1	A character here means the line is a comment
2–5	Jump address and format number
6	A character here is a continuation from the previous line
7–72	Program code
73–80	Comment

A.3.1 Data Types in Fortran

In earlier versions of Fortran, four basic data types were included. These were `INTEGER` and `REAL` numbers, a Boolean type called `LOGICAL`, and `CHARACTER`, which represents the alphabet and other special non-numeric types. In Fortran 90/95, the `REAL` data type is split into the `REAL` and `COMPLEX` data types. In addition to this, a derived data type can be used in Fortran 2003. A derived data type may contain one or more of the basic data types, other derived data types, and in addition, procedures which are part of the new OOP features in Fortran 2003.

INTEGER
An `INTEGER` data type is identified with the reserved word `INTEGER`. It has a valid range which varies with the way it is declared and the architecture of the computer it is compiled on. When nothing else is given, an `INTEGER` has a length of 32 bits on a typical workstation and can have a value from -2^{31} to 2^{30}, while a 64-bit `INTEGER` has a minimum value of -2^{63} and a maximum value of 2^{62}.

REAL and COMPLEX
In the same manner, a `REAL` number can be specified with various ranges and accuracies. A real number is identified with the reserved word `REAL` and can be declared with single or double precision. Table A.2 gives the number of bits and minimum and maximum values. A double precision real number is declared using the reserved words `DOUBLE PRECISION` or `REAL(KIND=8)`. The latter is the preferred declaration.

An extension of `REAL` numbers is `COMPLEX` numbers with their real and imaginary parts. A `COMPLEX` number is identified with the reserved word `COMPLEX`. The real part can be extracted by the function `REAL()` and the imaginary part by the function `AIMAG()`. There is no need to write explicit calculation functions for `COMPLEX` numbers, as one has to do in C/C++, which lacks the `COMPLEX` data type.

LOGICAL
The Boolean data type is identified by the reserved word `LOGICAL` and has only two values: "true" or "false". These values are identified with `.TRUE.` or `.FALSE.`. The dot (period mark) at the beginning and end of the declaration is a necessary part of the syntax. To omit one or more dots will give a compilation error.

CHARACTER
The `CHARACTER` data type is identified by the reserved word `CHARACTER` and contains letters and characters in order to represent data in a readable form. Legal characters are, among others, `a` to `z`, `A` to `Z`, and some special characters +, -, *, / and =.

Table A.2 The `REAL` numbers data type in Fortran

Precision	Sign	Exponent	Significand	Maximum value	Minimum value
Single	1	8	23	2^{128}	2^{-126}
Double	1	11	52	2^{1024}	2^{-1022}

Derived Data Types
These are data types which are defined for special purposes. A derived data type is built from components from one or more of the four basic data types, and also other derived data types. A derived data type is always identified by the reserved word `TYPE name` as prefix and `END TYPE name` as postfix.

A.4 Structure of Fortran

A.4.1 Declaration of Variables

In Fortran, there are two ways to declare a variable. The first is called *implicit* declaration and is inherited from the earliest versions of Fortran. The second is called *explicit* declaration and is in accordance with other programming languages. Explicit declaration means that all variables have to be declared *before* any instruction occurs.

Implicit declaration, on the other hand, means that a variable is declared when needed by giving it a value anywhere in the source code, that is, even within the instructions. The data type is determined by the first letter in the variable name. An `INTEGER` is recognized by starting with the letters `I` to `N` and a `REAL` variable by the rest of the alphabet. No special characters are allowed in a variable name. Only the letters `A–Z`, the numbers `0–9`, and the underscore character _ are allowed. A variable cannot start with a number. In addition, a `LOGICAL` variable is, in most compilers, identified by the letter `L`.

It should be emphasized, as a general rule, that an implicit declaration is not a good way to program. For one thing, it renders a code that is not easy to read. For another, it easily introduces errors into a program due to the risk of typing errors. It is therefore strongly recommended to use explicit declaration of variables. To ensure that all variables are declared, the keywords `IMPLICIT NONE` must be included in the second line of all programs, functions, and subroutines. This tells the compiler to check that all variables are declared. Finally, there are some variables that always have to be declared. These are arrays in one or more dimensions and character strings.

INTEGER Numbers
A good start is to show an example of how to declare an `INTEGER` in Fortran 95:

```
INTEGER                  :: i ! Declaration of an INTEGER
                              ! length(32 bit)
INTEGER(KIND=2)          :: j ! Declaration of an INTEGER (16bit)
INTEGER(KIND=4)          :: k ! Declaration of an INTEGER (32bit)
INTEGER(KIND=8)          :: m ! Declaration of an INTEGER (64bit)
INTEGER,DIMENSION(100):: n ! Declaration of an INTEGER array
                              ! (100 elements)
```

There are certain differences between the Fortran 77 and Fortran 95 ways of declaring variables. In Fortran 95 there is more to write, but this is offset by greater readability.

Note that in Fortran 95 a comment can start anywhere on the code line, but must always be preceded by an exclamation (!) mark.

REAL Numbers

In most compilers, the REAL data type now conforms to the IEEE standard for floating point numbers. Declarations of single and double precision are declared as in the following example:

```
REAL                :: x ! Declaration of REAL
                         ! defaultlength (32bit)
REAL(KIND=8)        :: y ! Declaration of REAL
                         ! double precision (64 bit)
REAL,DIMENSION(200) :: z ! Declaration of REAL array
                         ! (200 elements)
```

COMPLEX Numbers

Unlike C/C++, Fortran has an intrinsic data type of complex numbers. Declaration of a COMPLEX variable in Fortran is achieved as follows:

```
COMPLEX                :: a ! Complex number
COMPLEX,DIMENSION(100) :: b ! Array of complex numbers
                            ! (100 elements)
```

LOGICAL Variables

Unlike INTEGER and REAL numbers, a LOGICAL variable has only two values, .TRUE. or .FALSE., and therefore uses a minimum of space. The number of bits a LOGICAL variable uses depends on the architecture and the compiler. It is possible to declare a single LOGICAL variable or an array of them. The following example shows a Fortran 90/95 declaration. In other programming languages, the LOGICAL variable is often called a Boolean variable, after the mathematician George Boole.[3]

```
LOGICAL                :: l1 ! Single LOGICAL variable
LOGICAL,DIMENSION(100) :: l2 ! Array of LOGICAL variables
                             ! (100 elements)
```

CHARACTER Variables

Characters can either be declared as a single CHARACTER variable, a string of characters, or an array of single characters or character strings:

[3]George Boole (1815–1864) was an English mathematician, philosopher, and logician. He worked in the fields of differential equations and algebraic logic, and is now best known as the author of "The Laws of Thought", and as the inventor of the prototype of what is now called Boolean logic (source: Wikipedia).

```
CHARACTER                          :: c1 ! Single character
CHARACTER(LEN=80)                  :: c2 ! String of characters
CHARACTER,DIMENSION(10)            :: c3 ! Array of single
                                         ! characters
CHARACTER(LEN=80),DIMENSION(10)    :: c4 ! Array of character
                                         ! strings (10 elements)
```

Derived Data Types

The Fortran 95 syntax for the declaration of a derived data type can be like the one shown here:

```
TYPE derived
   ! Internal variables
   INTEGER            :: counter
   REAL               :: number
   LOGICAL            :: used
   CHARACTER(LEN=10)  :: string
END TYPE derived
! A declaration of a variable of
! the new derived data type
TYPE (derived)    :: my_type
```

The question arises as to why we should use derived data types. The answer is that it is sometimes desirable to group variables together to be able to refer to them under a common name. It is usually good practice to select a name of abstract data type to indicate the contents and area of use.

A.4.2 Instructions

There are two main types of instructions. One is for program control and the other is for assigning a value to a variable.

Instructions for Program Control

Instructions for program control can be split into three groups, one for loops, a second for tests (even though a loop usually involves an implicit test), and a third for assigning values to variables and performing mathematical operations on the variables.

In Fortran, all loops start with the reserved word `DO`. The following piece of code shows a short example of a simple loop:

```
DO i = 1, 100
  !// Here instructions are performed 100 times
  !// before the loop is finished
END DO
```

The next example shows a loop with instructions. This loop is a nonterminating loop, where an `IF`-test inside the loop is used to exit the loop when the result of the test is true:

```
....
do
  a = a * sqrt(b) + c
  if (a > z) then
    !// Jump out of the loop
    exit
  end if
end do
```

This piece of code instructs the computer to give the variable a a value equal to the sum of the square root of the variable b multiplied by the previous value of a and a third variable c. When the value of a becomes greater then the value of the variable z, the program transfers control to the next instruction following the loop. Note that it is assumed that all the variables are declared and initialized somewhere in the program *before* the loop, as indicated by the code line .... appearing before the loop. The various Fortran instructions will be described in the example in Sect. A.6.

A.5 Compiling a Program

In order to have an executable program, it must be compiled. This requires the sample program (or source code) to reside in a file on the computer, say daynr.f90, where the extension indicates that the file contains a Fortran program written in Fortran 90. The compilation process takes the file with the source code and creates a binary file linked in with the necessary system libraries, so that it may run the program on the computer. The binary file contains the program in machine-specific assembly language, i.e., instructions written in machine language.

The most common compiler is the open-source compiler called gfortran.[4] The command line for compiling the program daynr.f90 is simply

```
gfortran -o daynr daynr.f90
```

where gfortran is the name of the compiler. The argument -o means that the next argument to the compiler is the name of the executable program. The last argument is the name of the file containing the source code. One may also simply write

```
gfortran daynr.f90
```

In this case, the executable program by default is given the name a.out. To run the compiled program, use the command ./daynr (or ./a.out) in the terminal window.

[4] Open source means that the compiler may be downloaded and used free of charge.

A.6 Sample Programs

A.6.1 Daynumber Converter

The first sample program considered is a very simple one in which the task is to calculate the daynumber of a specific date in the year. It is assumed that the year is a non-leap year. We first write the program skeleton, and then fill in the necessary code to solve the problem:

```
PROGRAM daynumber
  implicit none

END PROGRAM daynumber
```

All Fortran programs begin with the reserved word PROGRAM, followed by the program name. In this case, the program name is daynumber. As mentioned above, the code line implicit none is almost mandatory, or at least good programming practice. It prevents the use of implicit declarations, which are otherwise the default behavior of the Fortran compiler.

Some variables and constants are then declared to calculate the daynumber:

```
PROGRAM daynumber
  implicit none
  integer                 :: counter
  integer,dimension(12)   :: months
  integer                 :: day, month
  integer                 :: daynr

END PROGRAM daynumber
```

Four integer variables are declared, namely, counter, day, month, and daynr, and one integer array months with 12 elements. The variable counter is used to traverse the array to select the number of days in the months before the given month. The variables day and month hold the day and month. The variable daynr contains the result of the calculations.

Next, numbers are specified for the constant integers day and month, and the variable daynr and the array months are initialized:

```
PROGRAM daynumber
  implicit none

  day = 16
  month = 9
  daynr = 0
  months(:) = 31

END PROGRAM daynumber
```

Initializing scalar arrays is not difficult, but each element of the array usually has to be initialized separately. Fortunately, Fortran 95 and 2003 have a built-in functionality which allows initialization of a whole array with one value.

However, not all months contain 31 days. Thus, the next step is to change the number of days in the months that differ from 31, that is, `months(2)` (February), `months(4)` (April) , `months(6)` (June), `months(9)` (September), and `months(11)` (November):

```
PROGRAM daynumber
  implicit none

  months(2)  = 28
  months(4)  = 30
  months(6)  = 30
  months(9)  = 30
  months(11)  = 30

END PROGRAM daynumber
```

The next step is to loop through all the elements in the array `months` up to the month minus one, summing up the number of days in each month in the `daynr` variable. To arrive at the result, the value from the variable day is added to `daynr`. To display the result, the command `PRINT *, daynr` writes the result on the terminal:

```
PROGRAM daynumber
  implicit none

  DO counter = 1, month - 1
    daynr = daynr + months(counter)
  END DO
  daynr = daynr + day
  PRINT*, daynr
END PROGRAM daynumber
```

The resulting output from this sample program with `month` = 9 and `day` = 16 is 259. You can use a calculator and perform the calculations by hand to check that the result is correct.

This simple program illustrates the rule that one should *never use implicit declarations of variables*. This is very important. There is a story from the 1970s about 10 implicit declarations where a typing error created an uninitialized variable that caused a NASA rocket launch to fail, and the rocket had to be destroyed before it could cause serious damage.

A.6.2 Temperature Converter

The next sample program we consider converts a temperature from degrees Fahrenheit to degrees Celsius (or centigrade). The two temperatures must appear side by side in the terminal window. The formulas for the conversions are

$$C = \frac{5}{9}(F - 32)\,, \qquad F = \frac{9}{5}C + 32\,, \tag{A.1}$$

where F represents the temperature in Fahrenheit and C the temperature in Celsius or centigrade.

To proceed, two floating point variables are needed, one to hold the temperature in Fahrenheit, say F, and a second to hold the temperature in Celsius, say C. To declare them as floating point variables, we use the REAL keyword. Thus, the following piece of code must be included:

```
!// The Fahrenheit variable
REAL :: F
!// The Centigrade variable
REAL :: C
```

Note that, in contrast to most other languages, Fortran is case insensitive. That means that a variable or function name is the same whether it is written with uppercase or lowercase letters. Moreover, beginning with the Fortran 90 version, a double colon is used to separate the variable type from the variable name, while an exclamation sign is used to tell the compiler that the rest of the line is a comment. In order to make the code more readable, it is recommended to include an additional double slash before writing the comment, as shown in the above example. A program that is easy to read is also easy to understand, and this makes it easier to find and correct errors.

The next step is to initialize the Fahrenheit variable and perform the conversion according to the formula. The whole program may be something like this:

```
PROGRAM f2c_simple
  IMPLICIT NONE
  !// Declare variables
  REAL :: F     ! Fahrenheit variable F (floating point)
  REAL :: C     ! Centigrade variable C (floating point)
  !// Assign a value to F as a constant number
  !// Note the decimal point
  F = 75.
  !// Perform the calculations
  C = (F - 32.)*5./9.
  !// Write the result to the terminal window
  PRINT *, C
END PROGRAM f2c_simple
```

The Fortran default of implicit declarations of variables is avoided by writing IMPLICIT NONE in the second code line. To tell the compiler that the constant value 75 is a real number, a decimal point is added as part of the number. If the dot is omitted the compiler will assume that it is an integer, and in some cases, the calculations will be wrong. Finally, parentheses are there to ensure that calculations are performed in the right order. The statement or command `PRINT *, C` writes the temperature in degrees Celsius in the terminal window.

As in the previous example, the program has to be compiled to obtain a binary executable program file. There are several commercial Fortran compilers, but the open-source GNU Fortran compiler `gfortran` is used in this example to compile the program. Assuming that the program is written into a file named `f2c_simple.f90`, it is compiled by the command

```
gfortran -o f2c f2c_simple.f90
```

where the `-o` option tells the compiler that the next argument is the name of the executable program and the last argument is the name of the file with the source code. In this case, the binary program file `f2c` is created. This may be run by typing `./f2c` in the terminal window. This is exactly as for the sample program in Sect. A.6.1. All Fortran programs are compiled and run this way.

A.6.3 A More User-Friendly Version of the Converter Program

The above program is not very user friendly. Every time a new temperature in degrees Fahrenheit is to be converted to degrees Celsius, the source code has to be changed. A new `F` has to be specified, and then the program has to be recompiled and rerun. To avoid this, a user interface asking for a temperature in degrees Fahrenheit may be added to the program.

This is accomplished by adding some code lines for communication in the form of text strings. In the example below, two text strings `prompt1` and `prompt2` are declared and assigned. The first text string, declared as `prompt1`, asks us whether our input is in Fahrenheit or Celsius, while the second, `prompt2`, asks us to enter the temperature that we, the user, would like to convert. To declare the text strings, the data type `CHARACTER` is used. Thus, the program begins with the following code lines:

```
!////////////////////////////////////////////////////////////
!//
!// f2c.f90
!//
!// Program to convert from Fahrenheit to
!// Celsius or vice versa
!////////////////////////////////////////////////////////////
PROGRAM f2c
  IMPLICIT NONE
  REAL :: F         !// Temperature in Fahrenheit
  REAL :: C         !// Temperature in degrees Celsius
  !// Character strings to hold the prompts for
  !// communicating with the user
```

```
CHARACTER(LEN=80) :: prompt1, prompt2
!// A single character to hold the answer
!// which is either F or C
CHARACTER :: answer
!// Assign a value to prompt1
prompt1 = 'Enter F for Fahrenheit or C for Celsius'
!// Assign a value to prompt2
prompt2 = 'Enter a temperature'
```

The declaration `CHARACTER(LEN=80)` means that a space for a text string up to 80 characters long is allocated. The character `answer` is unassigned, but is used to hold a single character variable that is the answer to the question `prompt1`, that is, `answer` is used to hold the characters (`F` or `C`) according to the answer to `prompt1`.

Note also that some comments are added above the `PROGRAM f2c` line. They give the name of the file containing the source code, and also a brief description of the purpose of the program. This makes the code more readable and easier to understand and is recommendable, even for short programs like this one.

The code continues:

```
!// Print the contents to the terminal window
!// without trailing blank characters
PRINT *, TRIM(prompt1)
!// Read the input from the keyboard
READ(*,*) answer
!// Print the contents to the terminal window
!// without trailing blank characters
PRINT *, TRIM(prompt2)
```

This part prints the assigned value of `prompt1` to the terminal window, and then reads the input and puts it in `answer`. It subsequently prints `prompt2` to the terminal window and waits for the next input. The keyword TRIM tells the compiler to print out the text without printing the trailing space characters. To read the answer to `prompt1`, the `READ` command is used. The construct `READ(*,*)` instructs the computer to read the input from the keyboard into the receiving variable using the declared format for that variable.

The next program step is:

```
!// Is the temperature given in Fahrenheit?
IF(answer .EQ. 'F') THEN
  !// Yes, read input into the F variable
  READ(*,*) F
  !// Convert from Fahrenheit to Celsius
  C = (F - 32) * (5. / 9.)
  !// Print the result to the screen
  PRINT *, C
ELSE
  !// No, read input into the C variable
  READ(*,*) C
```

```
      !// convert from Celsius to Fahrenheit
      F = (C * 9. / 5.) + 32
      !// Print the result to the screen
      PRINT *, F
    END IF
  END PROGRAM f2c
```

This reads the input, then branches out in accordance with the input `answer`, performs the conversion, and prints the result before ending the program properly using the code line `END PROGRAM f2c`.

A.6.4 Variable Types, Arrays, Loops, and Memory Allocation

More often than not, atmosphere and ocean variables are stored in large files where the variables are stored in multiple dimensional arrays. To illustrate this, it is best to begin by looking at the way data are stored in a one-dimensional array.[5]

For this purpose, we construct a program that calculates the Fibonacci sequence.[6] The sequence consists of a series of integers, the so-called Fibonacci numbers. These must be stored in a one-dimensional array or vector of integers. Recall that the formula for calculating the Fibonacci sequence is

$$F_{j-1} = F_{j-2} + F_{j-3} \,, \quad j = 3(1)n \,, \tag{A.2}$$

where the two first numbers are $F_0 = 0\,(j = 1)$ and $F_1 = 1\,(j = 2)$.

First, we require an array of integers to hold the numbers. The program will be constructed so that the user may choose the length of the sequence. Thus, we must include a user interface like the one in the temperature conversion program, asking for the length of the sequence. The length of this array (or vector) is not known in advance. We must therefore use what is known as an allocatable array, where the needed space is allocated at runtime. In addition, an index variable and a status variable are needed. The first is needed to access the various elements of the array, and the second to check the result of the allocation. Consequently, the first part of the code takes the form:

```
!/////////////////////////////////////////////////////////////////////
!//
!// fibonacci.f90
!//
```

[5] A one-dimensional array is often referred to as a vector.

[6] Leonardo Pisano Bigollo (ca. 1170–ca. 1250), also known as Fibonacci, Leonardo of Pisa, Leonardo Pisano, Leonardo Bonacci, and Leonardo Fibonacci, was an Italian mathematician, considered by some "the most talented Western mathematician of the Middle Ages". Fibonacci is best known to the modern world for spreading the Hindu–Arabic numeral system in Europe, primarily through his Liber Abaci (Book of Calculation), written in 1202, and for a numerical sequence called the Fibonacci sequence, which he did not discover, but used as an example in the Liber Abaci.

```
!// Program to display the Fibonacci
!// sequence from 1 to n
!//
!////////////////////////////////////////////////////////////////
PROGRAM fibonacci
  IMPLICIT NONE
!// Variable declarations
 !// Declaring integer variables
  !// first a counter variable
  INTEGER :: i
  !// then the length of the Fibonacci sequence
  INTEGER :: n
  !// and finally a status variable (if errors)
  INTEGER :: res
  !// Declare an array to hold the Fibonacci sequence
  !// with unknown length at compilation time
  INTEGER, ALLOCATABLE, DIMENSION(:) :: sequence
```

To declare an array with unknown length, in this case `sequence`, the keywords `INTEGER, ALLOCATABLE, DIMENSION(:)` are used, where the length of the array is replaced by a colon `:`. If the length of the array is known in advance, say an array containing 100 elements, the construct `INTEGER, DIMENSION(100)` is used.

The next step is to prompt for how many numbers of the infinite Fibonacci sequence are to be included in the calculation. Later this number is used to allocate space for the Fibonacci numbers in the sequence. Thus, a prompting character string of a certain length, say 80, has to be declared and assigned a value. As long as the text string is shorter than 80 characters, a line of code counting the actual number of characters is inserted. Furthermore, the prompt is pushed to the terminal window. Thus, the next part of the program takes the form

```
!// Declare the length of the prompt to ask
!// for the length of the Fibonacci sequence
CHARACTER(LEN=80) :: prompt
!// Assign value to the prompt
prompt = 'Enter the length of the sequence: '
!// Get the number of non blank characters in the prompt
i = LEN(TRIM(prompt))
!// Display the prompt asking for the length suppressing
!// the line feed
WRITE(*,FMT='(A)',ADVANCE='NO') prompt(1:i+1)
```

Since the keyword `PRINT *` always prints the text to the terminal window with a linefeed as the last operation, it is replaced by the command

```
WRITE(*,FMT='(A)',ADVANCE='NO')
```

The `WRITE` command makes it possible to avoid or suppress the linefeed by adding the formatting code in the `WRITE` command, as shown above.

To read in the input from the terminal window and then perform the calculations, the rest of the program takes the form:

```
    !// Read the keyboard input
    READ(*,*) n
    !// Allocate space for the sequence
    ALLOCATE(sequence(n), STAT=res)
    !// Test the value of the res variable for errors
    IF(res /= 0) THEN
      !// We have an error. Print a message and stop the program
      PRINT *, 'Error in allocating space, status: ', res
      STOP
    END IF
    !// Initialize the two first elements in the sequence
    sequence(1) = 0
    sequence(2) = 1
    !// Loop and calculate the Fibonacci numbers
    DO i = 3, n
      sequence(i) = sequence(i-1) + sequence(i-2)
    END DO
    !// Print the sequence to the screen
    PRINT *, sequence
  END PROGRAM fibonacci
```

Once the variable `n` has been read, space is allocated for the Fibonacci array `sequence` using the construct

```
ALLOCATE(sequence(n), STAT=res)
```

Then, using the integer `res`, an "if" test is made to check whether the allocation is acceptable. If different from zero, the allocation has failed and the program is stopped. Otherwise, the program continues by first specifying the first two numbers in the Fibonacci sequence. Next, a loop is started to calculate the following integers in the sequence using the formula given in (A.2). In accordance with (A.2), the loop starts with an index variable equal to 3. Finally, the code line `PRINT *, sequence` is added at the end to display the contents of the sequence in the terminal window (unformatted).

A.6.5 File Input/Output or I/O

The data input used in the programs is usually stored in files. These may be numbers generated through the output of another program, or observations produced by instrument sensors in one way or another. In either case, they are usually available to us on a file stored on a computer somewhere, or residing on a memory device of some sort.

In the example to follow, the reader will learn how to read data from a file into an array, perform some operation on the data set, and write the result to a new file. It is assumed that the length of the array, say seven elements long, is known in advance. The program starts like this:

```
!//////////////////////////////////////////////////////////////
!//
!// f2c_file.f90
!//
!// Program to calculate the degrees Celsius from a
!// file containing seven observations of temperatures
!// in Fahrenheit
!//
!//////////////////////////////////////////////////////////////
PROGRAM f2c_file
  IMPLICIT NONE
  !// Declare a static variable to hold the
  !// length of the arrays
  INTEGER, PARAMETER :: n = 7
  !// Declare the arrays for the temperatures,
  !// and a counter variable
  REAL,DIMENSION(n)   :: F            !// Fahrenheit
  REAL,DIMENSION(n)   :: C            !// Celsius
  INTEGER             :: j            !// Counter variable
```

To specify that `n` is constant or static, it is declared using the keyword `INTEGER, PARAMETER` to signal that it is unchanged at runtime. Next, the arrays that will hold the temperatures in Fahrenheit and Celsius are declared together with a counter variable. In addition to this, two character strings are needed to hold the names of the input and output files, as given by the following lines of code:

```
  !// Declare character strings to hold the filenames
  CHARACTER(LEN=80)   :: infile  !// Holds the Fahrenheit
                                 !// temperatures
  CHARACTER(LEN=80)   :: outfile !// Holds the results of
                                 !// the conversion to
                                 !// degrees Celsius
```

Furthermore, two parameters are needed to hold the unit numbers, which are used to reference the input and output files. The names of the input and output files are also needed, to recognize them in the directory once the operation is completed. The reference numbers are integers. In this regard, a status variable must be declared, as in the last example above. Thus, the next part of the program takes the form:

```
  !// Declare constant values for the Logical Unit Number for
  !// referencing the files for opening, reading, and writing
  INTEGER, PARAMETER :: ilun = 10
  INTEGER, PARAMETER :: olun = 11
  !// A status variable to hold the result of file
  !// operations
  INTEGER :: res
```

```
!// Assign an input filename for temperatures in Fahrenheit
infile = "fahrenheit.txt"
!// Assign an output filename for temperatures in Celsius
outfile = "celsius.txt"
```

To access the contents of the output file, it has to be opened using the declared unit number and its filename `infile`. We also add a test to see whether the opening was successful. To achieve this, the program continues with the following statements:

```
!// Open the input file
OPEN(UNIT=ilun,FILE=infile,FORM="FORMATTED",IOSTAT=res)
!// Test whether the operation was successful
IF(res /= 0) THEN
  !// No, an error occurred, print a message to
  !// the screen
  PRINT *, "Error in opening file, status: ", res
  !// Stop the program
  STOP
END IF
```

The next step is to read the contents of the input file into the array `F`. This is carried out using a loop that runs from `1` to the length of the array, in this example `n`. This is done using the `READ` statement, which is similar to the `OPEN` statement:

```
!// Loop and read each line of the file
DO j = 1, n
  !// Read the current line
  READ(UNIT=ilun,FMT='(F4.1,X,F4.1)',IOSTAT=res) F(j)
  !// Was the read successful?
  IF(res /= 0) THEN
    !// No, Test whether we have reached End Of File (EOF)
    IF(res /= -1) THEN
      !// No, an error has occurred, print a message
      PRINT *, "Error in reading file, status: ", res
      !// Close the file
      CLOSE(UNIT=ilun)
      !// Stop the program
      STOP
    END IF
  END IF
END DO
```

The `infile` is formatted, so we need to specify the format of the data contained in it. In this example, it is the file `fahrenheit.txt`. The argument specified in `FMT` must be an exact match of the format in `fahrenheit.txt`. The keyword/argument pair `FMT='(F4.1,X,F4.1)'` is used to tell the computer that the number is a floating point number with four digits, including the decimal point and one decimal, a space character, and then another floating point number like the first.

The conversion from Fahrenheit to Celsius may then proceed, storing the result in the second array, that is, C. To see the result in the terminal window, we also add a PRINT statement. This is accomplished by adding the code lines:

```
!// Loop and convert from Fahrenheit to Celsius
DO j = 1, n
  C(j) = (F(j) - 32) * 5. / 9.
END DO
!// Print the temperatures to the screen
DO j = 1, n
  PRINT *, " Degrees Fahrenheit ", F(j), &
           " Degrees Celsius ", C(j)
END DO
```

Once the conversion is completed, the contents of the array C may be written to the output file outfile, which was given the reference number olun = 11. To enable writing to the output file, it must be opened. This is carried out exactly as was done for the input file:

```
!// Open the output file
OPEN(UNIT=olun,FILE=outfile,FORM="FORMATTED",IOSTAT=res)
!// Test whether the operation was successful
IF(res /= 0) THEN
  !// No, an error occurred, print a message to
  !// the screen
  PRINT *, "Error in opening output file, status: ", res
  !// Stop the program
 STOP
END IF
```

To ensure that the outfile is formatted, the argument FORM="FORMATTED" is included in the OPEN statement.

The result of the conversion may now be written to the output file, by continuing with the following code lines:

```
DO j = 1, n
  WRITE(UNIT=olun,FMT='(F4.1,A1,F4.1)',IOSTAT=res) C(j)
  !// Test whether the operation was successful
  IF(res /= 0) THEN
    !// No, an error occurred, print a message to
    !// the screen
    PRINT *, "Error in writing file, status: ", res
    !// Exit the loop
    EXIT
  END IF
END DO
```

Finally, the input and output files must be closed before terminating the program:

```
    !// Close the input file
    CLOSE(UNIT=ilun)
    !// Close the output file
    CLOSE(UNIT=olun)
  END PROGRAM f2c_file
```

When using the `OPEN` function, use was made of the respective unit numbers (or logical unit numbers) `ilun = 10` and `olun = 11`. As part of the `FILE` argument, the filename was also provided. Finally, note that the result of the call to `OPEN` is returned in the `IOSTAT=res` keyword/argument pair. The same procedure may be used in reading from files.

In this example, formatted files were used. This means that the file content may be displayed in the terminal window. In contrast, writing binary files to the terminal is not very meaningful. To visualize, consider constructing a formatted file containing two pairs of seven temperatures in Fahrenheit, in two columns, formatted following the keyword/argument `FMT='FMT='(F4.1,X,F4.1)'`:

```
68.2 65.5
69.6 63.7
73.2 66.0
75.0 68.0
77.5 70.2
79.2 71.4
91.2 73.2
```

In contrast, the binary file looks like this:

```
The binary file looks like this:
.....
^@H<8D><BC>$<D0>^@^@^@<BE>
^@^@^@<B9><B8>_N^@H<C7>D$0P^@^@^@<BA>^C<FF><84>
^C3<C0>H<C7>D$8<80><DC>s^@L<8D>D$0H<C7>D$@
^@^@^@H<C7>D$H<E0>'N^@H<C7>D$P^D^@^@^@<E8>O<C3>
.....
This is the end of the binary file$
```

Clearly, the binary format is unreadable as is, unless a program is used to translate the binary information into a text file written in ASCII (American Standard Code for Information Interchange) format. ASCII is the format used by formatted files in Fortran.

A.6.6 Multidimensional Arrays

In the field of meteorology and oceanography, data sets are usually in four dimensions, three in space and one in time. Consequently, matrices in four dimensions have to be declared to be able to store the data. The next example shows how the entries in a two-dimensional matrix, consisting of temperatures in degrees Fahrenheit from two observing stations, are read in from a file, converted to degrees Celsius, and finally written to a file.

The program is of course very similar to the previous example, except that the operation is on a two-dimensional matrix. This is made possible using nested loops. The complete code is:

```
!///////////////////////////////////////////////////////////////
!//
!// f2c_advanced.f90
!//
!// A program calculating degrees Celsius from Fahrenheit
!// from two measuring stations
!//
!///////////////////////////////////////////////////////////////
PROGRAM f2c_advanced
  IMPLICIT NONE
  !// Declare a static variable to hold the
  !// dimension of the vectors or arrays
  INTEGER, PARAMETER :: m = 7
  INTEGER, PARAMETER :: n = 2
  !// Declare the arrays for the temperatures in
  !// Fahrenheit and Celsius, and counter variables
  REAL,DIMENSION(m,n) :: F
  REAL,DIMENSION(m,n) :: C
  INTEGER :: j, k
  !// Character strings to hold the filenames
  CHARACTER(LEN=80) :: F_file
  CHARACTER(LEN=80) :: C_file
  !// Constant values for the Logical Unit Number
  !// for referencing the files for opening, reading,
  !// and writing
  INTEGER, PARAMETER :: ilun = 10
  INTEGER, PARAMETER :: olun = 11
  !// A status variable to hold the result of file
  !// operations
  INTEGER :: res
  !// Assign the input filename for Fahrenheit temp.
  infile = "fahrenheit.txt"
  !// Assign the output filename for Centigrade temp.
  outfile = "celsius.txt"
  !// Open the Fahrenheit input file
  OPEN(UNIT=ilun,FILE=infile,FORM="FORMATTED",IOSTAT=res)
  !// Test whether the operation was successful
  IF(res /= 0) THEN
    !// No, an error occurred, print a message to
    !// the screen
    PRINT *, "Error in opening file, status: ", res
    !// Stop the program
    STOP
```

```
  END IF
  !// Loop and read each line of the file
  DO j = 1, m
    !// Read the current line
    READ(UNIT=ilun,FMT='(F4.1,X,F4.1)',IOSTAT=res) F(j,1), F(j,2)
    !// Successfully read?
    IF(res /= 0) THEN
      !// No, Test whether we have reached End Of File (EOF)
      IF(res /= -1) THEN
        !// No, an error has occurred, print a message
        PRINT *, "Error in reading file, status: ", res
        !// Close the file
        CLOSE(UNIT=ilun)
        !// Stop the program
        STOP
      END IF
    END IF
  END DO
  !// Close the input file
  CLOSE(UNIT=ilun)
  !// Loop and convert from Fahrenheit to Centigrade
  DO k = 1, n
    DO j = 1, m
      C(j,k) = (F(j,k) - 32.) * 5./9.
    END DO
  END DO
  !// Open the Centigrade output file
  OPEN(UNIT=olun,FILE=centigradefile,FORM="FORMATTED",IOSTAT=res)
  !// Test whether the operation was successful
  IF(res /= 0) THEN
    !// No, an error occurred, print a message to
    !// the screen
    PRINT *, "Error in opening output file, status: ", res
    !// Stop the program
    STOP
  END IF
  DO j = 1, m
    WRITE(UNIT=olun,FMT='(F4.1,A1,F4.1)',IOSTAT=res) C(i,1), &
    ' ', C(i,2)
    !// Test whether the operation was successful
    IF(res /= 0) THEN
      !// No, an error occurred, print a message to
      !// the terminal window
      PRINT *, "Error in writing file, status: ", res
      !// Exit the loop
      EXIT
    END IF
  END DO
  !// Close the output file
  CLOSE(UNIT=olun)
END PROGRAM f2c_advanced
```

It is important for the efficiency of the program to know how Fortran accesses a matrix. In fact, it does so columnwise.[7] Thus, all the row elements in a column are accessed before the row elements in the next column. Therefore, as a rule of thumb, *always let the first index be the innermost loop* in Fortran.

A.6.7 Functions and Subroutines

It is not a good programming practice to write one long program that includes all the necessary code in one single source file. It makes the program hard to understand, and difficult to maintain. Consequently, it is common to break the program into smaller parts, or subprograms, called into action by the main program. In Fortran, these subprograms are called `FUNCTION`s or `SUBROUTINE`s.

In what follows, the program in the previous section is broken into a main program and a subprogram. The task of the subprogram is simply to do the conversion from Fahrenheit to Celsius, or the other way around. This may be done either by a function or by a subroutine.

Functions

First the `FUNCTION` is used. In the former program, the conversion was either from Fahrenheit to Celsius or from Celsius to Fahrenheit. Two functions are therefore needed, one to convert from Fahrenheit to Celsius, here called `f2c(arg)`, and a second converting from Celsius to Fahrenheit, here called `c2f(arg)`. In this example `f2c(arg)` takes the form:

```
FUNCTION f2c(F) RESULT(C)
  IMPLICIT NONE
!// The input argument which is read only
  REAL(KIND=8), INTENT(IN) :: F
!// The result from the calculations
  REAL(KIND=8) :: C
!// Perform the calculation
  C = (F -32) * 5./9.
!// Return the result
  RETURN
END FUNCTION f2c
```

while `c2f(arg)` takes the form:

```
FUNCTION c2f(C) RESULT(F)
  IMPLICIT NONE
!// The input argument which is read only
  REAL(KIND=8), INTENT(IN) :: C
!// The result from the calculations
  REAL(KIND=8) :: F
```

[7] This is also true for Matlab.

```
!// Perform the calculation
  F = (C * 9. / 5.) + 32
!// Return the result
  RETURN
END FUNCTION c2f
```

In contrast to how mathematical functions are written, Fortran 90–2003 adds the keyword `RESULT(arg)`, where the data type of the argument (arg) states what kind of function it is.[8] Here, `INTENT(IN)` is used for the input argument to the functions to prevent accidental overwriting of the argument, since Fortran function and subroutine arguments are always called by reference and not by value. The only thing the main program needs to do, apart from providing the user interface, is to replace the formula for the conversion with a call to the respective functions. Using the Fortran intrinsic functions, they have to be declared as external functions of the correct type. This is taken care of by the code lines `REAL, EXTERNAL :: f2c` and `REAL, EXTERNAL :: c2f`. If omitted, the compiler will flag en error and the compilation will be aborted. The complete main program is as follows:

```
!////////////////////////////////////////////////////////////
!//
!// array2.f90
!//
!// A program converting from Fahrenheit to Celsius
!// and vice versa
!//
!////////////////////////////////////////////////////////////
PROGRAM array2
  IMPLICIT NONE
  !// Declare everything .....
  !// 1. Arrays for the temperatures
  REAL,DIMENSION(7,2) :: F    ! Fahrenheit
  REAL,DIMENSION(7,2) :: C    ! Celsius
  !// 2. Index variables
  INTEGER :: i, j
  !// 3. Character strings to hold the filenames
  CHARACTER(LEN=80) :: fahrenheitfile
  CHARACTER(LEN=80) :: centigradefile
  !// 3. Constant values for the Logical Unit Number (lun)
  !//    for referencing the files for opening, reading
  !//    and writing
  INTEGER, PARAMETER :: ilun = 10
  INTEGER, PARAMETER :: olun = 11
  !// 4. A status variable to hold the result of file
  !//    operations
  INTEGER :: res
  !// 5. External function(s)
  REAL, EXTERNAL :: f2c
```

[8] In Fortran 77 and older versions, the syntax was `REAL FUNCTION f2c(arg)`.

```
REAL, EXTERNAL :: c2f
!// 6. A character string for a prompt
CHARACTER(LEN=80) :: prompt
CHARACTER :: answer
!// ..... End declarations
!// Assign filenames for temperatures
fahrenheitfile = "fahrenheit.txt"
centigradefile = "centigrade.txt"
!// Ask if we are to convert from Fahrenheit to
!// Celsius or vice versa
prompt = 'Convert from Fahrenheit to Centigrade (F/C)?'
PRINT *, TRIM(prompt)
READ(*,*) answer
!// Check if the answer is F or f for Fahrenheit
IF(answer .EQ. 'F' .OR. answer .EQ. 'f') THEN
!// Yes, open the Fahrenheit input file
  OPEN(UNIT=ilun,FILE=fahrenheitfile,FORM="FORMATTED", &
  IOSTAT=res)
  !// Test whether the operation was successful
  IF(res /= 0) THEN
  !// No, an error occurred, print a message to
  !// the screen
    PRINT *, "Error in opening file, status: ", res
  !// Stop the program
    STOP
  END IF
  !// Loop and read each line of the file
  DO i = 1, 7
    !// Read the current line
    READ(UNIT=ilun,FMT='(F4.1,X,F4.1)',IOSTAT=res) &
    F(i,1), F(i,2)
    !// Was the read successful?
    IF(res /= 0) THEN
      !// No, Test whether we have reached End Of File (EOF)
      IF(res /= -1) THEN
        !// No, an error has occurred, print a message
        PRINT *, "Error in reading file, status: ", res
        !// Close the file
        CLOSE(UNIT=ilun)
        !// Stop the program
        STOP
      END IF
    END IF
  END DO
ELSE
  !// No, open the Celsius input file
  OPEN(UNIT=ilun,FILE=centigradefile,FORM="FORMATTED", &
  IOSTAT=res)
  !// Test whether the operation was successful
  IF(res /= 0) THEN
    !// No, an error occurred, print a message to
    !// the screen
```

```
      PRINT *, "Error in opening file, status: ", res
      !// Stop the program
      STOP
    END IF
    !// Loop and read each line of the file
    DO i = 1, 7
      !// Read the current line
      READ(UNIT=ilun,FMT='(F4.1,X,F4.1)',IOSTAT=res) &
      C(i,1), C(i,2)
      !// Was the read successful?
      IF(res /= 0) THEN
        !// No, Test whether we have reached End Of File (EOF)
        IF(res /= -1) THEN
          !// No, an error has occurred, print a message
          PRINT *, "Error in reading file, status: ", res
          !// Close the file
          CLOSE(UNIT=ilun)
          !// Stop the program
          STOP
        END IF
      END IF
    END DO
  END IF
  !// Close the input file
  CLOSE(UNIT=ilun)
  !// Which way to convert ?
  IF(answer .EQ. 'F' .OR. answer .EQ. 'f') THEN
    !// Loop and convert from Fahrenheit to Celsius
    DO j = 1, 2
      DO i = 1, 7
        C(i,j) = f2c(F(i,j))
      END DO
    END DO
  ELSE
  !// Loop and convert from Celsius to Fahrenheit
  DO j = 1, 2
    DO i = 1, 7
      F(i,j) = c2f(C(i,j))
    END DO
  END DO
  END IF
  !// Which file to write to ?
  IF(answer .EQ. 'F' .OR. answer .EQ. 'f') THEN
    !// Open the Centigrade output file
    OPEN(UNIT=olun,FILE=centigradefile,FORM="FORMATTED", &
    IOSTAT=res)
    !// Test whether the operation was successful
    IF(res /= 0) THEN
      !// No, an error occurred, print message to the screen
      PRINT *, "Error in opening output file, status: ", res
      !// Stop the program
      STOP
    END IF
```

```
      DO i = 1, 7
        WRITE(UNIT=olun,FMT='(F4.1,A1,F4.1)',IOSTAT=res) &
        C(i,1), ' ', C(i,2)
        !// Test whether the operation was successful
        IF(res /= 0) THEN
          !// No, an error occurred, print message to the screen
          PRINT *, "Error in writing file, status: ", res
          !// Exit the loop
          EXIT
        END IF
      END DO
    ELSE
      !// Open the Fahrenheit output file
      OPEN(UNIT=olun,FILE=fahrenheitfile,FORM="FORMATTED", &
      IOSTAT=res)
      !// Test whether the operation was successful
      IF(res /= 0) THEN
        !// No, an error occurred, print message to the screen
        PRINT *, "Error in opening output file, status: ", res
        !// Stop the program
        STOP
      END IF
      DO i = 1, 7
        WRITE(UNIT=olun,FMT='(F4.1,A1,F4.1)',IOSTAT=res) &
        F(i,1), ' ', F(i,2)
        !// Test whether the operation was successful
        IF(res /= 0) THEN
          !// No, an error occurred, print message to the screen
          PRINT *, "Error in writing file, status: ", res
          !// Exit the loop
          EXIT
        END IF
      END DO
    END IF
    !// Close the output file
    CLOSE(UNIT=olun)
  END PROGRAM array2
```

In contrast to functions, a subroutine does not return a value and is the same as a void function in other languages. The functions f2c() and c2f() may be replaced by corresponding subroutines, for instance, which can look like this:

```
!///////////////////////////////////////////////////////////
!//
!// SUBROUTINE f2c(F,C)
!//
!// Called from program array2.f90
!//
!///////////////////////////////////////////////////////////
SUBROUTINE f2c(F,C)
  IMPLICIT NONE
```

```
    !// The input argument which is read only
    REAL(KIND=8), INTENT(IN) :: F
    !// The result of the conversion which is
    !// write only
    REAL(KIND=8), INTENT(OUT) :: C
    !// Perform the conversion
    C = (F -32) * 5./9.
    !// Return the result through the second argument
    RETURN
  END SUBROUTINE f2c
```

and

```
  !////////////////////////////////////////////////////////////
  !//
  !// SUBROUTINE c2f(C,F)
  !//
  !// Called from program array2.f90
  !//
  !////////////////////////////////////////////////////////////
  SUBROUTINE c2f(C,F)
    IMPLICIT NONE
    !// The input argument which is read only
    REAL(KIND=8), INTENT(IN) :: C
    !// The result of the conversion which is
    REAL(KIND=8), INTENT(OUT) :: F
    !// Perform the conversion
    F = (C * 9. / 5.) + 32
    !// Return the result through the second argument
    RETURN
  END SUBROUTINE c2f
```

The call from the main program is like this:

```
    !// Loop and perform the conversion using
    !// nested loops
    DO j = 1, 2
      DO i = 1, 7
      !// Call the subroutine with the current element
      !// of the Fahrenheit array as the first argument to
      !// the subroutine and the current element of the
      !// centigrade array as the second argument.
        CALL f2c(Fahrenheit(i,j),centigrade(i,j))
      END DO
    END DO
```

and

```
    !// Loop and perform the conversion using
    !// nested loops
```

```
DO j = 1, 2
  DO i = 1, 7
  !// Call the subroutine with the current element
  !// of the Fahrenheit array as the first argument to
  !// the subroutine and the current element of the
  !// centigrade array as the second argument.
    CALL c2f(centigrade(i,j).Fahrenheit(i,j))
  END DO
END DO
```

Only non-intrinsic functions can be declared as external, subroutines cannot. Note also the use of `INTENT(OUT)`, which means that one can only give value to the argument (write only). Trying to read the value from the argument would flag a compilation error, just as it would if we were trying to give the argument a value when it had the attribute `INTENT(IN)` (read only).

A.7 Exercises

1. Use the code in Sect. A.6, fill in what is missing, and save the source code in a file. Compile the code and run it to check that, in a non-leap year, daynumber for September 16 is indeed 259.
2. Given the radius of a circle, write a program to calculate the length of the circumference of the circle. Compile and run the program to check that the result is correct.
3. Given the radius of a circle, write a program to calculate the area inside the circle. Compile and run the program to check that the result is correct.
4. Given the radius of a sphere, write a program to calculate the volume of the sphere. Compile and run the program to check that the result is correct.

Printed in the United States
By Bookmasters